# Epistemic Game Theory and Logic

Special Issue Editor

Paul Weirich

*Special Issue Editor*
Paul Weirich
University of Missouri
USA

*Editorial Office*
MDPI AG
St. Alban-Anlage 66
Basel, Switzerland

This edition is a reprint of the Special Issue published online in the open access journal *Games* (ISSN 2073-4336) from 2016–2017 (available at: http://www.mdpi.com/journal/games/special_issues/epistemic_game_theory_logic).

For citation purposes, cite each article independently as indicated on the article page online and as indicated below:

Author 1; Author 2; Author 3 etc. Article title. *Journal Name*. **Year**. Article number/page range.

ISBN 978-3-03842-422-2 (Pbk)
ISBN 978-3-03842-423-9 (PDF)

# Table of Contents

About the Guest Editor ........ v

Preface to "Epistemic Game Theory and Logic" ........ vii

## Section 1: Economics

**Andrés Perea and Willemien Kets**
When Do Types Induce the Same Belief Hierarchy?
Reprinted from: *Games* 2016, *7*(4), 28; doi: 10.3390/g7040028
http://www.mdpi.com/2073-4336/7/4/28 ........ 3

**Giacomo Bonanno**
Exploring the Gap between Perfect Bayesian Equilibrium and Sequential Equilibrium
Reprinted from: *Games* 2016, *7*(4), 35; doi: 10.3390/g7040035
http://www.mdpi.com/2073-4336/7/4/35 ........ 20

**Geir B. Asheim, Mark Voorneveld and Jörgen W. Weibull**
Epistemically Robust Strategy Subsets
Reprinted from: *Games* 2016, *7*(4), 37; doi: 10.3390/g7040037
http://www.mdpi.com/2073-4336/7/4/37 ........ 43

**Fan Yang and Ronald M. Harstad**
The Welfare Cost of Signaling
Reprinted from: *Games* 2017, *8*(1), 11; doi: doi:10.3390/g8010011
www.mdpi.com/2073-4336/8/1/11 ........ 59

## Section 2: Computer Science

**Jing Chen and Silvio Micali**
Leveraging Possibilistic Beliefs in Unrestricted Combinatorial Auctions
Reprinted from: *Games* 2016, *7*(4), 32; doi: 10.3390/g7040032
http://www.mdpi.com/2073-4336/7/4/32 ........ 83

## Section 3: Philosophy

**Cyril Hédoin**
Community-Based Reasoning in Games: Salience, Rule-Following, and Counterfactuals
Reprinted from: *Games* 2016, *7*(4), 36; doi: 10.3390/g7040036
http://www.mdpi.com/2073-4336/7/4/36 ........ 105

**Mikaël Cozic**
Probabilistic Unawareness
Reprinted from: *Games* 2016, *7*(4), 38; doi: 10.3390/g7040038
www.mdpi.com/2073-4336/7/4/38 ........ 122

**Ashton T. Sperry-Taylor**
Strategy Constrained by Cognitive Limits, and the Rationality of Belief-Revision Policies
Reprinted from: *Games* 2017, *8*(1), 3; doi: 10.3390/g8010003
www.mdpi.com/2073-4336/8/1/3 ........ 146

## Section 4: Psychology

**Ramzi Suleiman**
Economic Harmony: An Epistemic Theory of Economic Interactions
Reprinted from: *Games* 2017, *8*(1), 2; doi: 10.3390/g8010002
www.mdpi.com/2073-4336/8/1/2 ........ 161

# About the Guest Editor

**Paul Weirich** is a Curators' Distinguished Professor in the Philosophy Department at the University of Missouri. He is the author of Equilibrium and Rationality (Cambridge, 1998), Decision Space (Cambridge, 2001), Realistic Decision Theory (Oxford, 2004), Collective Rationality (Oxford, 2010), and Models of Decision-Making (Cambridge, 2015).

# Preface to "Epistemic Game Theory and Logic"

Epistemic game theory and the systems of logic that support it are crucial for understanding rational behavior in interactive situations in which the outcome for an agent depends, not just on her own behavior, but also on the behavior of those with whom she is interacting. Scholars in many fields study such interactive situations, that is, games of strategy.

Epistemic game theory presents the epistemic foundations of a game's solution, taken as a combination of strategies, one for each player in the game, such that each strategy is rational given the combination. It considers the beliefs of the players in a game and shows how, along with the players' goals, their beliefs guide their choices and settle the outcome of their game. Adopting the Bayesian account of probability, as rational degree of belief, it yields Bayesian game theory. Epistemic game theory, because it attends to how players reason strategically in games, contrasts with evolutionary game theory, which applies to non-reasoning organisms such as bacteria.

Logic advances rules of inference for strategic reasoning. It contributes not just standard rules of deductive logic, such as modus ponens, but also rules of epistemic logic, such as the rule going from knowledge of a set of propositions to knowledge of their deductive consequences, and rules of probabilistic reasoning such as Bayesian conditionalization, which uses probabilities conditional on receiving some new evidence to form new non-conditional probabilities after receiving exactly that new evidence.

Perea (2012) offers an overview, and Weirich (1998) shows how principles of choice support solutions to games of strategy.

The papers in the special issue came in response to the journal's call for papers. Diversity of perspectives was a goal. The papers include four by economists, one by computer scientists, three by philosophers, and one by a psychologist. They display a variety of approaches to epistemic game theory and logic. The following paragraphs briefly describe the topics of the papers, grouped according to discipline and within a discipline according to date of publication.

**1. Economics**

Perea and Kets (2016), "When Do Types Induce the Same Belief Hierarchy?" Higher-order beliefs, that is, beliefs about beliefs, play an important role in game theory. Whether one player decides to make a certain move may depend on the player's beliefs about another player's beliefs. A representation of a player's higher-order beliefs may use a type structure, that is, a set of player types that characterize a player's uncertainty. This paper investigates conditions under which two type structures may represent the same hierarchical beliefs.

Bonanno (2016), "Exploring the Gap between Perfect Bayesian Equilibrium and Sequential Equilibrium." This paper uses methods of belief revision to characterize types of equilibrium in a game with sequences of moves. It assumes that belief revision is conservative or minimal and then obtains two types of equilibrium that in a certain sense are intermediate between perfect Bayesian equilibrium and sequential equilibrium. It argues that refinements of subgame-perfect equilibrium in extensive-form games should attend to principles of belief revision, in particular, iterated belief revision.

Asheim et al. (2016), "Epistemically Robust Strategy Subsets." This paper examines the epistemic foundations of set-valued generalizations of strict Nash equilibrium. It explores, in particular, epistemic robustness, or support by epistemic considerations, of sets of strategy profiles taken as solutions to games in which players have doubts about the rationality of players, and so doubts about the strategy a player adopts in response to other players. It classifies a player according to her doxastic state, characterized by a probability distribution over the other players' strategies and types.

Yang and Harstad (2017), "The Welfare Cost of Signaling." This paper treats learning about differences among agents from observations of the agents' behavior. It investigates, in particular, separating workers according to skill using signals an employer elicits from the workers. A credible signal may involve a transfer of resources instead of consumption of resources, and so not impose a welfare cost on society. The paper shows specifically that an employer's charging a job application fee may generate such credible signals, and it explores the robustness of its results given variation in factors such as a job applicant's aversion to risk.

**2. Computer Science**

Chen and Micali (2016), "Leveraging Possibilistic Beliefs in Unrestricted Combinatorial Auctions." This paper addresses a topic in economics, namely, the design of auctions, with special attention to methods of computing a rational strategy given the design's adoption. The design of the auction aims to guarantee an amount of revenue, dependent on bidders' "possibilistic beliefs" about bidders' valuations of an item up for auction. A bidder's possibilistic beliefs settle the valuations of other bidders that are epistemically possible for the bidder, that is, may be correct for all the bidder knows. The paper describes the benefits of its auction design for a bidder's privacy concerning her valuations, for computationally simplifying a design's implementation, and for limiting collusion among bidders.

**3. Philosophy**

Hédoin (2016), "Community-Based Reasoning in Games: Salience, Rule-Following, and Counterfactuals." This paper argues for a representation of a game that includes, when the players know that they come from the same community, the players' knowledge of this fact. Their knowledge of their cultural ties assists their predictions of their behavior; it explains a focal point of their game, that is, a strategy profile that the players expect the players to realize. The paper's representation of a game includes an epistemic model grounding applications of epistemic logic that yield conclusions about the players' behavior. The model covers not only actual behavior, but also behavior in hypothetical situations that do not arise during a play of the game. The paper examines the literature on following rules because a culture's rules govern counterfactual as well as actual situations.

Cozic (2016), "Probabilistic Unawareness." This paper considers whether an agent in a game, for a proposition relevant to the game's solution, believes the proposition and whether, if the proposition is true, the agent is aware or unaware of its truth. An agent may be unaware of a proposition's truth because he fails to entertain it. The paper uses partial (and so probabilistic) belief as opposed to full belief in its representation of an agent's beliefs. It extends doxastic logic to obtain a version of probabilistic logic, namely, a formal system with an explicit representation of partial belief together with an explicit representation of awareness. Its main result is a soundness and completeness theorem, using its semantics of partial belief and awareness, for its probabilistic extension of doxastic logic.

Sperry-Taylor (2017), "Strategy Constrained by Cognitive Limits, and the Rationality of Belief-Revision Policies." This paper investigates the effect of cognitive limits on the strategic reasoning of the players in a game. It examines how as a game progresses, imperfect agents playing the game revise their beliefs (taken as probability assignments or assignments of credence to propositions), and how their methods of belief revision affect their choices and the outcome of their game. Cognitive limits may excuse an agent's failure to consider all the options in a decision problem and may excuse an agent's failure to consider all her strategies for playing a game. The paper advances principles of rationality that govern belief revision by cognitively limited agents who in games entertain some contingencies only as they arise.

**4. Psychology**

Suleiman (2017), "Economic Harmony: An Epistemic Theory of Economic Interactions." This paper maintains that some under-studied epistemic factors, in particular, an agent's aspirations, influence the agent's reasoning in a game of strategy. It defines an agent's subjective utility assignment to the possible outcomes of an option in a decision problem as a quantitative representation of the agent's mental attitudes to the outcomes, roughly, levels of satisfaction from the outcomes, so that the utilities of the outcomes form a ratio scale. Then it proposes that the value of an outcome for an agent is the ratio of the utility for the agent of the payoff the outcome yields divided by the utility for the agent of the payoff to which the agent aspires. As a solution to a game, it proposes a combination of strategies, with exactly one strategy for each player that yields an outcome with the same utility for all the players. The paper supports its proposals by noting their agreement with experimental data concerning behavior in games such as the Ultimatum Game and the Common Pool Resource Dilemma.

This Special Issue gathers recent scholarship in several disciplines treating epistemic game theory and logic. Some articles propose new ways of representing the mental states of the players in a game or new ways of making inferences about their mental states, some strengthen the logical foundations of epistemic game theory or extend its scope, and some present applications of epistemic game theory. The articles show the vitality and significance of this area of research.

Paul Weirich
*Guest Editor*

# Section 1: Economics

 games

Article

# When Do Types Induce the Same Belief Hierarchy?

**Andrés Perea [1] and Willemien Kets [2,*]**

[1] EpiCenter and Department of Quantitative Economics, Maastricht University, P.O. Box 616, 6200 MD Maastricht, The Netherlands; a.perea@maastrichtuniversity.nl
[2] MEDS, Kellogg School of Management, Northwestern University, Evanston, IL 60208-2001, USA
* Correspondence: willemien.kets@gmail.com

Academic Editor: Paul Weirich
Received: 22 August 2016 ; Accepted: 30 September 2016 ; Published: 9 October 2016

**Abstract:** Type structures are a simple device to describe higher-order beliefs. However, how can we check whether two types generate the same belief hierarchy? This paper generalizes the concept of a type morphism and shows that one type structure is contained in another if and only if the former can be mapped into the other using a generalized type morphism. Hence, every generalized type morphism is a hierarchy morphism and vice versa. Importantly, generalized type morphisms do not make reference to belief hierarchies. We use our results to characterize the conditions under which types generate the same belief hierarchy.

**Keywords:** types; belief hierarchies; epistemic game theory; morphisms

**JEL Classification:** C72

---

## 1. Introduction

Higher-order beliefs play a central role in game theory. Whether a player is willing to invest in a project, for example, may depend on what he or she thinks that his or her opponent thinks about the economic fundamentals, what he or she thinks that his or her opponent thinks that he or she thinks, and so on, up to arbitrarily high order (e.g., [1]). Higher-order beliefs can also affect economic conclusions in settings ranging from bargaining [2,3] and speculative trade [4] to mechanism design [5] . Higher-order beliefs about actions are central to epistemic characterizations, for example, of rationalizability [6,7], Nash equilibrium [8,9] and forward induction reasoning [10]. In principle, higher-order beliefs can be modeled explicitly, using belief hierarchies. For applications, the type structures introduced by Harsanyi [11] provide a simple, tractable modeling device to represent players' higher-order beliefs.

While type structures provide a convenient way to represent higher-order beliefs, it may be difficult to check whether types generate the same belief hierarchy. The literature has considered the following question: given two type structures, $\mathcal{T}$ and $\mathcal{T}'$, is it the case that for every type in $\mathcal{T}$, there is a type in $\mathcal{T}'$ that generates the same belief hierarchy? That is, is the type structure $\mathcal{T}$ contained in $\mathcal{T}'$?[1] The literature has considered two different tests to address this question, one based on hierarchy morphisms and one based on type morphisms. Hierarchy morphisms can be used to give a complete answer to this question: a type structure $\mathcal{T}$ is contained in $\mathcal{T}'$ if and only if there is a hierarchy morphism from the former to the latter. A problem with this test is that hierarchy morphisms make reference to belief hierarchies, as we shall see. Therefore, this test requires us to go outside the purview of type structures. The second test uses type morphisms. Type morphisms are defined solely in terms

[1] We follow the terminology of Friedenberg and Meier [12] here. A stronger condition for $\mathcal{T}$ to be contained in $\mathcal{T}'$ is that $\mathcal{T}$ can be embedded (using a type morphism) into $\mathcal{T}'$ as a belief-closed subset [13]. Our results can be used to characterize conditions under which $\mathcal{T}$ is contained in $\mathcal{T}'$ in this stronger sense in a straightforward way.

of the properties of the type structures. However, the test based on type morphisms only provides a sufficient condition: if there is a type morphism from $\mathcal{T}$ to $\mathcal{T}'$, then $\mathcal{T}$ is contained in $\mathcal{T}'$ [14]. However, as shown by Friedenberg and Meier [12], the condition is not necessary: it may be that $\mathcal{T}'$ contains $\mathcal{T}$, yet there is no type morphism from $\mathcal{T}$ to $\mathcal{T}'$. The work in [12] also provides a range of conditions under which the condition is both necessary and sufficient. However, they do not directly address the question of whether there might be an alternate test (which provides conditions that are both necessary and sufficient) that does not require us to describe the belief hierarchies explicitly.

This paper provides such a test, by generalizing the notion of a type morphism. We show that a type structure is contained in another if and only if there is a generalized type morphism from the former to the latter. Therefore, a generalized type morphism is a hierarchy morphism and vice versa. Unlike the definition of hierarchy morphisms, the definition of generalized type morphisms does not make reference to belief hierarchies. Therefore, this test can be carried out without leaving the purview of type structures. Using this result, it is straightforward to verify whether two types generate the same belief hierarchy, as we show.

Hierarchy morphisms are used in a number of different settings. For example, they can be used to check whether types have the same rationalizable actions [15] and play an important role in the literature on the robustness to misspecifying the parameter set more generally; see, e.g., Ely and Peski [16] and Liu [17]. Hierarchy morphisms are also used to study the robustness of Bayesian-Nash equilibria to misspecifications of players' belief hierarchies [18,19] and in epistemic game theory. The current results make it possible to study these issues without describing players' belief hierarchies explicitly, using that every hierarchy morphism is a generalized type morphism and conversely.

A critical ingredient in the definition of a generalized type morphism is the $\sigma$-algebra on a player's type set, which separates his or her types if and only if they differ in the belief hierarchy that they generate. Mertens and Zamir ([13], p. 6) use this $\sigma$-algebra to define non-redundant type structures, and this $\sigma$-algebra also plays an important role in the work of Friedenberg and Meier [12], where it is used to characterize the conditions under which hierarchy morphisms and type morphisms coincide. The work in [13] provides a nonconstructive definition of this $\sigma$-algebra, and [12] show that the $\sigma$-algebra defined by [13] is the $\sigma$-algebra generated by the functions that map types into belief hierarchies. We provide a constructive definition of this $\sigma$-algebra, by means of a type partitioning procedure that does not make reference to belief hierarchies.

While many of the ingredients that underlie our results are known in some form or another, we view the contribution of this paper as combining these ideas in a new way to generalize the concept of a type morphism, so that it provides a necessary and sufficient condition for a type structure to be contained in another that does not refer to belief hierarchies.

A number of papers has shown that the measurable structure associated with type structures can impose restrictions on reasoning [12,20–23]. This paper contributes to that literature in two ways. First, we elucidate the connection by constructing the measurable structure on type sets that is generated by players' higher-order beliefs. Second, we provide tools to easily go from the domain of type structures to the domain of belief hierarchies and vice versa.

The outline of this paper is as follows. The next section introduces basic concepts. Section 3 discusses type morphisms and hierarchy morphisms. Section 4 defines our generalization of a type morphism and proves the main result. Section 5 applies this result to characterize the conditions under which types generate the same belief hierarchy. Section 6 considers the special case where players have finitely many types. Proofs are relegated to the Appendix A.

## 2. Belief Hierarchies and Types

In this section, we show how belief hierarchies can be encoded by means of a type structure. The original idea behind this construction goes back to Harsanyi (1967). We first provide the definition of a type structure and subsequently explain how to derive a belief hierarchy from a type in a type structure. We conclude the section with an example of two type structures that are equivalent, in the sense that they produce exactly the same sets of belief hierarchies for the players. This example thus shows that the same belief hierarchy can be encoded within different type structures.

### 2.1. Type Structures

Consider a finite set of players $I$. Assume that each player $i$ faces a basic space of uncertainty $(X_i, \Sigma_i)$, where $X_i$ is a set and $\Sigma_i$ a $\sigma$-algebra on $X_i$. That is, $\mathcal{X}_i = (X_i, \Sigma_i)$ is a measurable space. The combination $\mathcal{X} = (X_i, \Sigma_i)_{i \in I}$ of basic uncertainty spaces is called a multi-agent uncertainty space. The basic space of uncertainty for player $i$ could, for instance, be the set of opponents' choice combinations, or the set of parameters determining the utility functions of the players, or even a combination of the two.

A belief hierarchy for player $i$ specifies a probability measure on $\mathcal{X}_i$, the first-order belief, a probability measure on $\mathcal{X}_i$ and the opponents' possible first-order beliefs, the second-order belief, and so on. As is standard, we encode such infinite belief hierarchies by means of type structures.

For any measurable space $(Y, \hat{\Sigma})$, we denote by $\Delta(Y, \hat{\Sigma})$ the set of probability measures on $(Y, \hat{\Sigma})$. We endow $\Delta(Y, \hat{\Sigma})$ with the coarsest $\sigma$-algebra that contains the sets:

$$\{\mu \in \Delta(Y, \hat{\Sigma}) \mid \mu(E) \geq p\}: \quad E \in \hat{\Sigma}, p \in [0,1].$$

This is the $\sigma$-algebra used in Heifetz and Samet [14] and many subsequent papers; it coincides with the Borel $\sigma$-algebra on $\Delta(Y, \hat{\Sigma})$ (induced by the weak convergence topology) if $Y$ is metrizable and $\hat{\Sigma}$ is the Borel $\sigma$-algebra. Product spaces are endowed with the product $\sigma$-algebra. Given a collection of measurable spaces $(Y_i, \mathcal{Y}_i)$, $i \in I$, write $\mathcal{Y}$ for the product $\sigma$-algebra $\bigotimes_{j \in I} \mathcal{Y}_j$ and $\mathcal{Y}_{-i}$ for the product $\sigma$-algebra $\bigotimes_{j \neq i} \mathcal{Y}_j$, where $i \in I$.

**Definition 1.** *(Type structure) Consider a multi-agent uncertainty space $\mathcal{X} = (X_i, \Sigma_i)_{i \in I}$. A type structure for $\mathcal{X}$ is a tuple $\mathcal{T} = (T_i, \Sigma_i^{\mathcal{T}}, b_i)_{i \in I}$ where, for every player $i$,*

*(a) $T_i$ is a set of types for player $i$, endowed with a $\sigma$-algebra $\Sigma_i^{\mathcal{T}}$, and*

*(b) $b_i : T_i \to \Delta(X_i \times T_{-i}, \hat{\Sigma}_i)$ is a measurable mapping that assigns to every type $t_i$ a probabilistic belief $b_i(t_i) \in \Delta(X_i \times T_{-i}, \hat{\Sigma}_i)$ on its basic uncertainty space and the opponents' type combinations, where $\hat{\Sigma}_i = \Sigma_i \otimes \Sigma_{-i}^{\mathcal{T}}$ is the product $\sigma$-algebra on $X_i \times T_{-i}$.*

Finally, if $f : Y \to (Y', \Sigma')$ is a function from $Y$ to the measurable space $(Y', \Sigma')$, then $\sigma(f)$ is the $\sigma$-algebra on $Y$ generated by $f$, that is, it is the coarsest $\sigma$-algebra that contains the sets $\{y \in Y : f(y) \in E\}$ for $E \in \Sigma'$.

### 2.2. From Type Structures to Belief Hierarchies

In the previous subsection, we have introduced the formal definition of a type structure. We now show how to "decode" a type within a type structure, by deriving the full belief hierarchy that it induces.

Consider a type structure $\mathcal{T} = (T_i, \Sigma_i^{\mathcal{T}}, b_i)_{i \in I}$ for $\mathcal{X}$. Then, every type $t_i$ within $\mathcal{T}$ induces an infinite belief hierarchy:

$$h_i^{\mathcal{T}}(t_i) = (\mu_i^{\mathcal{T},1}(t_i), \mu_i^{\mathcal{T},2}(t_i), \ldots),$$

where $\mu_i^{\mathcal{T},1}(t_i)$ is the induced first-order belief, $\mu_i^{\mathcal{T},2}(t_i)$ is the induced second-order belief, and so on. We will inductively define, for every $n$, the $n$-th order beliefs induced by types $t_i$ in $\mathcal{T}$, building upon the $(n-1)$-th order beliefs that have been defined in the preceding step.

We start by defining the first-order beliefs. For each player $i$, define:

$$H_i^1 := \Delta(X_i, \Sigma_i)$$

to be the set of beliefs about $X_i$, and for every type $t_i \in T_i$, define its first-order belief $\mu_i^{\mathcal{T},1}(t_i)$ by:

$$\mu_i^{\mathcal{T},1}(t_i)(E_i) := b_i(t_i)(E_i \times T_{-i}) \text{ for all } E_i \in \Sigma_i.$$

Clearly, $\mu_i^{\mathcal{T},1}(t_i) \in \Delta(X_i, \Sigma_i)$ for every type $t_i$. Define $h_i^{\mathcal{T},1}(t_i) := \mu_i^{\mathcal{T},1}(t_i)$. The mapping $\mu_i^{\mathcal{T},1}$ from $T_i$ to $H_i^1$ is measurable by standard arguments. For $n > 1$, suppose the set $H_i^{n-1}$ has been defined and that the function $h_i^{\mathcal{T},n-1}$ from $T_i$ to $H_i^{n-1}$ is measurable. Let $\hat{\Sigma}_i^{n-1}$ be the product $\sigma$-algebra on $X_i \times \times_{j \neq i} H_j^{n-1}$, and define:

$$H_i^n := H_i^{n-1} \times \Delta(X_i \times H_{-i}^{n-1}, \hat{\Sigma}_i^{n-1}).$$

For every type $t_i$, define its $n$-th-order belief $\mu_i^{\mathcal{T},n}(t_i)$ by:

$$\text{for all } E \in \hat{\Sigma}_i^{n-1}: \qquad \mu_i^{\mathcal{T},n}(t_i)(E) = b_i(t_i)(\{(x_i, t_{-i}) \in X_i \times T_{-i} \mid (x_i, h_{-i}^{\mathcal{T},n-1}(t_{-i})) \in E\}),$$

with $h_{-i}^{\mathcal{T},n-1}(t_{-i}) = (h_j^{\mathcal{T},n-1}(t_j))_{j \neq i}$. Since $h_j^{\mathcal{T},n-1}$ is measurable for every player $j$, $\mu_i^{\mathcal{T},n}$ is indeed a probability measure on $(X_i \times H_{-i}^{n-1}, \hat{\Sigma}_i^{n-1})$. Define $h_i^{\mathcal{T},n}(t_i) := (h_i^{\mathcal{T},n-1}(t_i), \mu_i^{\mathcal{T},n}(t_i))$. It follows that $h_i^{\mathcal{T},n}(t_i) \in H_i^n$. Moreover, $h_i^{\mathcal{T},n}$ is measurable.

Note that, formally speaking, the $n$-th-order belief $\mu_i^{\mathcal{T},n}(t_i)$ is a belief about $X_i$ and the opponents' first-order until $(n-1)$-th order beliefs. Moreover, by construction, the $n$-th and $(n+1)$-th order beliefs $\mu_i^{\mathcal{T},n}(t_i)$ and $\mu_i^{\mathcal{T},n+1}(t_i)$ are coherent in the sense that they induce the same belief on $X_i$ and the opponents' first-order until $(n-1)$-th order beliefs.

Finally, for every type $t_i \in T_i$, we denote by:

$$h_i^{\mathcal{T}}(t_i) := (\mu_i^{\mathcal{T},n}(t_i))_{n \in \mathbb{N}}$$

the belief hierarchy induced by type $t_i$ in $\mathcal{T}$. Furthermore, define $H_i$ to be the set $\Delta(X_i) \times \times_{n \geq 1} \Delta(X_i \times H_{-i}^n)$ of all belief hierarchies. We say that two types, $t_i$ and $t_i'$, of player $i$ generate the same belief hierarchy if $h_i^{\mathcal{T}}(t_i) = h_i^{\mathcal{T}}(t_i')$. Types $t_i$ and $t_i'$ generate the same $n$-th-order belief if $\mu_i^{\mathcal{T},n}(t_i) = \mu_i^{\mathcal{T},n}(t_i')$.[2]

## 2.3. Example

Consider a multi-agent uncertainty space $\mathcal{X} = (X_i, \Sigma_i)_{i \in I}$ where $I = \{1, 2\}$, $X_1 = \{c, d\}$, $X_2 = \{e, f\}$ and $\Sigma_1, \Sigma_2$ are the discrete $\sigma$-algebras on $X_1$ and $X_2$, respectively. Consider the type structures $\mathcal{T} = (T_1, T_2, \Sigma_1, \Sigma_2, b_1, b_2)$ and $\mathcal{T}' = (R_1, R_2, \Sigma_1, \Sigma_2, \beta_1, \beta_2)$ in Table 1.

Then, it can be verified that the types $t_1, t_1', r_1'$ and $r_1''$ generate the same belief hierarchy, and so do the types $t_1''$ and $r_1$, the types $t_2, t_2''$ and $r_2'$ and the types $t_2', r_2$ and $r_2''$. In particular, for every type in $\mathcal{T}$, there is another type in $\mathcal{T}'$ generating the same belief hierarchy, and vice versa. In this sense, the two type structures $\mathcal{T}$ and $\mathcal{T}'$ are equivalent.

[2] Clearly, $t_i$ and $t_i'$ generate the same $n$-th-order belief if and only if $h_i^{\mathcal{T},n}(t_i) = h_i^{\mathcal{T},n}(t_i')$.

Table 1. Two equivalent type structures.

| Type Structure $\mathcal{T}$ |
| --- |
| $T_1 = \{t_1, t_1', t_1''\}, \quad T_2 = \{t_2, t_2', t_2''\}$ |
| $b_1(t_1) = \frac{1}{2}(c, t_2) + \frac{1}{2}(d, t_2')$<br>$b_1(t_1') = \frac{1}{6}(c, t_2) + \frac{1}{3}(c, t_2'') + \frac{1}{2}(d, t_2')$<br>$b_1(t_1'') = \frac{1}{2}(c, t_2') + \frac{1}{2}(d, t_2'')$ |
| $b_2(t_2) = \frac{1}{4}(e, t_1) + \frac{1}{2}(e, t_1') + \frac{1}{4}(f, t_1'')$<br>$b_2(t_2') = \frac{1}{8}(e, t_1) + \frac{1}{8}(e, t_1') + \frac{3}{4}(f, t_1'')$<br>$b_2(t_2'') = \frac{3}{8}(e, t_1) + \frac{3}{8}(e, t_1') + \frac{1}{4}(f, t_1'')$ |
| **Type Structure $\mathcal{T}'$** |
| $R_1 = \{r_1, r_1', r_1''\}, \quad R_2 = \{r_2, r_2', r_2''\}$ |
| $\beta_1(r_1) = \frac{1}{4}(c, r_2) + \frac{1}{4}(c, r_2'') + \frac{1}{2}(d, r_2')$<br>$\beta_1(r_1') = \frac{1}{2}(c, r_2') + \frac{1}{8}(d, r_2) + \frac{3}{8}(d, r_2'')$<br>$\beta_1(r_1'') = \frac{1}{2}(c, r_2') + \frac{3}{8}(d, r_2) + \frac{1}{8}(d, r_2'')$ |
| $\beta_2(r_2) = \frac{1}{4}(e, r_1') + \frac{3}{4}(f, r_1)$<br>$\beta_2(r_2') = \frac{3}{4}(e, r_1') + \frac{1}{4}(f, r_1)$<br>$\beta_2(r_2'') = \frac{1}{8}(e, r_1') + \frac{1}{8}(e, r_1'') + \frac{3}{4}(f, r_1)$ |

## 3. Hierarchy and Type Morphisms

The literature has considered two concepts that map type structures into each other, type morphisms and hierarchy morphisms. Throughout the remainder of the paper, fix two type structures, $\mathcal{T} = (T_i, \Sigma_i^{\mathcal{T}}, b_i)_{i \in I}$ and $\mathcal{T}' = (T_i', \Sigma_i^{\mathcal{T}'}, b_i')_{i \in I}$ on $\mathcal{X}$. The functions that map types from $\mathcal{T}$ and $\mathcal{T}'$ into belief hierarchies are denoted by $h_i^{\mathcal{T}}$ and $h_i^{\mathcal{T}'}$, respectively.

**Definition 2.** *(Hierarchy morphism) For each player $i \in I$, let $\varphi_i$ be a function from $T_i$ to $T_i'$, such that for every type $t_i \in T_i$, $h_i^{\mathcal{T}'}(\varphi_i(t_i)) = h_i^{\mathcal{T}}(t_i)$. Then, $\varphi_i$ is a hierarchy morphism (from $\mathcal{T}$ to $\mathcal{T}'$). With some abuse of notation, we refer to the profile $(\varphi_i)_{i \in I}$ as a hierarchy morphism.*

Therefore, if there is a hierarchy morphism between $\mathcal{T}$ and $\mathcal{T}'$, then every type in $\mathcal{T}$ can be mapped into a type in $\mathcal{T}'$ in a way that preserves belief hierarchies. We say that the type structure $\mathcal{T}'$ contains $\mathcal{T}$ if, and only if, there is a hierarchy morphism from $\mathcal{T}$ to $\mathcal{T}'$.

Type morphisms are mappings between type structures that preserve beliefs.

**Definition 3.** *(Type morphism) For each player $i \in I$, let $\varphi_i$ be a function from $T_i$ to $T_i'$ that is measurable with respect to $\Sigma_i^{\mathcal{T}}$ and $\Sigma_i^{\mathcal{T}'}$.*[3] *Suppose that for each player $i$, type $t_i \in T_i$ and $E \in \Sigma_i \otimes \Sigma_{-i}^{\mathcal{T}'}$,*

$$b_i(t_i)(\{(x_i, t_{-i}) \in X_i \times T_{-i} \mid (x_i, \varphi_{-i}(t_{-i})) \in E\}) = b_i'(\varphi_i(t_i))(E). \tag{1}$$

*Then, $\varphi := (\varphi_i)_{i \in I}$ is a type morphism (from $\mathcal{T}$ to $\mathcal{T}'$).*

Heifetz and Samet [14] have shown that one type structure is contained in another whenever there is a type morphism from the former to the latter.

**Proposition 1.** *([14], Prop. 5.1) If $\varphi$ is a type morphism from $\mathcal{T}$ to $\mathcal{T}'$, then it is a hierarchy morphism. Therefore, if there is a type morphism from $\mathcal{T}$ to $\mathcal{T}'$, then $\mathcal{T}'$ contains $\mathcal{T}$.*

[3] That is, for each $E \in \Sigma_i^{\mathcal{T}'}$, we have $\{t_i \in T_i \mid \varphi_i(t_i) \in E\} \in \Sigma_i^{\mathcal{T}}$.

Unlike hierarchy morphisms, type morphisms do not make reference to belief hierarchies. Therefore, to check whether there is a type morphism from one type structure to another, we need to consider only the type structures. However, the condition that there be a type morphism from one type structure to another provides only a sufficient condition for the former to be contained in the latter. Indeed, Friedenberg and Meier [12] show that the condition is not necessary: there are type structures such that one is contained in the other, yet there is no type morphism between the two.

## 4. Generalized Type Morphisms

Type morphisms require beliefs to be preserved for every event in the types' $\sigma$-algebra. However, for two types to generate the same belief hierarchy, it suffices that their beliefs are preserved only for events that can be described in terms of players' belief hierarchies. We use this insight to define generalized type morphisms and show that a type structure contains another if and only if there is a generalized type morphism from the latter to the former.

The first step is to define the relevant $\sigma$-algebra. Mertens and Zamir ([13], p. 6) provide the relevant condition. We follow the presentation of Friedenberg and Meier [12].

**Definition 4.** *([12], Def. 5.1) Fix a type structure $\mathcal{T}$ and fix a sub-$\sigma$ algebra $\widetilde{\Sigma}_i^{\mathcal{T}} \subseteq \Sigma_i^{\mathcal{T}}$ for each player $i \in I$. Then, the product $\sigma$-algebra $\widetilde{\Sigma}^{\mathcal{T}}$ is closed under $\mathcal{T}$ if for each player $i$,*

$$\{t_i \in T_i \mid b_i(t_i)(E) \geq p\} \in \widetilde{\Sigma}_i^{\mathcal{T}}$$

*for all $E \in \Sigma_i \otimes \widetilde{\Sigma}_{-i}^{\mathcal{T}}$ and $p \in [0,1]$.*

The coarsest (sub-)$\sigma$ algebra that is closed under $\mathcal{T}$ is of special interest, and we denote it by $\mathcal{F}^{\mathcal{T}} = \bigotimes_{i \in I} \mathcal{F}_i^{\mathcal{T}}$. It is the intersection of all $\sigma$-algebras that are closed under $\mathcal{T}$.[4] The work in [13] uses this $\sigma$-algebra to define non-redundant type spaces, and [12] use it to characterize the condition under which a hierarchy morphism is a type morphism.

Friedenberg and Meier [12] provide a characterization of the $\sigma$-algebra $\mathcal{F}^{\mathcal{T}}$ in terms of the hierarchy mappings. Recall that $\sigma(h_i^{\mathcal{T}})$ is the $\sigma$-algebra on $T_i$ generated by the mapping $h_i^{\mathcal{T}}$. That is, $\sigma(h_i^{\mathcal{T}})$ is the coarsest $\sigma$-algebra that contains the sets:

$$\{t_i \in T_i \mid h_i^{\mathcal{T}}(t_i) \in E\} : \qquad E \subseteq H_i \text{ measurable.}$$

**Lemma 1.** *([12], Lemma 6.4) Let the product $\sigma$-algebra $\mathcal{F}^{\mathcal{T}}$ be the coarsest $\sigma$-algebra that is closed under $\mathcal{T}$. Then, for each player $i$, $\mathcal{F}_i^{\mathcal{T}} = \sigma(h_i^{\mathcal{T}})$.*

We are now ready to define generalized type morphisms.

**Definition 5.** *(Generalized type morphism) For each player $i \in I$, let $\varphi_i$ be a function from $T_i$ to $T_i'$ that is measurable with respect to $\Sigma_i^{\mathcal{T}}$ and $\mathcal{F}_i^{\mathcal{T}'}$.[5] Suppose that for each player $i$, type $t_i \in T_i$ and $E \in \Sigma_i \otimes \mathcal{F}_{-i}^{\mathcal{T}'}$,*

$$b_i(t_i)(\{(x_i, t_{-i}) \in X_i \times T_{-i} \mid (x_i, \varphi_{-i}(t_{-i})) \in E\}) = b_i'(\varphi_i(t_i))(E).$$

*Then, $\varphi := (\varphi_i)_{i \in I}$ is a generalized type morphism (from $\mathcal{T}$ to $\mathcal{T}'$).*

Note that a type morphism is always a generalized type morphism, but not vice versa. Like type morphisms, generalized type morphisms are defined using the language of type structures alone;

---

4 Since $\Sigma^{\mathcal{T}}$ is closed under $\mathcal{T}$ (by measurability of the belief maps $b_i$), the intersection is nonempty. It is easy to verify that the intersection is a $\sigma$-algebra.

5 That is, for each $E \in \mathcal{F}_i^{\mathcal{T}'}$, we have $\{t_i \in T_i \mid \varphi_i(t_i) \in E\} \in \Sigma_i^{\mathcal{T}}$.

the definition does not make reference to belief hierarchies. The difference between type morphisms and generalized type morphisms is that the former requires beliefs to be preserved for all events in the $\sigma$-algebra $\Sigma_i \otimes \Sigma_{-i}^{\mathcal{T}'}$ for player $i$, while the latter requires beliefs to be preserved only for events in the $\sigma$-algebra $\Sigma_i \otimes \mathcal{F}_{-i}^{\mathcal{T}'}$, and this $\sigma$-algebra is a coarsening of $\Sigma_i \otimes \Sigma_{-i}^{\mathcal{T}'}$ (Definition 4 and Lemma 1).

Our main result states that one structure is contained in another if and only if there is a generalized type morphism from the former to the latter.

**Theorem 1.** *A mapping $\varphi$ is a hierarchy morphism from $\mathcal{T}$ to $\mathcal{T}'$ if and only if it is a generalized type morphism from $\mathcal{T}$ to $\mathcal{T}'$. Hence, a type structure $\mathcal{T}'$ contains $\mathcal{T}$ if and only if there is a generalized type morphism from $\mathcal{T}$ to $\mathcal{T}'$.*

This result establishes an equivalence between generalized type morphisms and hierarchy morphisms. It thus provides a test that can be used to verify whether one type structure is contained in the other that does not refer to belief hierarchies.

While the characterization in Theorem 1 does not make reference to belief hierarchies, the result may not be easy to apply directly. The $\sigma$-algebras $\mathcal{F}_i^{\mathcal{T}}$ are defined as the intersection of $\sigma$-algebras that are closed under $\mathcal{T}$, and there can be (uncountably) many of those. We next define a simple procedure to construct this $\sigma$-algebra.

**Procedure 1.** *(Type partitioning procedure) Consider a multi-agent uncertainty space $\mathcal{X} = (X_i, \Sigma_i)_{i \in I}$ and a type structure $\mathcal{T} = (T_i, \Sigma_i^{\mathcal{T}}, b_i)_{i \in I}$ for $\mathcal{X}$.*

*Initial step: For every player i, let $\mathcal{S}_i^{\mathcal{T},0} = \{T_i, \varnothing\}$ be the trivial $\sigma$-algebra of his or her set of types $T_i$.*

*Inductive step: Suppose that $n \geq 1$ and that the sub-$\sigma$ algebra $\mathcal{S}_i^{\mathcal{T},n-1}$ on $T_i$ has been defined for every player i. Then, for every player i, let $\mathcal{S}_i^{\mathcal{T},n}$ be the coarsest $\sigma$-algebra that contains the sets:*

$$\{t_i \in T_i \mid b_i(t_i)(E) \geq p\}$$

*for all $E \in \Sigma_i \otimes \mathcal{S}_{-i}^{\mathcal{T},n-1}$ and all $p \in [0,1]$. Furthermore, let $\mathcal{S}_i^{\mathcal{T},\infty}$ be the $\sigma$-algebra generated by the union $\bigcup_n \mathcal{S}_i^{\mathcal{T},n}$.*

A simple inductive argument shows that $\mathcal{S}_i^{\mathcal{T},n}$ refines $\mathcal{S}_i^{\mathcal{T},n-1}$ for all players $i$ and all $n$; clearly, $\mathcal{S}_i^{\mathcal{T},\infty}$ refines $\mathcal{S}_i^{\mathcal{T},n}$ for any $n$. The next result shows that the type partitioning procedure delivers the $\sigma$-algebras that are generated by the hierarchy mappings.

**Proposition 2.** *Fix a type structure $\mathcal{T}$, and let $i \in I$. Then, $\mathcal{S}_i^{\mathcal{T},\infty} = \sigma(h_i^{\mathcal{T}})$ and $\mathcal{S}_i^{\mathcal{T},n} = \sigma(h_i^{\mathcal{T},n})$ for all $n \geq 1$. Therefore, $\mathcal{S}_i^{\mathcal{T},\infty} = \mathcal{F}_i^{\mathcal{T}}$.*

Hence, we can use the type partitioning procedure to construct the $\sigma$-algebras, which we need for our characterization result (Theorem 1). Heifetz and Samet [24] consider a similar procedure in the context of knowledge spaces to show that a universal space does not exist for that setting. The procedure also has connections with the construction in Kets [22] of type structures that describe the beliefs of players with a finite depth of reasoning. In the next section, we use Theorem 1 and the type partitioning procedure to characterize the types that generate the same belief hierarchies.

## 5. Characterizing Types with the Same Belief Hierarchy

We can use the results in the previous section to provide simple tests to determine whether two types, from the same type structure or from different structures, generate the same belief hierarchy. We assume in this section that $\mathcal{X}_i$ is countably generated: there is a countable collection of subsets $E_i^n \subseteq X_i$, $n = 1, 2, \ldots$, such that $\Sigma_i$ is the coarsest $\sigma$-algebra that contains these subsets. Examples of countably-generated $\sigma$-algebras include the discrete $\sigma$-algebra on a finite or countable set and the Borel

$\sigma$-algebra on a finite-dimensional Euclidean space. Recall that an atom of a $\sigma$-algebra $\Sigma$ on a set $Y$ is a set $a \in \Sigma$, such that $\Sigma$ does not contain a nonempty proper subset of $a$. That is, for any $a' \in \Sigma$, such that $a' \subseteq a$, we have $a' = a$ or $a' = \emptyset$.[6]

**Lemma 2.** *Let $i \in I$ and $n \geq 1$. The $\sigma$-algebras $\mathcal{S}_i^{\mathcal{T},n}$ and $\mathcal{S}_i^{\mathcal{T},\infty}$ are atomic. That is, for each $t_i \in T_i$, there are atoms $a_i^n(t_i)$ and $a_i^\infty(t_i)$ in $\mathcal{S}_i^{\mathcal{T},n}$ and $\mathcal{S}_i^{\mathcal{T},\infty}$, respectively, such that $t_i \in a_i^n(t_i)$ and $t_i \in a_i^\infty(t_i)$.*

This result motivates the name "type partitioning procedure": the procedure constructs a $\sigma$-algebra that partitions the type sets into atoms. Proposition 3 shows that these atoms contain precisely the types that generate the same higher-order beliefs.

**Proposition 3.** *For every player $i$, every $n \geq 1$ and every two types $t_i, t_i' \in T_i$, we have that*

*(a) for every $n \geq 0$, types $t_i$ and $t_i'$ generate the same $n$-th-order belief if and only if there is an atom $a_i^n \in \mathcal{S}_i^{\mathcal{T},n}$, such that $t_i, t_i' \in a_i^n$;*

*(b) types $t_i$ and $t_i'$ generate the same belief hierarchy if and only if there is an atom $a_i^\infty \in \mathcal{S}_i^{\mathcal{T},\infty}$, such that $t_i, t_i' \in a_i^\infty$.*

There is a connection between Proposition 3 and the work of Mertens and Zamir [13]. The work in [13] defines a type structure $\mathcal{T}$ to be non-redundant if for every player $i$, the $\sigma$-algebra $\mathcal{F}_i^{\mathcal{T}}$ separates types; see Liu ([17], Prop. 2) for a result that shows that this definition is equivalent to the requirement that there are no two types that generate the same belief hierarchy. Therefore, [13] already note the connection between the separating properties of $\mathcal{F}^{\mathcal{T}}$ and the question of whether types generate the same belief hierarchy. The contribution of Proposition 3 is to provide a simple procedure to construct the $\sigma$-algebra $\mathcal{F}^{\mathcal{T}}$ and to show that the separating sets can be taken to be atoms (as long as the $\sigma$-algebra on $X_i$ is countably generated).

Proposition 3 can also be used to verify whether two types from different type structures generate the same higher-order beliefs, by merging the two structures. Specifically, consider two different type structures, $\mathcal{T}^1 = (T_i^1, \Sigma_i^1, b_i^1)_{i \in I}$ and $\mathcal{T}^2 = (T_i^2, \Sigma_i^2, b_i^2)_{i \in I}$, for the same multi-agent uncertainty space $\mathcal{X} = (X_i, \Sigma_i)_{i \in I}$. To check whether two types $t_i^1 \in T_i^1$ and $t_i^2 \in T_i^2$ induce the same belief hierarchy, we can merge the two type structures into one large type structure and then run the type partitioning procedure on this larger type structure. That is, define the type structure $\mathcal{T}^* = (T_i^*, \Sigma_i^*, b_i^*)_{i \in I}$ as follows. For each player $i$, let $T_i^*$ be the union of $T_i^1$ and $T_i^2$ (possibly made disjoint by replacing $T_i^1$ or $T_i^2$ with a homeomorphic copy), and define the $\sigma$-algebra $\Sigma_i^*$ on $T_i^*$ by:

$$E \in \Sigma_i^* \text{ if and only if } E \cap T_i^1 \in \Sigma_i^1 \text{ and } E \cap T_i^2 \in \Sigma_i^2.$$

Furthermore, define $b_i^*$ by:

$$b_i^*(t_i) := \begin{cases} b_i^1(t_i), & \text{if } t_i \in T_i^1 \\ b_i^2(t_i), & \text{if } t_i \in T_i^2 \end{cases}$$

for all types $t_i \in T_i$.[7] Applying the type partitioning procedure on $\mathcal{T}^*$ gives a $\sigma$-algebra $\mathcal{S}_i^{*,\infty}$ on $T_i^*$ for each player $i$. If $t_i^1 \in T_i^1$ and $t_i^2 \in T_i^2$ belong to the same atom of $\mathcal{S}_i^{*,\infty}$, then $t_i^1$ and $t_i^2$ induce the same belief hierarchy. The converse also holds, and hence, we obtain the following result.

---

[6] Clearly, for any $y \in Y$, if there is an atom $a$ that contains $y$ (i.e., $y \in a$), then this atom is unique.

[7] This is with some abuse of notation, since $b_i^*$ is defined on $X_i \times T_{-i}^*$, while $b_i^1$ and $b_i^2$ are defined on $X_i \times T_{-i}^1$ and $X_i \times T_{-i}^2$, respectively. By defining the $\sigma$-algebra $\Sigma_j^{\mathcal{T}^*}$ on $T_j^*$ as above, the extension of $b_i^1$ and $b_i^2$ to the larger domain is well defined.

**Proposition 4.** *Consider two type structures $T^1 = (T_i^1, \Sigma_i^1, b_i^1)_{i \in I}$ and $T^2 = (T_i^2, \Sigma_i^2, b_i^2)_{i \in I}$. Let $T^* = (T_i^*, \Sigma_i^*, b_i^*)_{i \in I}$ be the large type structure defined above, obtained by merging the two type structures, and let $\mathcal{S}_i^{*,\infty}$, for a given player $i$, be the $\sigma$-algebra on $T_i^*$ generated by the type partitioning procedure. Then, two types $t_i^1 \in T_i^1$ and $t_i^2 \in T_i^2$ induce the same belief hierarchy, if and only if, $t_i^1$ and $t_i^2$ belong to the same atom of $S_i^{*,\infty}$.*

The type partitioning procedure is thus an easy and effective way to check whether two types, from possibly different type structures, generate the same belief hierarchy or not.

We expect our main results to apply more broadly. The proofs can easily be modified so that the main results extend to conditional probability systems in dynamic games [25], lexicographic beliefs [26], beliefs of players with a finite depth of reasoning [22,27] and the $\Delta$-hierarchies introduced by Ely and Peski [16].

## 6. Finite Type Structures

When type structures are finite, our results take on a particularly simple and intuitive form. Say that a type structure $\mathcal{T}$ is finite if the type set $T_i$ is finite for every player $i$. For finite type structures, we can replace $\sigma$-algebras by partitions.

We first define the type partitioning procedure for the case of finite type structures. A finite partition of a set $A$ is a finite collection $\mathcal{P} = \{P_1, \ldots, P_K\}$ of nonempty subsets $P_k \subseteq A$, such that $\bigcup_{k=1}^K P_k = A$ and $P_k \cap P_m = \emptyset$ whenever $k \neq m$. We refer to the sets $P_k$ as equivalence classes. For an element $a \in A$, we denote by $\mathcal{P}(a)$ the equivalence class $P_k$ to which $a$ belongs. The trivial partition of $A$ is the partition $\mathcal{P} = \{A\}$ containing a single set; the full set $A$. For two partitions $\mathcal{P}^1$ and $\mathcal{P}^2$ on $A$, we say that $\mathcal{P}^1$ is a refinement of $\mathcal{P}^2$ if for every set $P^1 \in \mathcal{P}^1$, there is a set $P^2 \in \mathcal{P}^2$, such that $P^1 \subseteq P^2$.

In the procedure, we recursively partition the set of types of an agent into equivalence classes, starting from the trivial partition and refining the previous partition with every step, until these partitions cannot be refined any further. We show that the equivalence classes produced in round $n$ contain exactly the types that induce the same $n$-th order belief. In particular, the equivalence classes produced at the end contain precisely those types that induce the same (infinite) belief hierarchy.

**Procedure 2** (Type partitioning procedure (finite type structures)). *Consider a multi-agent uncertainty space $\mathcal{X} = (X_i, \Sigma_i)_{i \in I}$ and a finite type structure $\mathcal{T} = (T_i, \Sigma_i^{\mathcal{T}}, b_i)_{i \in I}$ for $\mathcal{X}$.*

*Initial step: For every agent $i$, let $\mathcal{P}_i^0$ be the trivial partition of his or her set of types $T_i$.*

*Inductive step: Suppose that $n \geq 1$ and that the partitions $\mathcal{P}_i^{n-1}$ have been defined for every agent $i$. Then, for every agent $i$, and every $t_i \in T_i$,*

$$\mathcal{P}_i^n(t_i) = \{t_i' \in T_i \mid b_i(t_i')(E_i \times P_{-i}^{n-1}) = b_i(t_i)(E_i \times P_{-i}^{n-1}) \text{ for all } E_i \in \Sigma_i, \text{ and all } P_{-i}^{n-1} \in \mathcal{P}_{-i}^{n-1}\}. \tag{2}$$

*The procedure terminates at round $n$ whenever $\mathcal{P}_i^n = \mathcal{P}_i^{n-1}$ for every agent $i$.*

In this procedure, $\mathcal{P}_{-i}^{n-1}$ is the partition of the set $T_{-i}$ induced by the partitions $\mathcal{P}_j^{n-1}$ on $T_j$. Again, it follows from a simple inductive argument that $\mathcal{P}_i^n$ is a refinement of $\mathcal{P}_i^{n-1}$ for every player $i$ and every $n$. Note that if the total number of types, viz., $\sum_{i \in I} |T_i|$, equals $N$, then the procedure terminates in at most $N - |I|$ steps. We now illustrate the procedure by means of an example.

**Example 1.** *Consider the first type structure $\mathcal{T} = (T_1, T_2, \Sigma_1, \Sigma_2, b_1, b_2)$ from Table 1.*

*Initial step: Let $\mathcal{P}_1^0$ be the trivial partition of the set of types $T_1$, and let $\mathcal{P}_2^0$ be the trivial partition of the set of types $T_2$. That is,*

$$\mathcal{P}_1^0 = \{\{t_1, t_1', t_1''\}\} \text{ and } \mathcal{P}_2^0 = \{\{t_2, t_2', t_2''\}\}.$$

*Round 1: By Equation* (2),

$$\begin{aligned}\mathcal{P}_1^1(t_1) = \{\tau_1 \in T_1 \mid \\ b_1(\tau_1)(\{c\} \times T_2) = b_1(t_1)(\{c\} \times T_2) = \tfrac{1}{2}, \\ b_1(\tau_1)(\{d\} \times T_2) = b_1(t_1)(\{d\} \times T_2) = \tfrac{1}{2}\} \\ = \{t_1, t_1', t_1''\},\end{aligned}$$

*which implies that:*

$$\mathcal{P}_1^1 = \mathcal{P}_1^0 = \{\{t_1, t_1', t_1''\}\}.$$

*At the same time,*

$$\begin{aligned}\mathcal{P}_2^1(t_2) = \{\tau_2 \in T_2 \mid \\ b_2(\tau_2)(\{e\} \times T_1) = b_2(t_2)(\{e\} \times T_1) = \tfrac{3}{4}, \\ b_2(\tau_2)(\{f\} \times T_1) = b_2(t_2)(\{f\} \times T_1) = \tfrac{1}{4}\} \\ = \{t_2, t_2''\}\end{aligned}$$

*which implies that* $\mathcal{P}_2^1(t_2') = \{t_2'\}$, *and hence:*

$$\mathcal{P}_2^1 = \{\{t_2, t_2''\}, \{t_2'\}\}.$$

*Round 2: By Equation* (2),

$$\begin{aligned}\mathcal{P}_1^2(t_1) = \{\tau_1 \in T_1 \mid \\ b_1(\tau_1)(\{c\} \times \{t_2, t_2''\}) = b_1(t_1)(\{c\} \times \{t_2, t_2''\}) = \tfrac{1}{2}, \\ b_1(\tau_1)(\{c\} \times \{t_2'\}) = b_1(t_1)(\{c\} \times \{t_2'\}) = 0, \\ b_1(\tau_1)(\{d\} \times \{t_2, t_2''\}) = b_1(t_1)(\{d\} \times \{t_2, t_2''\}) = 0, \\ b_1(\tau_1)(\{d\} \times \{t_2'\}) = b_1(t_1)(\{d\} \times \{t_2'\}) = \tfrac{1}{2}\} \\ = \{t_1, t_1'\},\end{aligned}$$

*which implies that* $\mathcal{P}_1^2(t_1'') = \{t_1''\}$, *and hence:*

$$\mathcal{P}_1^2 = \{\{t_1, t_1'\}, \{t_1''\}\}.$$

*Since* $\mathcal{P}_1^1 = \mathcal{P}_1^0$, *we may immediately conclude that:*

$$\mathcal{P}_2^2 = \mathcal{P}_2^1 = \{\{t_2, t_2''\}, \{t_2'\}\}.$$

*Round 3: As* $\mathcal{P}_2^2 = \mathcal{P}_2^1$, *we may immediately conclude that:*

$$\mathcal{P}_1^3 = \mathcal{P}_1^2 = \{\{t_1, t_1'\}, \{t_1''\}\}.$$

*By Equation* (2)*,*

$$\begin{aligned}\mathcal{P}_2^3(t_2) &= \{\tau_2 \in T_2 \mid \\ & b_2(\tau_2)(\{e\} \times \{t_1, t_1'\}) = b_2(t_2)(\{e\} \times \{t_1, t_1'\}) = \tfrac{3}{4}, \\ & b_2(\tau_2)(\{e\} \times \{t_1''\}) = b_2(t_2)(\{e\} \times \{t_1''\}) = 0, \\ & b_2(\tau_2)(\{f\} \times \{t_1, t_1'\}) = b_2(t_2)(\{f\} \times \{t_1, t_1'\}) = 0, \\ & b_2(\tau_2)(\{f\} \times \{t_1''\}) = b_2(t_2)(\{f\} \times \{t_1''\}) = \tfrac{1}{4}\} \\ &= \{t_2, t_2''\},\end{aligned}$$

*which implies that* $\mathcal{P}_2^3(t_2') = \{t_2'\}$*, and hence,*

$$\mathcal{P}_2^3 = \{\{t_2, t_2''\}, \{t_2'\}\} = \mathcal{P}_2^2.$$

*As* $\mathcal{P}_1^3 = \mathcal{P}_1^2$ *and* $\mathcal{P}_2^3 = \mathcal{P}_2^2$*, the procedure terminates at Round 3. The final partitions of the types are thus given by:*

$$\mathcal{P}_1^\infty = \{\{t_1, t_1'\}, \{t_1''\}\} \text{ and } \mathcal{P}_2^\infty = \{\{t_2, t_2''\}, \{t_2'\}\}.$$

*The reader may check that all types within the same equivalence class indeed induce the same belief hierarchy. That is,* $t_1$ *induces the same belief hierarchy as* $t_1'$*, and* $t_2$ *induces the same belief hierarchy as* $t_2''$*. Moreover,* $t_1$ *and* $t_1''$ *induce different belief hierarchies, and so do* $t_2$ *and* $t_2'$*.*

Our characterization result for the case of finite type structures states that the type partitioning procedure characterizes precisely those groups of types that induce the same belief hierarchy. We actually prove a little more: we show that the partitions generated in round $n$ of the procedure characterize exactly those types that yield the same $n$-th order belief.

**Proposition 5** (Characterization result (finite type structures))**.** *Consider a finite type structure* $\mathcal{T} = (T_i, \Sigma_i, b_i)_{i \in I}$*, where* $\Sigma_i$ *is the discrete* $\sigma$*-algebra on* $T_i$ *for every player* $i$*. For every agent* $i$*, every* $n \geq 1$ *and every two types* $t_i, t_i' \in T_i$*, we have that*

*(a)* $h_i^{\mathcal{T},n}(t_i) = h_i^{\mathcal{T},n}(t_i')$*, if and only if,* $t_i' \in \mathcal{P}_i^n(t_i)$*;*

*(b)* $h_i^{\mathcal{T}}(t_i) = h_i^{\mathcal{T}}(t_i')$*, if and only if,* $t_i' \in \mathcal{P}_i^\infty(t_i)$*.*

The proof follows directly from Proposition 3 and is therefore omitted. As before, this result can be used to verify whether two types from different type structures generate the same belief hierarchies, by first merging the two type structures and then running the type partitioning procedure on this "large" type structure.

**Acknowledgments:** This paper is a substantially revised version of an earlier paper with the same title by Andrés Perea. We would like to thank Pierpaolo Battigalli, Eddie Dekel, Amanda Friedenberg, Christian Nauerz, Miklós Pintér, Elias Tsakas and two anonymous reviewers for very useful comments.

**Author Contributions:** Both authors have contributed equally to this paper.

**Conflicts of Interest:** The authors declare no conflict of interest.

## Appendix A. Proofs

### *Appendix A.1. Proof of Theorem 1*

By definition, $\mathcal{T}'$ contains $\mathcal{T}$ if and only if there is a hierarchy morphism from $\mathcal{T}$ to $\mathcal{T}'$. Therefore, it suffices to show that every generalized type morphism is a hierarchy morphism and vice versa.

Appendix A.1.1. Every Hierarchy Morphism Is a Generalized Type Morphism

To show that every hierarchy morphism is a generalized type morphism, we need to show two things. First, we need to show that any hierarchy morphism is measurable with respect to the appropriate $\sigma$-algebra. Second, we need to show that beliefs are preserved for the relevant events.

Let us start with the measurability condition. Suppose $\varphi$ is a hierarchy morphism. Let $i \in I$ and $E \in \mathcal{F}_i^{\mathcal{T}'}$. We need to show that:

$$\{t_i \in T_i \mid \varphi_i(t_i) \in E\} \in \Sigma_i^{\mathcal{T}}.$$

Recall that $\mathcal{F}_i^{\mathcal{T}'} = \sigma(h_i^{\mathcal{T}'})$ (Lemma 1). Therefore, there is a measurable subset $B$ of the set $H_i$ of belief hierarchies, such that:

$$E = \{t_i' \in T_i' \mid h_i^{\mathcal{T}'}(t_i') \in B\}.$$

Hence,

$$\begin{aligned} \{t_i \in T_i \mid \varphi_i(t_i) \in E\} &= \{t_i \in T_i \mid h_i^{\mathcal{T}'}(\varphi_i(t_i)) \in B\} \\ &= \{t_i \in T_i \mid h_i^{\mathcal{T}}(t_i) \in B\}, \end{aligned}$$

where the second equality follows from the assumption that $\varphi$ is a hierarchy morphism. By Lemma 1, we have:

$$\{t_i \in T_i \mid h_i^{\mathcal{T}}(t_i) \in B\} \in \mathcal{F}_i^{\mathcal{T}}.$$

Since $\Sigma_i^{\mathcal{T}} \supseteq \mathcal{F}_i^{\mathcal{T}}$ (Definition 4 and Lemma 1), the result follows.

We next ask whether hierarchy morphisms preserve beliefs for the relevant events. Again, let $\varphi$ be a hierarchy morphism. Let $i \in I$, $t_i \in T_i$ and $E' \in \Sigma_i \otimes \mathcal{F}_{-i}^{\mathcal{T}'}$. We need to show that:

$$b_i(t_i) \circ (\mathrm{Id}_{X_i}, \varphi_{-i})^{-1}(E') = b_i'(\varphi_i(t_i))(E'),$$

where $\mathrm{Id}_{X_i}$ is the identity function on $X_i$ and where we have used the notation $(f_1, \ldots, f_m)$ for the induced function that maps $(x_1, \ldots, x_m)$ into $(f_1(x_1), \ldots, f_m(x_m))$, so that $b_i(t_i) \circ (\mathrm{Id}_{X_i}, \varphi_{-i})^{-1}$ is the image measure induced by $(\mathrm{Id}_{X_i}, \varphi_{-i})$. By a similar argument as before, there is a measurable subset $B'$ of the set $X_i \times H_{-i}$, such that:

$$E' = \{(x_i, t_{-i}') \in X_i \times T_{-i}' \mid (x_i, h_{-i}^{\mathcal{T}'}(t_{-i}')) \in B'\}.$$

If $E'$ is an element of $\Sigma_i \otimes \bigotimes_{j \neq i} \{T_j', \emptyset\}$, then the result follows directly from the definitions. Therefore, suppose $E' \notin \Sigma_i \otimes \bigotimes_{j \neq i} \{T_j', \emptyset\}$. Then, for every $n \geq 1$, define:

$$\begin{aligned} B^n := \{(x_i, \mu_{-i}^1, \ldots, \mu_{-i}^n) \in X_i \times H_{-i}^n \mid (x_i, \mu_{-i}^1, \ldots, \mu_{-i}^n, \mu_{-i}^{n+1}, \ldots) \in B' \\ \text{for some } (\mu_{-i}^{n+1}, \mu_{-i}^{n+2}, \ldots)\} \end{aligned}$$

and:

$$E^n := \{(x_i, t_{-i}') \in X_i \times T_{-i}' \mid (x_i, h_{-i}^{\mathcal{T}', n}(t_{-i}')) \in B^n\}.$$

Then, $E^n \supseteq E'$ and $E^n \downarrow E'$. Furthermore, we have $E^n \in \Sigma_i \otimes \bigotimes_{j\neq i}\sigma(h_j^{\mathcal{T}',n})$, and thus, $E^n \in \Sigma_i \otimes \mathcal{F}_{-i}^{\mathcal{T}'}$ (Lemma 1). For every $n$,

$$\begin{aligned} b_i(t_i) \circ (\mathrm{Id}_{X_i}, \varphi_{-i})^{-1}(E^{n-1}) &= b_i(t_i) \circ (\mathrm{Id}_{X_i}, \varphi_{-i})^{-1} \circ (\mathrm{Id}_{X_i}, h_{-i}^{\mathcal{T}',n-1})^{-1}(B^{n-1}) \\ &= b_i(t_i) \circ (\mathrm{Id}_{X_i}, h_{-i}^{\mathcal{T}',n-1} \circ \varphi_{-i})^{-1}(B^{n-1}) \\ &= \mu_i^{\mathcal{T},n}(t_i)(B^{n-1}) \\ &= b_i'(\varphi(t_i)) \circ (\mathrm{Id}_{X_i}, h_{-i}^{\mathcal{T}',n-1})^{-1}(B^{n-1}) \\ &= b_i'(\varphi(t_i))(E^{n-1}), \end{aligned}$$

where the penultimate equality uses the definition of a hierarchy morphism. By the continuity of the probability measures $b_i(t_i)$ and $b_i'(\varphi_i(t_i))$ (e.g., [28], Thm. 10.8), we have $b_i(t_i) \circ (\mathrm{Id}_{X_i}, \varphi_{-i})^{-1}(E') = b_i'(\varphi_i(t_i))(E')$, and the result follows.

Appendix A.1.2. Every Generalized Type Morphism Is a Hierarchy Morphism

For the other direction, that is to show that every generalized type morphism is a hierarchy morphism, suppose that $\varphi$ is a generalized type morphism from $\mathcal{T} = (T_i, \Sigma_i^{\mathcal{T}}, b_i)_{i\in I}$ to $\mathcal{T}' = (T_i', \Sigma_i^{\mathcal{T}'}, b_i')_{i\in I}$. We can use an inductive argument to show that it is a hierarchy morphism. Let $i \in I$ and $t_i \in T_i$. Then, for all $E \in \Sigma_i$,

$$\begin{aligned} \mu_i^{\mathcal{T}',1}(\varphi_i(t_i))(E) &= b_i'(\varphi_i(t_i))(E \times T_{-i}') \\ &= b_i(t_i)(E \times T_{-i}) \\ &= \mu_i^{\mathcal{T},1}(t_i), \end{aligned}$$

where the first and the last equality use the definition of a first-order belief induced by a type and the second uses the definition of a generalized type morphism. Therefore, $\mu_i^{\mathcal{T},1}(t_i) = \mu_i^{\mathcal{T}',1}(\varphi_i(t_i))$, and thus, $h_i^{\mathcal{T},1}(t_i) = h_i^{\mathcal{T}',1}(\varphi_i(t_i))$ for each player $i$ and every type $t_i \in T_i$.

For $n > 1$, suppose that for each player $i$ and every type $t_i \in T_i$, we have $h_i^{\mathcal{T},n-1}(t_i) = h_i^{\mathcal{T}',n-1}(\varphi_i(t_i))$. We will use the notation $(f_1, \ldots, f_m)$ for the induced function that maps $(x_1, \ldots, x_m)$ into $(f_1(x_1), \ldots, f_m(x_m))$, so that $\mu \circ (f_1, \ldots, f_m)^{-1}$ is the image measure induced by a probability measure $\mu$ and $(f_1, \ldots, f_m)$.

Let $E$ be a measurable subset of $X_i \times H_{-i}^n$. By Lemma 1, we have $\mathcal{F}_j^{\mathcal{T}'} = \sigma(h_j^{\mathcal{T}'})$; and clearly, $\sigma(h_j^{\mathcal{T}'}) \supseteq \sigma(h_j^{\mathcal{T}',n-1})$. Therefore, if we write $\mathrm{Id}_{X_i}$ for the identity function on $X_i$, we have $(\mathrm{Id}_{X_i}, h_{-i}^{\mathcal{T}',n-1})^{-1}(E) \in \Sigma_i \otimes \mathcal{F}_{-i}^{\mathcal{T}'}$. Then, for every player $i$ and type $t_i \in T_i$,

$$\begin{aligned} \mu_i^{\mathcal{T}',n}(\varphi_i(t_i))(E) &= b_i'(\varphi_i(t_i)) \circ (\mathrm{Id}_{X_i}, h_{-i}^{\mathcal{T}',n-1})^{-1}(E) \\ &= b_i(t_i) \circ (\mathrm{Id}_{X_i}, h_{-i}^{\mathcal{T}',n-1})^{-1} \circ (\mathrm{Id}_{X_i}, \varphi_{-i})^{-1}(E) \\ &= b_i(t_i) \circ (\mathrm{Id}_{X_i}, \varphi_{-i} \circ h_{-i}^{\mathcal{T}',n-1})^{-1}(E) \\ &= b_i(t_i) \circ (\mathrm{Id}_{X_i}, h_{-i}^{\mathcal{T},n-1})^{-1}(E) \\ &= \mu_i^{\mathcal{T},n}(t_i)(E), \end{aligned}$$

where the first equality uses the definition of an $n$-th-order belief, the second uses the definition of a generalized type morphism, the third uses the definition of the composition operator, the fourth uses the induction hypothesis and the fifth uses the definition of an $n$-th-order belief again. Conclude that $\mu_i^{\mathcal{T},n}(t_i) = \mu_i^{\mathcal{T}',n}(\varphi_i(t_i))$ and thus $h_i^{\mathcal{T},n}(t_i) = h_i^{\mathcal{T}',n}(\varphi_i(t_i))$ for each player $i$ and every type $t_i \in T_i$.

Therefore, for each player $i \in I$ and each type $t_i \in T_i$, we have $h_i^{\mathcal{T}}(t_i) = h_i^{\mathcal{T}'}(\varphi_i(t_i))$, which shows that $\varphi$ is a hierarchy morphism. □

*Appendix A.2. Proof of Proposition 2*

Let $i \in I$. It will be convenient to define $h_i^{\mathcal{T},0}$ to be the trivial function from $T_i$ into some singleton $\{v_i\}$. Therefore, the $\sigma$-algebra $\sigma(h_i^{\mathcal{T},0})$ generated by $h_i^{\mathcal{T},0}$ is just the trivial $\sigma$-algebra $\mathcal{S}_i^0 = \{T_i, \emptyset\}$. Next, consider $n = 1$. Fix player $i \in I$. By definition, $\sigma(h_i^{\mathcal{T},1})$ is the coarsest $\sigma$-algebra that contains the sets:

$$\{t_i \in T_i \mid h_i^{\mathcal{T},1}(t_i) \in E\}: \quad E \subseteq H_i^1 \text{ measurable.}$$

It suffices to restrict attention to the generating sets $E$ of the $\sigma$-algebra on $H_i^1 = \Delta(X_i)$ (e.g., [28]). Therefore, $\sigma(h_i^{\mathcal{T},1})$ is the coarsest $\sigma$-algebra that contains the sets:

$$\{t_i \in T_i \mid h_i^{\mathcal{T},1}(t_i) \in E\}$$

where $E$ is of the form $\{\mu \in \Delta(X_i) \mid \mu(F) \geq p\}$ for $F \in \Sigma_i$ and $p \in [0,1]$. Using that for each type $t_i$, $h_i^{\mathcal{T},1}(t_i)$ is the marginal on $X_i$ of $b_i(t_i)$, we have that $\sigma(h_i^{\mathcal{T},1})$ is the coarsest $\sigma$-algebra that contains the sets:

$$\{t_i \in T_i \mid b_i(t_i)(E) \geq p\}: \quad E \in \Sigma_i \otimes \mathcal{S}_{-i}^0, p \in [0,1].$$

That is, $\sigma(h_i^{\mathcal{T},1}) = \mathcal{S}_i^1$. In particular, $h_i^{\mathcal{T},1}$ is measurable with respect to $\mathcal{S}_i^{\mathcal{T},1}$.

For $n > 1$, suppose, inductively, that for each player $i \in I$, $\sigma(h_i^{\mathcal{T},n-1}) = \mathcal{S}_i^{\mathcal{T},n-1}$, so that $h_i^{\mathcal{T},n-1}$ is measurable with respect to $\mathcal{S}_i^{\mathcal{T},n-1}$. Fix $i \in I$. By definition, $\sigma(h_i^{\mathcal{T},n})$ is the coarsest $\sigma$-algebra that contains the sets in $\sigma(h_i^{\mathcal{T},n-1})$ and the sets:

$$\{t_i \in T_i \mid \mu_i^{\mathcal{T},n}(t_i) \in E\}: \quad E \subseteq \Delta(X_i \times H_{-i}^{n-1}) \text{ measurable.}$$

Again, it suffices to consider the generating sets of the $\sigma$-algebra on $\Delta(X_i \times H_{-i}^{n-1})$. Hence, $\sigma(h_i^{\mathcal{T},n})$ is the coarsest $\sigma$-algebra that contains the sets:

$$\{t_i \in T_i \mid \mu_i^{\mathcal{T},n}(t_i)(F) \geq p\}: \qquad F \subseteq X_i \times H_{-i}^{n-1} \text{ measurable and } p \in [0,1].$$

(Note that this includes the generating sets of $\sigma(h_i^{\mathcal{T},n-1})$, given that the $n$-th-order belief induced by a type is consistent with its $(n-1)$-th-order belief.) Using the definition of $\mu_i^{\mathcal{T},n}$ and the induction assumption that $\mathcal{S}_{-i}^{\mathcal{T},n-1} = \sigma(h_{-i}^{\mathcal{T},n-1})$, we see that $\sigma(h_i^{\mathcal{T},n})$ is the coarsest $\sigma$-algebra on $T_i$ that contains the sets:

$$\{t_i \in T_i \mid b_i(t_i)(F) \geq p\}: \qquad F \in \Sigma_i \otimes \mathcal{S}_{-i}^{\mathcal{T},n-1}, p \in [0,1].$$

That is, $\sigma(h_i^{\mathcal{T},n}) = \mathcal{S}_i^{\mathcal{T},n}$, and $h_i^{\mathcal{T},n}$ is measurable with respect to $\mathcal{S}_i^{\mathcal{T},n}$.

Therefore, for each player $i$ and $n \geq 1$, $\sigma(h_i^{\mathcal{T},n}) = \mathcal{S}_i^{\mathcal{T},n}$. It follows immediately that $\sigma(h_i^{\mathcal{T}})$, as the $\sigma$-algebra on $T_i$ generated by the "cylinders" $\sigma(h_i^{\mathcal{T},n})$, is equal to $\mathcal{S}_i^{\mathcal{T},\infty}$. □

*Appendix A.3. Proof of Lemma 2*

Let $i \in I$. Recall that $\mathcal{X}_i$ is countably generated, that is there is a countable subset $\mathcal{D}_i^0$ of $\Sigma_i$, such that $\mathcal{D}_i^0$ generates $\Sigma_i$ (i.e., $\Sigma_i$ is the coarsest $\sigma$-algebra that contains $\mathcal{D}_i^0$). Throughout this proof, we write $\sigma(D)$ for the $\sigma$-algebra on a set $Y$ generated by a collection $D$ of subsets of $Y$.

The following result says that a countable collection of subsets of a set $Y$ generates a countable algebra on $Y$. For a collection $D$ of subsets of a set $Y$, denote the algebra generated by $D$ by $A(D)$. Therefore, $A(D)$ is the coarsest algebra on $Y$ that contains $D$.

**Lemma A1.** *Let $D$ be a countable collection of subsets of a set $Y$. Then, the algebra $A(D)$ generated by $D$ is countable.*

**Proof.** We can construct the algebra generated by $D$. Denote the elements of $D$ by $D_\lambda$, $\lambda \in \Lambda$, where $\Lambda$ is a countable index set. Define:

$$A(D) = \Big\{ \bigcup_{m \in F} \bigcap_{\ell \in L_m} D_\ell, F \text{ a finite subset of } \mathbb{N}, L_m \text{ a finite subset of } \Lambda, \\ D_\ell = A_k \text{ or } D_\ell = A_k^c \text{ for some } k \Big\}, \tag{A1}$$

where $E^c$ is the complement of a set $E$. That is, $A(D)$ is the collection of finite unions of finite intersection of elements of $D$ and their complements. We check that $A(D)$ is an algebra. Clearly, $A(D)$ is nonempty (it contains $D$) and $\emptyset \in A(D)$. We next show that $A(D)$ is closed under finite intersections. Let:

$$A_1 := \bigcup_{m_1 \in F_1} \bigcap_{\ell \in L^1_{m_1}} D_\ell, \quad A_2 := \bigcup_{m_2 \in F_2} \bigcap_{\ell \in L^2_{m_2}} D_\ell,$$

be elements of $A(D)$. Then,

$$A_1 \cap A_2 = \bigcup_{(m_1, m_2) \in F_1 \times F_2} \bigcap_{\ell_1 \in L^1_{m_1}} \bigcap_{\ell_2 \in L^2_{m_2}} D_{\ell_1} \cap D_{\ell_2}.$$

Clearly, $F_1 \times F_2$ is finite and so are the sets $L^1_m$ and $L^2_m$. We can thus rewrite $A_1 \cap A_2$ so that it is of the form as the elements in (A1). We can likewise show that $A(D)$ is closed under complements: let $A := \bigcup_{m \in F} \bigcap_{\ell \in L_m} D_\ell \in A(D)$, so that $A^c = \bigcap_{m \in F} \bigcup_{\ell \in L_m} D^c_\ell$; then, since $\bigcup_{\ell \in L_m} D^c_\ell \in A(D)$ for every $m$, we have $A^c \in A(D)$. Therefore, $A(D)$ is an algebra that contains $D$, and it is in fact the coarsest such one (by construction, any proper subset of $A(D)$ does not contain all finite intersections of the sets in $D$ and their complements). As $D$ is countable, so is the collection of the elements in $D$ and their complements; the collections of the finite intersections of such sets are also countable. Hence, $A(D)$ is countable. □

Note that for any $p \in [0,1]$, the set $\{t_i \in T_i : b_i(t_i)(E) \geq p\}$ can be written as the countable intersection of sets $\{t_i \in T_i : b_i(t_i)(E) \geq p_\ell\}$ for some rational $p_\ell$, $\ell = 1, 2, \ldots$. Therefore, by Proposition 2, the $\sigma$-algebra $\sigma(h_i^{\mathcal{T},n})$, $n = 1, 2, \ldots$, on the type set $T_i$, $i \in I$, is the coarsest $\sigma$-algebra that contains the sets:

$$\{t_i \in T_i : b_i(t_i)(E) \geq p\} : \quad E \in \Sigma_i \otimes \bigotimes_{j \neq i} \sigma(h_j^{\mathcal{T},n-1}), p \in \mathbb{Q}$$

We are now ready to prove Lemma 2. Fix $i \in I$. By Lemma A1, the set $\mathcal{D}_i^0$ generates a countable algebra $A(\mathcal{D}_i^0)$ on $X_i$. Then, by Proposition 2 and by Lemma 4.5 of Heifetz and Samet [14], we have that the $\sigma$-algebra $\sigma(h_i^{\mathcal{T},1})$ is generated by the sets:

$$\{t_i \in T_i : b_i(t_i)(E) \geq p\} : \qquad E \in \mathcal{D}_i^0 \otimes \bigotimes_{j \neq i} \sigma(h_j^{\mathcal{T},0}), p \in \mathbb{Q}.$$

Denote this collection of these sets by $\mathcal{D}_i^1$, so that $\sigma(h_i^{\mathcal{T},1}) = \sigma(\mathcal{D}_i^1)$; clearly, $\mathcal{D}_i^1$ is countable and $A(\mathcal{D}_i^1) \subseteq \sigma(h_i^{\mathcal{T},1})$ (so that $\sigma(h_i^{\mathcal{T},1}) = \sigma(A(\mathcal{D}_i^1))$).

For $m > 1$, suppose that for every $i \in I$, the $\sigma$-algebra $\sigma(h_i^{\mathcal{T},m-1})$ on $T_i$ is generated by a countable collection $\mathcal{D}_i^{m-1}$ of subsets of $T_i$, such that $A(\mathcal{D}_i^{m-1}) \subseteq \sigma(h_i^{\mathcal{T},m-1})$. Fix $i \in I$. By Proposition 2 and Lemma 4.5 of Heifetz and Samet [14], the $\sigma$-algebra $\sigma(h_i^{\mathcal{T},m})$ is generated by the sets:

$$\{t_i \in T_i : b_i(t_i)(E) \geq p\} : \qquad E \in \mathcal{D}_i^0 \otimes \bigotimes_{j \neq i} A(\mathcal{D}_j^{m-1}), p \in \mathbb{Q}.$$

Denote this collection of these sets by $\mathcal{D}_i^m$; as before, $\mathcal{D}_i^m$ is clearly countable and $A(\mathcal{D}_i^m) \subseteq \sigma(h_i^{\mathcal{T},m})$. Again, we have $\sigma(h_i^{\mathcal{T},m}) = \sigma(\mathcal{D}_i^m) = \sigma(A(\mathcal{D}_i^m))$.

Therefore, we have shown that for every $i \in I$ and $m = 1, 2, \ldots$, the $\sigma$-algebra $\sigma(h_i^{\mathcal{T},m})$ is generated by a countable collection $\mathcal{D}_i^m$ of subsets of $T_i$. The $\sigma$-algebra $\sigma(h_i^{\mathcal{T}})$ is generated by the algebra $\bigcup_m \sigma(h_i^{\mathcal{T},m}) = \bigcup_m \sigma(\mathcal{D}_i^m)$ or, equivalently, by the union $\bigcup_m \mathcal{D}_i^m$ (Proposition 2). Since the latter set, as the countable union of countable sets, is countable, the $\sigma$-algebra $\sigma(h_i^{\mathcal{T}})$ is countably generated.

It now follows from Theorem V.2.1 of Parthasarathy [29] that for each player $i$, the $\sigma$-algebras $\sigma(h_i^{\mathcal{T},m})$, $m = 1, 2, \ldots$ are atomic in the sense that for each $t_i \in T_i$, there is a unique atom $a_{t_i}^m$ in $\sigma(h_i^{\mathcal{T},m})$ containing $t_i$; the analogous statement holds for $\sigma(h_i^{\mathcal{T}})$. □

*Appendix A.4. Proof of Proposition 3*

Fix a player $i \in I$. By Proposition 2, we have $\mathcal{S}_i^{\mathcal{T},\infty} = \sigma(h_i^{\mathcal{T}})$ and $\mathcal{S}_i^{\mathcal{T},n} = \sigma(h_i^{\mathcal{T},n})$ for each $n \geq 1$. By Lemma 2, the $\sigma$-algebras $\sigma(h_i^{\mathcal{T},\infty})$ and $\sigma(h_i^{\mathcal{T},n})$ are atomic for every $n \geq 1$. Let $n \geq 1$. Let $t_i, t_i' \in T_i$. Since $\sigma(h_i^{\mathcal{T},n})$ is atomic, there exist a unique atom $a_i^n(t_i) \in \sigma(h_i^{\mathcal{T},n})$, such that $t_i \in a_i^n(t_i)$, and a unique atom $a_i^n(t_i') \in \sigma(h_i^{\mathcal{T},n})$, such that $t_i' \in a_i^n(t_i')$. Suppose $h_i^{\mathcal{T},n}(t_i) = h_i^{\mathcal{T},n}(t_i')$. Then, for every generating set $E$ of the $\sigma$-algebra $\sigma(h_i^{\mathcal{T},n})$, either $t_i, t_i' \in E$ or $t_i, t_i' \notin E$. Therefore, $a_i^n(t_i) = a_i^n(t_i')$. Suppose $h_i^{\mathcal{T}}(t_i) \neq h_i^{\mathcal{T}}(t_i')$. Then, there is a generating set $E$ of $\sigma(h_i^{\mathcal{T},n})$ that separates $t_i$ and $t_i'$, that is, $t_i \in E$, $t_i' \notin E$. Therefore, $a_i^n(t_i) \neq a_i^n(t_i')$. The proof of the claim that there is a unique atom $a_i^\infty$ in $\sigma(h_i^{\mathcal{T}})$ that contains both $t_i$ and $t_i'$ if and only if $h_i^{\mathcal{T}}(t_i) = h_i^{\mathcal{T}}(t_i')$ is analogous and therefore omitted. □

## References


1. Carlsson, H.; van Damme, E. Global Games and Equilibrium Selection. *Econometrica* **1993**, *61*, 989–1018.
2. Feinberg, Y.; Skrzypacz, A. Uncertainty about uncertainty and delay in bargaining. *Econometrica* **2005**, *73*, 69–91.
3. Friedenberg, A. *Bargaining under Strategic Uncertainty*; Working Paper; Arizona State University: Tempe, AZ, USA, 2014.
4. Geanakoplos, J.D.; Polemarchakis, H.M. We can't disagree forever. *J. Econ. Theory* **1982**, *28*, 192–200.
5. Neeman, Z. The relevance of private information in mechanism design. *J. Econ. Theory* **2004**, *117*, 55–77.
6. Brandenburger, A.; Dekel, E. Rationalizability and Correlated Equilibria. *Econometrica* **1987**, *55*, 1391–1402.
7. Tan, T.; Werlang, S. The Bayesian foundations of solution concepts in games. *J. Econ. Theory* **1988**, *45*, 370–391.
8. Aumann, R.; Brandenburger, A. Epistemic conditions for Nash equilibrium. *Econometrica* **1995**, *63*, 1161–1180.
9. Perea, A. A one-person doxastic characterization of Nash strategies. *Synthese* **2007**, *158*, 251–271.
10. Battigalli, P.; Siniscalchi, M. Strong belief and forward-induction reasoning. *J. Econ. Theory* **2002**, *106*, 356–391.
11. Harsanyi, J.C. Games on incomplete information played by Bayesian players. Part I. *Manag. Sci.* **1967**, *14*, 159–182.
12. Friedenberg, A.; Meier, M. On the relationship between hierarchy and type morphisms. *Econ. Theory* **2011**, *46*, 377–399.
13. Mertens, J.F.; Zamir, S. Formulation of Bayesian analysis for games with incomplete information. *Int. J. Game Theory* **1985**, *14*, 1–29.
14. Heifetz, A.; Samet, D. Topology-Free Typology of Beliefs. *J. Econ. Theory* **1998**, *82*, 324–341.
15. Dekel, E.; Fudenberg, D.; Morris, S. Interim Correlated Rationalizability. *Theor. Econ.* **2007**, *2*, 15–40.
16. Ely, J.; Peski, M. Hierarchies of beliefs and interim rationalizability. *Theor. Econ.* **2006**, *1*, 19–65.
17. Liu, Q. On redundant types and Bayesian formulation of incomplete information. *J. Econ. Theory* **2009**, *144*, 2115–2145.
18. Friedenberg, A.; Meier, M. The context of the game. *Econ. Theory* **2016**, in press.
19. Yildiz, M. Invariance to Representation of Information. *Games Econ. Behav.* **2015**, *94*, 142–156.
20. Brandenburger, A.; Keisler, H.J. An Impossibility Theorem on Beliefs in Games. *Stud. Log.* **2006**, *84*, 211–240.
21. Friedenberg, A. When Do Type Structures Contain All Hierarchies of Beliefs? *Games Econ. Behav.* **2010**, *68*, 108–129.
22. Kets, W. *Bounded Reasoning and Higher-Order Uncertainty*; Working Paper; Northwestern University: Evanston, IL, USA, 2011.
23. Friedenberg, A.; Keisler, H.J. *Iterated Dominance Revisited*; Working Paper; Arizona State University: Tempe, AZ, USA; University of Wisconsin-Madison: Madison, WI, USA, 2011.

24. Heifetz, A.; Samet, D. Knowledge Spaces with Arbitrarily High Rank. *Games Econ. Behav.* **1998**, *22*, 260–273.
25. Battigalli, P.; Siniscalchi, M. Hierarchies of conditional beliefs and interactive epistemology in dynamic games. *J. Econ. Theory* **1999**, *88*, 188–230.
26. Blume, L.; Brandenburger, A.; Dekel, E. Lexicographic Probabilities and Choice Under Uncertainty. *Econometrica* **1991**, *59*, 61–79.
27. Heifetz, A.; Kets, W. *Robust Multiplicity with a Grain of Naiveté*; Working Paper; Northwestern University: Evanston, IL, USA, 2013.
28. Aliprantis, C.D.; Border, K.C. *Infinite Dimensional Analysis: A Hitchhiker's Guide*, 3rd ed.; Springer: Berlin, Germany, 2005.
29. Parthasarathy, K. *Probability Measures on Metric Spaces*; AMS Chelsea Publishing: Providence, RI, USA, 2005.

*Article*

# Exploring the Gap between Perfect Bayesian Equilibrium and Sequential Equilibrium †

**Giacomo Bonanno**

Department of Economics, University of California, Davis, CA 95616-8578, USA; gfbonanno@ucdavis.edu
† I am grateful to two anonymous reviewers for helpful comments and suggestions.

Academic Editor: Paul Weirich
Received: 9 August 2016; Accepted: 4 November 2016; Published: 10 November 2016

**Abstract:** In (Bonanno, 2013), a solution concept for extensive-form games, called perfect Bayesian equilibrium (PBE), was introduced and shown to be a strict refinement of subgame-perfect equilibrium; it was also shown that, in turn, sequential equilibrium (SE) is a strict refinement of PBE. In (Bonanno, 2016), the notion of PBE was used to provide a characterization of SE in terms of a strengthening of the two defining components of PBE (besides sequential rationality), namely AGM consistency and Bayes consistency. In this paper we explore the gap between PBE and SE by identifying solution concepts that lie strictly between PBE and SE; these solution concepts embody a notion of "conservative" belief revision. Furthermore, we provide a method for determining if a plausibility order on the set of histories is choice measurable, which is a necessary condition for a PBE to be a SE.

**Keywords:** plausibility order; minimal belief revision; Bayesian updating; independence; sequential equilibrium

---

## 1. Introduction

Since its introduction in 1982 [1], sequential equilibrium has been the most widely used solution concept for extensive-form games. In applications, however, checking the "consistency" requirement for beliefs has proved to be rather difficult; thus, similarly motivated—but simpler—notions of equilibrium have been sought. The simplest solution concept is "weak sequential equilibrium" [2,3] which is defined as an assessment that is sequentially rational and satisfies Bayes' rule at information sets that are reached with positive probability by the strategy profile (while no restrictions are imposed on the beliefs at information sets that have zero probability of being reached). However, this solution concept is too weak in that it is possible for an assessment $(\sigma, \mu)$ (where $\sigma$ is a strategy profile and $\mu$ is a system of beliefs) to be a weak sequential equilibrium without $\sigma$ being a subgame-perfect equilibrium [4]. Hence the search in the literature for a "simple" (yet sufficiently strong) solution concept that lies in the gap between subgame-perfect equilibrium and sequential equilibrium. The minimal desired properties of such a solution concept, which is usually referred to as "perfect Bayesian equilibrium" (PBE), are sequential rationality and the "persistent" application of Bayes' rule. The exact meaning of the latter requirement has not been easy to formalize.

Several attempts have been made in the literature to provide a satisfactory definition of PBE; they are reviewed in Section 5. In this paper we continue the study of one such notion, introduced in [5], where it is shown that (a) the proposed solution concept is a strict refinement of subgame-perfect equilibrium; and (b) in general, the set of sequential equilibria is a proper subset of the set of perfect Bayesian equilibria. This definition of PBE is based on two notions (besides sequential rationality): (1) the qualitative property of AGM-consistency relative to a plausibility order[1]; and (2) the quantitative property of Bayes consistency. This notion of PBE was further used in [8] to provide a new characterization of sequential equilibrium, in terms of a strengthening of both AGM consistency and Bayes consistency. In this paper we explore the gap between PBE and sequential equilibrium, by identifying solution concepts that lie strictly between PBE and sequential equilibrium. These solution concepts capture the notion of revising one's beliefs in a "conservative" or "minimal" way.

The paper is organized as follows. Section 2 reviews the notation, definitions and main results of [5,8]. The new material is contained in Sections 3 and 4. In Section 3 we introduce properties of the plausibility order that can be used to define solution concepts that lie between PBE and sequential equilibrium; the main result in this section is Proposition 2. In Section 4 we offer a method (Proposition 3) for determining whether a plausibility order satisfies the property of "choice measurability", which is one of the two conditions that, together, are necessary and sufficient for a PBE to be a sequential equilibrium. Section 5 discusses related literature and Section 6 concludes. The proofs are given in Appendix A.

## 2. Perfect Bayesian Equilibrium and Sequential Equilibrium

In this section we review the notation and the main definitions and results of [5,8].

We adopt the history-based definition of extensive-form game (see, for example, [9]). If $A$ is a set, we denote by $A^*$ the set of finite sequences in $A$. If $h = \langle a_1, ..., a_k \rangle \in A^*$ and $1 \leq j \leq k$, the sequence $h' = \langle a_1, ..., a_j \rangle$ is called a *prefix* of $h$; if $j < k$ then we say that $h'$ is a *proper* prefix of $h$. If $h = \langle a_1, ..., a_k \rangle \in A^*$ and $a \in A$, we denote the sequence $\langle a_1, ..., a_k, a \rangle \in A^*$ by $ha$.

A *finite extensive form* is a tuple $\langle A, H, N, \iota, \{\approx_i\}_{i \in N} \rangle$ whose elements are:

- A finite set of actions $A$.
- A finite set of histories $H \subseteq A^*$ which is closed under prefixes (that is, if $h \in H$ and $h' \in A^*$ is a prefix of $h$, then $h' \in H$). The null history $\langle\rangle$, denoted by $\emptyset$, is an element of $H$ and is a prefix of every history. A history $h \in H$ such that, for every $a \in A$, $ha \notin H$, is called a *terminal history*. The set of terminal histories is denoted by $Z$. $D = H \backslash Z$ denotes the set of non-terminal or *decision* histories. For every history $h \in H$, we denote by $A(h)$ the set of actions available at $h$, that is, $A(h) = \{a \in A : ha \in H\}$. Thus $A(h) \neq \varnothing$ if and only if $h \in D$. We assume that $A = \bigcup_{h \in D} A(h)$ (that is, we restrict attention to actions that are available at some decision history).
- A finite set $N = \{1, ..., n\}$ of players. In some cases there is also an additional, fictitious, player called *chance*.
- A function $\iota : D \to N \cup \{chance\}$ that assigns a player to each decision history. Thus $\iota(h)$ is the player who moves at history $h$. A game is said to be *without chance moves* if $\iota(h) \in N$ for every $h \in D$. For every $i \in N \cup \{chance\}$, let $D_i = \iota^{-1}(i)$ be the set of histories assigned to player $i$. Thus $\{D_{chance}, D_1, ..., D_n\}$ is a partition of $D$. If history $h$ is assigned to chance, then a probability distribution over $A(h)$ is given that assigns positive probability to every $a \in A(h)$.
- For every player $i \in N$, $\approx_i$ is an equivalence relation on $D_i$. The interpretation of $h \approx_i h'$ is that, when choosing an action at history $h$, player $i$ does not know whether she is moving at $h$ or at $h'$. The equivalence class of $h \in D_i$ is denoted by $I_i(h)$ and is called an *information set of player i*; thus $I_i(h) = \{h' \in D_i : h' \approx_i h\}$. The following restriction applies: if $h' \in I_i(h)$ then $A(h') = A(h)$, that is, the set of actions available to a player is the same at any two histories that belong to the same information set of that player.

[1] The acronym 'AGM' stands for 'Alchourrón-Gärdenfors-Makinson' who pioneered the literature on belief revision: see [6]. As shown in [7], AGM-consistency can be derived from the primitive concept of a player's epistemic state, which encodes the player's initial beliefs and her disposition to revise those beliefs upon receiving (possibly unexpected) information. The existence of a plausibility order that rationalizes the epistemic state of each player guarantees that the belief revision policy of each player satisfies the so-called AGM axioms for rational belief revision, which were introduced in [6].

- The following property, known as *perfect recall*, is assumed: for every player $i \in N$, if $h_1, h_2 \in D_i$, $a \in A(h_1)$ and $h_1 a$ is a prefix of $h_2$ then for every $h' \in I_i(h_2)$ there exists an $h \in I_i(h_1)$ such that $ha$ is a prefix of $h'$. Intuitively, perfect recall requires a player to remember what she knew in the past and what actions she took previously.

Given an extensive form, one obtains an *extensive game* by adding, for every player $i \in N$, a *utility (or payoff) function* $U_i : Z \to \mathbb{R}$ (where $\mathbb{R}$ denotes the set of real numbers).

A total pre-order on the set of histories $H$ is a binary relation $\precsim$ which is complete[2] and transitive[3]. We write $h \sim h'$ as a short-hand for the conjunction: $h \precsim h'$ and $h' \precsim h$, and write $h \prec h'$ as a short-hand for the conjunction: $h \precsim h'$ and not $h' \precsim h$.

**Definition 1.** *Given an extensive form, a* plausibility order *is a total pre-order $\precsim$ on $H$ that satisfies the following properties: $\forall h \in D$,*

*PL1.* $h \precsim ha, \ \forall a \in A(h)$,

*PL2.* *(i)* $\exists a \in A(h)$ *such that* $h \sim ha$,
*(ii)* $\forall a \in A(h)$, *if* $h \sim ha$ *then*, $\forall h' \in I(h), h' \sim h'a$,

*PL3.* *if history $h$ is assigned to chance, then* $h \sim ha, \forall a \in A(h)$.

The interpretation of $h \precsim h'$ is that history $h$ is *at least as plausible* as history $h'$; thus $h \prec h'$ means that $h$ is *more plausible* than $h'$ and $h \sim h'$ means that $h$ is *just as plausible* as $h'$[4]. Property $PL1$ says that adding an action to a decision history $h$ cannot yield a more plausible history than $h$ itself. Property $PL2$ says that at every decision history $h$ there is at least one action $a$ which is "plausibility preserving" in the sense that adding $a$ to $h$ yields a history which is as plausible as $h$; furthermore, any such action $a$ performs the same role with any other history that belongs to the same information set as $h$. Property $PL3$ says that all the actions at a history assigned to chance are plausibility preserving.

An *assessment* is a pair $(\sigma, \mu)$ where $\sigma$ is a behavior strategy profile and $\mu$ is a system of beliefs[5].

**Definition 2.** *Given an extensive-form, an assessment $(\sigma, \mu)$ is AGM-consistent if there exists a plausibility order $\precsim$ on the set of histories $H$ such that:*

*(i)* *the actions that are assigned positive probability by $\sigma$ are precisely the plausibility-preserving actions: $\forall h \in D, \forall a \in A(h)$,*

$$\sigma(a) > 0 \text{ if and only if } h \sim ha, \tag{P1}$$

*(ii)* *the histories that are assigned positive probability by $\mu$ are precisely those that are most plausible within the corresponding information set: $\forall h \in D$,*

$$\mu(h) > 0 \text{ if and only if } h \precsim h', \forall h' \in I(h). \tag{P2}$$

*If $\precsim$ satisfies properties P1 and P2 with respect to $(\sigma, \mu)$, we say that $\precsim$* rationalizes *$(\sigma, \mu)$.*

An assessment $(\sigma, \mu)$ is sequentially rational if, for every player $i$ and every information set $I$ of hers, player $i$'s expected payoff—given the strategy profile $\sigma$ and her beliefs at $I$ (as specified by

---

2 $\forall h, h' \in H$, either $h \precsim h'$ or $h' \precsim h$.
3 $\forall h, h', h'' \in H$, if $h \precsim h'$ and $h' \precsim h''$ then $h \precsim h''$.
4 As in [5] we use the notation $h \precsim h'$ rather than the, perhaps more natural, notation $h \succsim h'$, for two reasons: (1) it is the standard notation in the extensive literature that deals with AGM belief revision (for a recent survey of this literature see the special issue of the *Journal of Philosophical Logic*, Vol. 40 (2), April 2011); and (2) when representing the order $\precsim$ numerically it is convenient to assign lower values to more plausible histories. An alternative reading of $h \precsim h'$ is "history $h$ (weakly) precedes $h'$ in terms of plausibility".
5 A behavior strategy profile is a list of probability distributions, one for every information set, over the actions available at that information set. A system of beliefs is a collection of probability distributions, one for every information set, over the histories in that information set.

$\mu$)—cannot be increased by unilaterally changing her choice at $I$ and possibly at information sets of hers that follow $I$[6].

Consider the extensive-form game shown in Figure 1[7] and the assessment $(\sigma, \mu)$ where $\sigma = (d, e, g)$ and $\mu$ is the following system of beliefs: $\mu(a) = 0$, $\mu(b) = \frac{1}{3}$, $\mu(c) = \frac{2}{3}$ and $\mu(af) = \mu(bf) = \frac{1}{2}$. This assessment is AGM-consistent, since it is rationalized by the following plausibility order[8]:

$$\begin{pmatrix} \text{most plausible} & \varnothing, d \\ & b, c, be, ce \\ & a, ae \\ & af, bf, cf, afg, bfg \\ \text{least plausible} & afk, bfk \end{pmatrix} \tag{1}$$

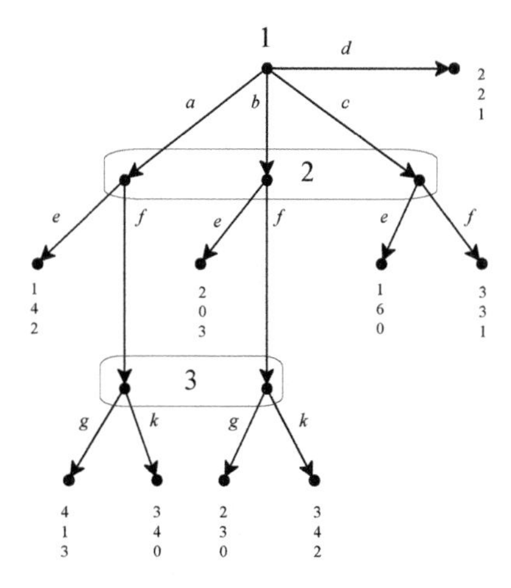


**Figure 1.** An extensive-form game.

---

6 The precise definition is as follows. Let $Z$ denote the set of terminal histories and, for every player $i$, let $U_i : Z \to \mathbb{R}$ be player $i$'s von Neumann-Morgenstern utility function. Given a decision history $h$, let $Z(h)$ be the set of terminal histories that have $h$ as a prefix. Let $\mathbb{P}_{h,\sigma}$ be the probability distribution over $Z(h)$ induced by the strategy profile $\sigma$, starting from history $h$ (that is, if $z$ is a terminal history and $z = ha_1...a_m$ then $\mathbb{P}_{h,\sigma}(z) = \prod_{j=1}^{m} \sigma(a_j)$). Let $I$ be an information set of player $i$ and let $u_i(I|\sigma, \mu) = \sum_{h \in I} \mu(h) \sum_{z \in Z(h)} \mathbb{P}_{h,\sigma}(z) U_i(z)$ be player $i$'s expected utility at $I$ if $\sigma$ is played, given her beliefs at $I$ (as specified by $\mu$). We say that player $i$'s strategy $\sigma_i$ is *sequentially rational at I* if $u_i(I|(\sigma_i, \sigma_{-i}), \mu) \geq u_i(I|(\tau_i, \sigma_{-i}), \mu)$ for every strategy $\tau_i$ of player $i$ (where $\sigma_{-i}$ denotes the strategy profile of the players other than $i$). An assessment $(\sigma, \mu)$ is *sequentially rational* if, for every player $i$ and for every information set $I$ of player $i$, $\sigma_i$ is sequentially rational at $I$. Note that there are two definitions of sequential rationality: the *weakly local* one—which is the one adopted here—according to which at an information set a player can contemplate changing her choice not only there but possibly also at subsequent information sets of hers, and a *strictly local* one, according to which at an information set a player contemplates changing her choice only there. If the definition of perfect Bayesian equilibrium (Definition 4 below) is modified by using the strictly local definition of sequential rationality, then an extra condition needs to be added, namely the "pre-consistency" condition identified in [10,11] as being necessary and sufficient for the equivalence of the two notions. For simplicity we have chosen the weakly local definition.

7 Rounded rectangles represent information sets and the payoffs are listed in the following order: Player 1's payoff at the top, Player 2's payoff in the middle and Player 3's payoff at the bottom.

8 We use the following convention to represent a total pre-order: if the row to which history $h$ belongs is above the row to which $h'$ belongs, then $h \prec h'$ ($h$ is more plausible than $h'$) and if $h$ and $h'$ belong to the same row then $h \sim h'$ ($h$ is as plausible as $h'$). $\varnothing$ denotes the empty history, which corresponds to the root of the tree. In (1) the plausibility-preserving actions are $d$, $e$ and $g$; the most plausible histories in the information set $\{a, b, c\}$ are $b$ and $c$ and the two histories in the information set $\{af, bf\}$ are equally plausible.

Furthermore $(\sigma, \mu)$ is sequentially rational[9]. The property of AGM-consistency imposes restrictions on the support of the behavior strategy $\sigma$ and on the support of the system of beliefs $\mu$. The following property imposes constraints on how probabilities can be distributed over those supports.

**Definition 3.** *Given an extensive form, let $\precsim$ be a plausibility order that rationalizes the assessment $(\sigma, \mu)$. We say that $(\sigma, \mu)$ is Bayes consistent (or Bayesian) relative to $\precsim$ if, for every equivalence class E of $\precsim$ that contains some decision history h with $\mu(h) > 0$ [that is, $E \cap D_\mu^+ \neq \varnothing$, where $D_\mu^+ = \{h \in D : \mu(h) > 0\}$], there exists a probability density function $\nu_E : H \to [0, 1]$ (recall that H is a finite set) such that:*

*B1. $\nu_E(h) > 0$ if and only if $h \in E \cap D_\mu^+$.*

*B2. If $h, h' \in E \cap D_\mu^+$ and $h' = ha_1...a_m$ (that is, h is a prefix of h′) then $\nu_E(h') = \nu_E(h) \times \sigma(a_1) \times \ldots \times \sigma(a_m)$.*

*B3. If $h \in E \cap D_\mu^+$, then, $\forall h' \in I(h)$, $\mu(h') = \nu_E(h' \mid I(h)) \overset{def}{=} \frac{\nu_E(h')}{\sum\limits_{h'' \in I(h)} \nu_E(h'')}$.*

Property $B1$ requires that $\nu_E(h) > 0$ if and only if $h \in E$ and $\mu(h) > 0$. Property $B2$ requires $\nu_E$ to be consistent with the strategy profile $\sigma$ in the sense that if $h, h' \in E$, $\mu(h) > 0$, $\mu(h') > 0$ and $h' = ha_1...a_m$ then the probability that $\nu_E$ assigns to $h'$ is equal to the probability that $\nu_E$ assigns to $h$ multiplied by the probabilities (according to $\sigma$) of the actions that lead from $h$ to $h'$[10]. Property $B3$ requires the system of beliefs $\mu$ to satisfy Bayes' rule in the sense that if $h \in E$ and $\mu(h) > 0$ (so that $E$ is the equivalence class of the most plausible elements of $I(h)$) then, for every history $h' \in I(h)$, $\mu(h')$ (the probability assigned to $h'$ by $\mu$) coincides with the probability of $h'$ conditional on $I(h)$ using the probability density function $\nu_E$[11].

Consider again the game of Figure 1, and the assessment $(\sigma, \mu)$ where $\sigma = (d, e, g)$ and $\mu(a) = 0$, $\mu(b) = \frac{1}{3}$, $\mu(c) = \frac{2}{3}$ and $\mu(af) = \mu(bf) = \frac{1}{2}$. Let $\precsim$ be the plausibility order (1) given above, which rationalizes $(\sigma, \mu)$. Then $(\sigma, \mu)$ is Bayes consistent relative to $\precsim$. In fact, we have that $D_\mu^+ = \{\varnothing, b, c, af, bf\}$ and the equivalence classes of $\precsim$ that have a non-empty intersection with $D_\mu^+$ are $E_1 = \{\varnothing, d\}$, $E_2 = \{b, c, be, ce\}$ and $E_3 = \{af, bf, cf, afg, bfg\}$. Let $\nu_{E_1}(\varnothing) = 1$, $\nu_{E_2}(b) = \frac{1}{3}$, $\nu_{E_2}(c) = \frac{2}{3}$ and $\nu_{E_3}(af) = \nu_{E_3}(bf) = \frac{1}{2}$. Then the three probability density functions $\nu_{E_1}$, $\nu_{E_2}$ and $\nu_{E_3}$ satisfy the properties of Definition 3 and hence $(\sigma, \mu)$ is Bayes consistent relative to $\precsim$.

**Definition 4.** *An assessment $(\sigma, \mu)$ is a perfect Bayesian equilibrium (PBE) if it is sequentially rational, it is rationalized by a plausibility order on the set of histories and is Bayes consistent relative to it.*

We saw above that, for the game illustrated in Figure 1, the assessment $(\sigma, \mu)$ where $\sigma = (d, e, g)$ and $\mu(a) = 0$, $\mu(b) = \frac{1}{3}$, $\mu(c) = \frac{2}{3}$ and $\mu(af) = \mu(bf) = \frac{1}{2}$ is sequentially rational, it is rationalized by the plausibility order (1) and is Bayes consistent relative to it. Thus it is a perfect Bayesian equilibrium.

**Remark 1.** *It is proved in [5] that if $(\sigma, \mu)$ is a perfect Bayesian equilibrium then $\sigma$ is a subgame-perfect equilibrium and that every sequential equilibrium is a perfect Bayesian equilibrium. Furthermore, the notion of PBE is a strict refinement of subgame-perfect equilibrium and sequential equilibrium is a strict refinement of PBE.*

---

9 Given $\sigma$, for Player 1 $d$ yields a payoff of 2 while $a$ and $c$ yield 1 and $b$ yields 2; thus $d$ is sequentially rational. Given $\sigma$ and $\mu$, at her information set $\{a, b, c\}$ with $e$ Player 2 obtains an expected payoff of 4 while with $f$ her expected payoff is 3; thus $e$ is sequentially rational. Given $\mu$, at his information set $\{af, bf\}$, Player 3's expected payoff from playing with $g$ is 1.5 while his expected payoff from playing with $k$ is 1; thus $g$ is sequentially rational.

10 Note that if $h, h' \in E$ and $h' = ha_1...a_m$, then $\sigma(a_j) > 0$, for all $j = 1, ..., m$. In fact, since $h' \sim h$, every action $a_j$ is plausibility preserving and therefore, by Property $P1$ of Definition 2, $\sigma(a_j) > 0$.

11 For an interpretation of the probabilities $\nu_E(h)$ see [8].

Next we recall the definition of sequential equilibrium [1]. An assessment $(\sigma, \mu)$ is KW-consistent (KW stands for 'Kreps-Wilson') if there is an infinite sequence $\langle \sigma^1, ..., \sigma^m, ... \rangle$ of completely mixed behavior strategy profiles such that, letting $\mu^m$ be the unique system of beliefs obtained from $\sigma^m$ by applying Bayes' rule[12], $\lim_{m \to \infty}(\sigma^m, \mu^m) = (\sigma, \mu)$. An assessment $(\sigma, \mu)$ is a *sequential equilibrium* if it is KW-consistent and sequentially rational. In [8] it is shown that sequential equilibrium can be characterized as a strengthening of PBE based on two properties: (1) a property of the plausibility order that constrains the supports of the belief system; and (2) a strengthening of the notion of Bayes consistency, that imposes constraints on how the probabilities can be distributed over those supports. The details are given below.

Given a plausibility order $\precsim$ on the finite set of histories $H$, a function $F : H \to \mathbb{N}$ (where $\mathbb{N}$ denotes the set of non-negative integers) is said to be an *ordinal integer-valued representation* of $\precsim$ if, for every $h, h' \in H$,

$$F(h) \leq F(h') \text{ if and only if } h \precsim h'. \tag{2}$$

Since $H$ is finite, the set of ordinal integer-valued representations is non-empty. A particular ordinal integer-valued representation, which we will call *canonical* and denote by $\rho$, is defined as follows.

**Definition 5.** *Let $H_0 = \{h \in H : h \precsim x, \ \forall x \in H\}$, $H_1 = \{h \in H \setminus H_0 : \ h \precsim x, \ \forall x \in H \setminus H_0\}$ and, in general, for every integer $k \geq 1$, $H_k = \{h \in H \setminus H_0 \cup ... \cup H_{k-1} : h \precsim x, \ \forall x \in H \setminus H_0 \cup ... \cup H_{k-1}\}$. Thus $H_0$ is the equivalence class of $\precsim$ containing the most plausible histories, $H_1$ is the equivalence class containing the most plausible among the histories left after removing those in $H_0$, etc.[13] The canonical ordinal integer-valued representation of $\precsim$, $\rho : H \to \mathbb{N}$, is given by*

$$\rho(h) = k \text{ if and only if } h \in H_k. \tag{3}$$

*We call $\rho(h)$ the* rank *of history $h$.*

Instead of an ordinal integer-valued representation of the plausibility order one could seek a *cardinal* representation which, besides (2), satisfies the following property: if $h$ and $h'$ belong to the same information set $\big(\text{that is, } h' \in I(h)\big)$ and $a \in A(h)$, then

$$F(h') - F(h) = F(h'a) - F(ha). \tag{CM}$$

If we think of $F$ as measuring the "plausibility distance" between histories, then we can interpret $CM$ as a distance-preserving condition: the plausibility distance between two histories in the same information set is preserved by the addition of the same action.

**Definition 6.** *A plausibility order $\precsim$ on the set of histories $H$ is choice measurable if it has at least one integer-valued representation that satisfies property $CM$.*

For example, the plausibility order (1) is not choice measurable, since any integer-valued representation $F$ of it must be such that $F(a) - F(b) > 0$ and $F(af) - F(bf) = 0$.

Let $(\sigma, \mu)$ be an assessment which is rationalized by a plausibility order $\precsim$. As before, let $D_\mu^+$ be the set of decision histories to which $\mu$ assigns positive probability: $D_\mu^+ = \{h \in D : \mu(h) > 0\}$. Let $\mathcal{E}_\mu^+$

[12] That is, for every $h \in D \setminus \{\varnothing\}$, $\mu^m(h) = \dfrac{\prod_{a \in A_h} \sigma^m(a)}{\sum_{h' \in I(h)} \prod_{a \in A_{h'}} \sigma^m(a)}$ (where $A_h$ is the set of actions that occur in history $h$). Since $\sigma^m$ is completely mixed, $\sigma^m(a) > 0$ for every $a \in A$ and thus $\mu^m(h) > 0$ for all $h \in D \setminus \{\varnothing\}$.

[13] Since $H$ is finite, there is an $m \in \mathbb{N}$ such that $\{H_0, ..., H_m\}$ is a partition of $H$ and, for every $j, k \in \mathbb{N}$, with $j < k \leq m$, and for every $h, h' \in H$, if $h \in H_j$ and $h' \in H_k$ then $h \prec h'$.

be the set of equivalence classes of $\precsim$ that have a non-empty intersection with $D_\mu^+$. Clearly $\mathcal{E}_\mu^+$ is a non-empty, finite set. Suppose that $(\sigma, \mu)$ is Bayesian relative to $\precsim$ and let $\{\nu_E\}_{E \in \mathcal{E}_\mu^+}$ be a collection of probability density functions that satisfy the properties of Definition 3. We call a probability density function $\nu : D \to (0,1]$ a *full-support common prior* of $\{\nu_E\}_{E \in \mathcal{E}_\mu^+}$ if, for every $E \in \mathcal{E}_\mu^+$, $\nu_E(\cdot) = \nu(\cdot \mid E \cap D_\mu^+)$, that is, for all $h \in E \cap D_\mu^+$, $\nu_E(h) = \frac{\nu(h)}{\sum\limits_{h' \in E \cap D_\mu^+} \nu(h')}$. Note that a full support common prior assigns positive probability to *all* decision histories, not only to those in $D_\mu^+$.

**Definition 7.** *Consider an extensive form. Let $(\sigma, \mu)$ be an assessment which is rationalized by the plausibility order $\precsim$ and is Bayesian relative to it and let $\{\nu_E\}_{E \in \mathcal{E}_\mu^+}$ be a collection of probability density functions that satisfy the properties of Definition 3. We say that $(\sigma, \mu)$ is* uniformly Bayesian relative to $\precsim$ *if there exists a full-support common prior $\nu : D \to (0,1]$ of $\{\nu_E\}_{E \in \mathcal{E}_\mu^+}$ that satisfies the following properties.*

- *UB1.* *If $a \in A(h)$ and $ha \in D$, then*
  *(i) $\nu(ha) \leq \nu(h)$ and, (ii) if $\sigma(a) > 0$ then $\nu(ha) = \nu(h) \times \sigma(a)$.*
- *UB2.* *If $a \in A(h)$, $h$ and $h'$ belong to the same information set and $ha, h'a \in D$ then $\frac{\nu(h)}{\nu(h')} = \frac{\nu(ha)}{\nu(h'a)}$.*

*We call such a function $\nu$ a* uniform full-support common prior *of $\{\nu_E\}_{E \in \mathcal{E}_\mu^+}$.*

*UB1* requires that the common prior $\nu$ be consistent with the strategy profile $\sigma$, in the sense that if $\sigma(a) > 0$ then $\nu(ha) = \nu(h) \times \sigma(a)$ (thus extending Property *B2* of Definition 3 from $D_\mu^+$ to $D$). *UB2* requires that the relative probability, according to the common prior $\nu$, of any two histories that belong to the same information set remain unchanged by the addition of the same action.

It is shown in [8] that choice measurability and uniform Bayesian consistency are independent properties. The following proposition is proved in [8].

**Proposition 1.** *(I) and (II) below are equivalent:*

- *(I) $(\sigma, \mu)$ is a perfect Bayesian equilibrium which is rationalized by a choice measurable plausibility order and is uniformly Bayesian relative to it.*
- *(II) $(\sigma, \mu)$ is a sequential equilibrium.*

## 3. Exploring the Gap between PBE and Sequential Equilibrium

The notion of perfect Bayesian equilibrium (Definition 4) incorporates—through the property of AGM-consistency—a belief revision policy which can be interpreted either as the epistemic state of an external observer[14] or as a belief revision policy which is shared by all the players[15]. For example, the perfect Bayesian equilibrium considered in Section 2 for the game of Figure 1, namely $\sigma = (d, e, g)$ and $\mu(a) = 0$, $\mu(b) = \frac{1}{3}$, $\mu(c) = \frac{2}{3}$, $\mu(af) = \mu(bf) = \frac{1}{2}$ reflects the following belief revision policy: the initial beliefs are that Player 1 will play $d$; conditional on learning that Player 1 did not play $d$, the observer would become convinced that Player 1 played either $b$ or $c$ (that is, she would judge $a$ to be less plausible than $b$ and she would consider $c$ to be as plausible as $b$) and would expect Player 2 to play $e$; upon learning that (Player 1 did not play $d$ and) Player 2 played $f$, the observer would become convinced that Player 1 played either $a$ or $b$, hence judging $af$ to be as plausible as $bf$, thereby modifying her earlier judgment that $a$ was less plausible than $b$. Although such a belief revision policy does not violate the rationality constraints introduced in [6], it does involve a belief change that is not "minimal"or "conservative". Such "non-minimal" belief changes can be ruled out by

---

[14] For example, [12] adopts this interpretation.
[15] For such an interpretation see [7].

imposing the following restriction on the plausibility order: if $h$ and $h'$ belong to the same information set (that is, $h' \in I(h)$) and $a$ is an action available at $h$ $(a \in A(h))$, then

$$h \precsim h' \text{ if and only if } ha \precsim h'a. \qquad (IND_1)$$

$IND_1$ says that if $h$ is deemed to be at least as plausible as $h'$ then the addition of any available action $a$ must preserve this judgment, in the sense that $ha$ must be deemed to be at least as plausible as $h'a$, and *vice versa;* it can also be viewed as an "independence" condition, in the sense that observation of a new action cannot lead to a change in the relative plausibility of previous histories[16]. Any plausibility order that rationalizes the assessment $\sigma = (d,e,g)$ and $\mu(a) = 0$, $\mu(b) = \frac{1}{3}$, $\mu(c) = \frac{2}{3}$, $\mu(af) = \mu(bf) = \frac{1}{2}$ for the game of Figure 1 must violate $IND_1$ (since $b \prec a$ while $bf \sim af$).

We can obtain a strengthening of the notion of perfect Bayesian equilibrium (Definition 4) by (1) adding property $IND_1$; and (2) strengthening Bayes consistency (Definition 3) to uniform Bayesian consistency (Definition 7).

**Definition 8.** *Given an extensive-form game, an assessment $(\sigma,\mu)$ is a weakly independent perfect Bayesian equilibrium if it is sequentially rational, it is rationalized by a plausibility order that satisfies $IND_1$ and is uniformly Bayesian relative to that plausibility order.*

As an example of a weakly independent PBE consider the game of Figure 2 and the assessment $(\sigma,\mu)$ where $\sigma = (c,d,g,\ell)$ (highlighted by double edges in Figure 2) and $\mu(b) = \mu(ae) = \mu(bf) = 1$ (thus $\mu(a) = \mu(af) = \mu(be) = 0$) (the decision histories $x$ such that $\mu(x) > 0$ are shown as black nodes and the decision histories $x$ such that $\mu(x) = 0$ are shown as gray nodes)). This assessment is sequentially rational and is rationalized by the following plausibility order:

$$\begin{pmatrix} \text{most plausible} & \emptyset, c \\ & b, bd \\ & a, ad \\ & bf, bf\ell \\ & be, be\ell \\ & ae, aeg \\ & af, afg \\ & bfm \\ & bem \\ & aek \\ \text{least plausible} & afk \end{pmatrix} \qquad (4)$$

It is straightforward to check that plausibility order (4) satisfies $IND_1$[17]. To see that $(\sigma,\mu)$ is uniformly Bayesian relative to plausibility order (4), note that $D_\mu^+ = \{\emptyset, b, ae, bf\}$ and thus the only equivalence classes that have a non-empty intersection with $D_\mu^+$ are $E_1 = \{\emptyset, c\}$, $E_2 = \{b, bd\}$, $E_3 = \{ae, aeg\}$ and $E_4 = \{bf, bf\ell\}$. Letting $\nu_{E_1}(\emptyset) = 1$, $\nu_{E_2}(b) = 1$, $\nu_{E_3}(ae) = 1$ and $\nu_{E_4}(bf) = 1$, this collection of probability distributions satisfies the Properties of Definition 3. Let $\nu$ be the uniform distribution over the set of decision histories $D = \{\emptyset, a, b, ae, af, be, bf\}$ (thus $\nu(h) = \frac{1}{7}$ for every $h \in D$). Then $\nu$ is a full support common prior of the collection $\{\nu_{E_i}\}_{i\in\{1,2,3,4\}}$ and satisfies Properties $UB1$ and $UB2$ of Definition 7.

[16] Note, however, that $IND_1$ is compatible with the following: $a \prec b$ (with $b \in I(a)$) and $bc \prec ad$ (with $bc \in I(ad)$, $c,d \in A(a)$, $c \neq d$).

[17] We have that (1) $b \prec a$, $bd \prec ad$, $be \prec ae$ and $bf \prec af$, (2) $ae \prec af$, $aeg \prec afg$ and $aek \prec afk$, (3) $bf \prec be$, $bf\ell \prec be\ell$ and $bfm \prec bem$.

Note, however, that $(\sigma,\mu)$ is not a sequential equilibrium. This can be established by showing that $(\sigma,\mu)$ is not KW-consistent; however, we will show it by appealing to the following lemma (proved in Appendix A) which highlights a property that will motivate a further restriction on belief revision (property $IND_2$ below).

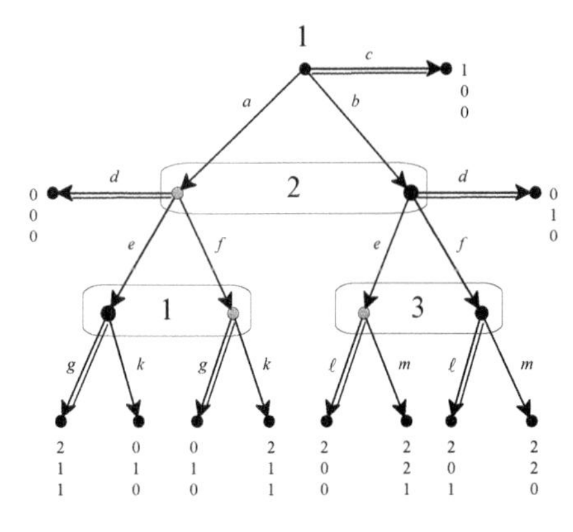


**Figure 2**

**Lemma 1.** *Let $\precsim$ be a plausibility order over the set $H$ of histories of an extensive-form game and let $F : H \to \mathbb{N}$ be an integer-valued representation of $\precsim$ (that is, for all $h, h' \in H$, $F(h) \leq F(h')$ if and only if $h \precsim h'$). Then the following are equivalent:*

*(A) $F$ satisfies Property $CM$ (Definition 6)*
*(B) $F$ satisfies the following property: for all $h, h' \in H$ and $a, b \in A(h)$, if $h' \in I(h)$ then*

$$F(hb) - F(ha) = F(h'b) - F(h'a). \qquad (CM')$$

Using Lemma 1 we can prove that the assessment $(\sigma,\mu)$ where $\sigma = (c, d, g, \ell)$ and $\mu(b) = \mu(ae) = \mu(bf) = 1$, for the game of Figure 2, is not a sequential equilibrium. By Proposition 1 it will be sufficient to show that $(\sigma,\mu)$ cannot be rationalized by a choice measurable plausibility order (Definition 6). Let $\precsim$ be a plausibility order that rationalizes $(\sigma,\mu)$ and let $F$ be an integer-valued representation of $\precsim$. Then, by (P2) of Definition 2, it must be that $ae \prec af$ (because $\mu(ae) > 0$ and $\mu(af) = 0$) and $bf \prec be$ (because $\mu(bf) > 0$ and $\mu(be) = 0$); thus $F(ae) - F(af) < 0$ and $F(be) - F(bf) > 0$, so that $F$ violates property $CM'$; hence, by Lemma 1, $F$ violates property $CM$ and thus $\precsim$ is not choice measurable.

The ordinal counterpart to Property $CM'$ is Property $IND_2$ below, which can be viewed as another "independence" condition: it says that if action $a$ is implicitly judged to be at least as plausible as action $b$, conditional on history $h$ (that is, $ha \precsim hb$), then the same judgment must be made conditional on any other history that belongs to the same information set as $h$: if $h' \in I(h)$ and $a, b \in A(h)$, then

$$ha \precsim hb \text{ if and only if } h'a \precsim h'b. \qquad (IND_2)$$

Note that Properties $IND_1$ and $IND_2$ are independent. An example of a plausibility order that violates $IND_1$ but satisfies $IND_2$ is plausibility order (1) for the game of Figure 1: $IND_1$ is violated because $b \prec a$ but $bf \sim af$ and $IND_2$ is satisfied because at every non-singleton information set there are only two choices, one of which is plausibility preserving and the other is not. An example of a plausibility order that satisfies $IND_1$ but violates $IND_2$ is plausibility order (4) for the game of

Figure 2[18]. Adding Property $IND_2$ to the properties given in Definition 8 we obtain a refinement of the notion of weakly independent perfect Bayesian equilibrium.

**Definition 9.** *Given an extensive-form game, an assessment $(\sigma,\mu)$ is a strongly independent perfect Bayesian equilibrium if it is sequentially rational, it is rationalized by a plausibility order that satisfies Properties $IND_1$ and $IND_2$, and is uniformly Bayesian relative to that plausibility order.*

The following proposition states that the notions of weakly/strongly independent PBE identify two (nested) solution concepts that lie strictly in the gap between PBE and sequential equilibrium. The proof of the first part of Proposition 2 is given in Appendix A, while the example of Figure 3 establishes the second part.

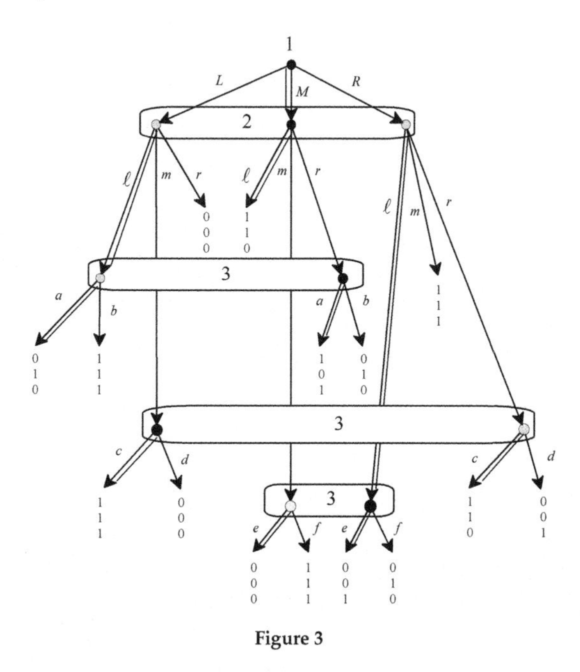


**Figure 3**

**Proposition 2.** *Consider an extensive-form game and an assessment $(\sigma,\mu)$. If $(\sigma,\mu)$ is a sequential equilibrium then it is a strongly independent perfect Bayesian equilibrium (PBE). Furthermore, there are games where there is a strongly independent PBE which is not a sequential equilibrium.*

To see that the notion of strongly independent PBE is weaker than sequential equilibrium, consider the game of Figure 3 (which is based on an example discussed in [12–14]) and the assessment $(\sigma,\mu)$ where $\sigma = (M,\ell,a,c,e)$ (highlighted by double edges), $\mu(x) = 1$ for $x \in \{\emptyset, M, Mr, Lm, R\ell\}$ and $\mu(x) = 0$ for every other decision history $x$ (the decision histories $x$ such that $\mu(x) > 0$ are

[18] That $IND_1$ is satisfied was shown in Footnote 17. $IND_2$ is violated because $b \in I(a)$ and $bf \prec be$ but $ae \prec af$.

shown as black nodes and the decision histories $x$ such that $\mu(x) = 0$ are shown as gray nodes). This assessment is rationalized by the following plausibility order:

$$\begin{pmatrix} \text{most plausible} & \emptyset, M, M\ell \\ & R, R\ell, R\ell e \\ & Mm, Mme \\ & Mr, Mra \\ & L, L\ell, L\ell a \\ & Rm \\ & Lm, Lmc \\ & Rr, Rrc \\ & Lr \\ & R\ell f \\ & Mmf \\ & Lmd \\ & Rrd \\ & Mrb \\ \text{least plausible} & L\ell b \end{pmatrix} \tag{5}$$

It is straightforward to check that plausibility order (5) satisfies Properties $IND_1$[19] and $IND_2$[20]. Furthermore $(\sigma,\mu)$ is trivially uniformly Bayesian relative to plausibility order (5)[21]. Thus $(\sigma,\mu)$ is a strongly independent PBE. Next we show that $(\sigma,\mu)$ is not a sequential equilibrium, by appealing to Proposition 1 and showing that *any* plausibility order that rationalizes $(\sigma,\mu)$ is not choice measurable[22]. Let $\precsim$ be a plausibility order that rationalizes $(\sigma,\mu)$; then it must satisfy the following properties:

- $Lm \prec Rr$ (because they belong to the same information set and $\mu(Lm) > 0$ while $\mu(Rr) = 0$). Thus if $F$ is any integer-valued representation of $\precsim$ it must be that

$$F(Lm) < F(Rr). \tag{6}$$

- $Mr \prec L\ell \sim L$ (because $Mr$ and $L\ell$ belong to the same information set and $\mu(Mr) > 0$ while $\mu(L\ell) = 0$; furthermore, $\ell$ is a plausibility-preserving action since $\sigma(\ell) > 0$). Thus if $F$ is any integer-valued representation of $\precsim$ it must be that

$$F(Mr) < F(L). \tag{7}$$

- $R \sim R\ell \prec Mm$ (because $\ell$ is a plausibility-preserving action, $R\ell$ and $Mm$ belong to the same information set and $\mu(R\ell) > 0$ while $\mu(Mm) = 0$). Thus if $F$ is any integer-valued representation of $\precsim$ it must be that

$$F(R) < F(Mm). \tag{8}$$

---

19 In fact, (1) $M \prec L$ and $Mx \prec Lx$ for every $x \in \{\ell, m, r\}$; (2) $M \prec R$ and $Mx \prec Rx$ for every $x \in \{\ell, m, r\}$; (3) $R \prec L$ and $Rx \prec Lx$ for every $x \in \{\ell, m, r\}$; (4) $Mr \prec L\ell$ and $Mrx \prec L\ell x$ for every $x \in \{a, b\}$; (5) $Lm \prec Rr$ and $Lmx \prec Rrx$ for every $x \in \{c, d\}$; and (6) $R\ell \prec Mm$ and $R\ell x \prec Mmx$ for every $x \in \{e, f\}$.

20 This is easily verified: the important observation is that $Mm \prec Mr$ and $Lm \prec Lr$ and $Rm \prec Rr$. The other comparisons involve a plausibility-preserving action *versus* a non-plausibility-preserving action and thus $IND_2$ is trivially satisfied.

21 As uniform full support common prior one can take, for example, the uniform distribution over the set of decision histories. Note that, for every equivalence class $E$ of the order, $E \cap D_\mu^+$ is either empty or a singleton.

22 To prove that $(\sigma,\mu)$ is not a sequential equilibrium it is not sufficient to show that plausibility order (5) is not choice measurable, because there could be *another* plausibility order which is choice measurable and rationalizes $(\sigma,\mu)$.

Suppose that $\precsim$ is choice measurable and let $F$ be an integer-valued representation of it that satisfies Property $CM$. From (6) and (7) we get that

$$F(Lm) - F(L) < F(Rr) - F(Mr) \tag{9}$$

and by Property $CM$ it must be that

$$F(Rr) - F(Mr) = F(R) - F(M). \tag{10}$$

It follows from (9) and (10) that

$$F(Lm) - F(L) < F(R) - F(M). \tag{11}$$

Subtracting $F(M)$ from both sides of (8) we obtain

$$F(R) - F(M) < F(Mm) - F(M). \tag{12}$$

It follows from (11) and (12) that $F(Lm) - F(L) < F(Mm) - F(M)$, which can be written as $F(M) - F(L) < F(Mm) - F(Lm)$, yielding a contradiction, because Property $CM$ requires that $F(M) - F(L) = F(Mm) - F(Lm)$.

Are the notions of weakly/strongly independent PBE "better" or "more natural" than the basic notion of PBE? This issue will be discussed briefly in Section 6.

## 4. How to Determine if a Plausibility Order Is Choice Measurable

In this section we provide a method for determining if a plausibility order is choice measurable. More generally, we provide a necessary and sufficient condition that applies not only to plausibility orders over sets of histories in a game but to a more general class of structures.

Let $S$ be an arbitrary finite set and let $\precsim$ be a total pre-order on $S$. Let $S\backslash_{\sim}$ be the set of $\precsim$-equivalence classes of $S$. If $s \in S$, the equivalence class of $s$ is denoted by $[s] = \{t \in S : s \sim t\}$ (where, as before, $s \sim t$ is a short-hand for "$s \precsim t$ and $t \precsim s$"); thus $S\backslash_{\sim} = \{[s] : s \in S\}$. Let $\doteq$ be an equivalence relation on $S\backslash_{\sim} \times S\backslash_{\sim}$. The interpretation of $([s_1], [s_2]) \doteq ([t_1], [t_2])$ is that the distance between the equivalence classes $[s_1]$ and $[s_2]$ is required to be equal to the distance between the equivalence classes $[t_1]$ and $[t_2]$.

**Remark 2.** *In the special case of a plausibility order $\precsim$ on the set of histories $H$ of a game, we shall be interested in the following equivalence relation $\doteq$ on $H\backslash_{\sim} \times H\backslash_{\sim}$, which is meant to capture property $CM$ above: if $E_1$, $E_2$, $F_1$ and $F_2$ are equivalence classes of $\precsim$ then $(E_1, E_2) \doteq (F_1, F_2)$ if and only if there exist two decision histories $h, h' \in H$ that belong to the same information set [$h' \in I(h)$] and a non-plausibility-preserving action $a \in A(h)$ such that $h \in E_1, h' \in E_2, ha \in F_1$ and $h'a \in F_2$ (or $ha \in E_1, h'a \in E_2, h \in F_1$ and $h' \in F_2$).*

The general problem that we are addressing is the following.

**Problem 1.** *Given a pair $(\precsim, \doteq)$, where $\precsim$ is a total pre-order on a finite set $S$ and $\doteq$ is an equivalence relation on the set of pairs of equivalence classes of $\precsim$, determine whether there exists a function $F : S \to \mathbb{N}$ such that, for all $s, t, x, y \in S$, (1) $F(s) \leq F(t)$ if and only if $s \precsim t$ and (2) if $([s], [t]) \doteq ([x], [y])$, with $s \prec t$ and $x \prec y$, then $F(t) - F(s) = F(y) - F(x)$.*

Instead of expressing the equivalence relation $\doteq$ in terms of pairs of elements of $S\backslash_{\sim}$, we shall express it in terms of pairs of numbers $(j, k)$ obtained by using the canonical ordinal representation

$\rho$ of $\precsim$[23]. That is, if $s_1, s_2, t_1, t_2 \in S$ and $([s_1],[s_2]) \doteq ([t_1],[t_2])$ then we shall write this as $(\rho(s_1), \rho(s_2)) \doteq (\rho(t_1), \rho(t_2))$. For example, let $S = \{a, b, c, d, e, f, g, h, \ell, m\}$ and let $\precsim$ be as shown in (13) below, together with the corresponding canonical representation $\rho$[24]:

$$\begin{pmatrix} \precsim: & \rho: \\ a & 0 \\ b,c & 1 \\ d & 2 \\ e & 3 \\ f,g & 4 \\ h,\ell & 5 \\ m & 6 \end{pmatrix} \tag{13}$$

If the equivalence relation $\doteq$ contains the following pairs[25]:

$$\begin{array}{lcl} ([a],[b]) \doteq ([h],[m]) & & (0,1) \doteq (5,6) \\ ([a],[b]) \doteq ([e],[f]) & & (0,1) \doteq (3,4) \\ ([a],[d]) \doteq ([f],[m]) & \text{then we express them (using } \rho \text{) as} & (0,2) \doteq (4,6) \\ ([b],[e]) \doteq ([e],[f]) & & (1,3) \doteq (3,4) \\ ([b],[e]) \doteq ([f],[m]) & & (1,3) \doteq (4,6) \end{array} \tag{14}$$

A *bag* (or *multiset*) is a generalization of the notion of set in which members are allowed to appear more than once. An example of a bag is $\{1,2,2,3,4,4,5,6\}$. Given two bags $B_1$ and $B_2$ their *union*, denoted by $B_1 \uplus B_2$, is the bag that contains those elements that occur in either $B_1$ or $B_2$ and, furthermore, the number of times that each element occurs in $B_1 \uplus B_2$ is equal to the number of times it occurs in $B_1$ plus the number of times it occurs in $B_2$. For instance, if $B_1 = \{1,2,2,3,4,4,5,6\}$ and $B_2 = \{2,3,6,6\}$ then $B_1 \uplus B_2 = \{1,2,2,2,3,3,4,4,5,6,6,6\}$. We say that $B_1$ is a *proper sub-bag* of $B_2$, denoted by $B_1 \sqsubset B_2$, if $B_1 \neq B_2$ and each element that occurs in $B_1$ occurs also, and at least as many times, in $B_2$. For example, $\{1,2,4,4,5,6\} \sqsubset \{1,1,2,4,4,5,5,6\}$.

Given a pair $(i,j)$ with $i < j$, we associate with it the set $B_{(i,j)} = \{i+1, i+2, ..., j\}$. For example, $B_{(2,5)} = \{3,4,5\}$. Given a set of pairs $P = \{(i_1, j_1), (i_2, j_2), ..., (i_m, j_m)\}$ (with $i_k < j_k$, for every $k = 1, ..., m$) we associate with it the bag $B_P = B_{(i_1,j_1)} \uplus B_{(i_2,j_2)} \uplus ... \uplus B_{(i_m,j_m)}$. For example, if $P = \{(0,2),(1,4),(2,5)\}$ then $B_P = \{1,2\} \uplus \{2,3,4\} \uplus \{3,4,5\} = \{1,2,2,3,3,4,4,5\}$.

**Definition 10.** *For every element of $\doteq$, expressed (using the canonical representation $\rho$) as $(i,j) \doteq (k,\ell)$ (with $i < j$ and $k < \ell$), the* equation corresponding to it *is $x_{i+1} + x_{i+2} + ... + x_j = x_{k+1} + x_{k+2} + ... + x_\ell$. By the system of equations corresponding to $\doteq$ we mean the set of all such equations*[26].

For example, consider the total pre-order given in (13) and the following equivalence relation $\doteq$ (expressed in terms of $\rho$ and omitting the reflexive pairs):

$$\{(0,3) \doteq (2,4),\ (2,4) \doteq (0,3),\ (2,4) \doteq (3,5),\ (3,5) \doteq (2,4),\ (0,3) \doteq (3,5),\ (3,5) \doteq (0,3)\}$$

---

23 As in Definition 5, let $S_0 = \{s \in S : s \precsim t,\ \forall t \in S\}$, and, for every integer $k \geq 1$, $S_k = \{h \in S \setminus S_0 \cup ... \cup S_{k-1} : s \precsim t,\ \forall t \in S \setminus S_0 \cup ... \cup S_{k-1}\}$. The canonical ordinal integer-valued representation of $\precsim$, $\rho : S \to \mathbb{N}$, is given by $\rho(s) = k$ if and only if $s \in S_k$.

24 Thus $a \prec x$ for every $x \in S \setminus \{a\}$, $[b] = \{b, c\}$, $b \prec d$, etc.

25 For example, $\doteq$ is the smallest reflexive, symmetric and transitive relation that contains the pairs given in (14).

26 The system of linear equations of Definition 10 is somewhat related to the system of multiplicative equations considered in [13] (Theorem 5.1). A direct comparison is beyond the scope of this paper and is not straightforward, because the structures considered in Definition 10 are more general than those considered in [13].

Then the corresponding system of equations is given by:

$$\begin{aligned}
x_1 + x_2 + x_3 &= x_3 + x_4 \\
x_3 + x_4 &= x_1 + x_2 + x_3 \\
x_3 + x_4 &= x_4 + x_5 \\
x_4 + x_5 &= x_3 + x_4 \\
x_1 + x_2 + x_3 &= x_4 + x_5 \\
x_4 + x_5 &= x_1 + x_2 + x_3
\end{aligned} \tag{15}$$

We are now ready to state the solution to Problem 1. The proof is given in Appendix A.

**Proposition 3.** *Given a pair $(\precsim, \doteq)$, where $\precsim$ is a total pre-order on a finite set $S$ and $\doteq$ is an equivalence relation on the set of pairs of equivalence classes of $\precsim$, (A), (B) and (C) below are equivalent.*

- *(A) There is a function $F : S \to \mathbb{N}$ such that, for all $s, t, x, y \in S$, (1) $F(s) \leq F(t)$ if and only if $s \precsim t$; and (2) if $([s], [t]) \doteq ([x], [y])$, with $s \prec t$ and $x \prec y$, then $F(t) - F(s) = F(y) - F(x)$,*
- *(B) The system of equations corresponding to $\doteq$ (Definition 10) has a solution consisting of positive integers.*
- *(C) There is no sequence $\langle((i_1, j_1) \doteq (k_1, \ell_1)), ..., ((i_m, j_m) \doteq (k_m, \ell_m))\rangle$ in $\doteq$ (expressed in terms of the canonical representation $\rho$ of $\precsim$ ) such that $B_{left} \sqsubset B_{right}$ where $B_{left} = B_{(i_1,j_1)} \uplus ... \uplus B_{(i_m,j_m)}$ and $B_{right} = B_{(k_1,\ell_1)} \uplus ... \uplus B_{(k_m,\ell_m)}$.*

As an application of Proposition 3 consider again the game of Figure 3 and plausibility order (5) which rationalizes the assessment $\sigma = (M, \ell, a, c, e)$, $\mu(x) = 1$ for $x \in \{\emptyset, M, Mr, Lm, R\ell\}$ and $\mu(x) = 0$ for every other decision history $x$; the order is reproduced below together with the canonical integer-valued representation $\rho$:

$$\begin{pmatrix}
 & \precsim : & \rho : \\
\text{most plausible} & \emptyset, M, M\ell & 0 \\
 & R, R\ell, R\ell e & 1 \\
 & Mm, Mme & 2 \\
 & Mr, Mra & 3 \\
 & L, L\ell, L\ell a & 4 \\
 & Rm & 5 \\
 & Lm, Lmc & 6 \\
 & Rr, Rrc & 7 \\
 & Lr & 8 \\
 & R\ell f & 9 \\
 & Mmf & 10 \\
 & Lmd & 11 \\
 & Rrd & 12 \\
 & Mrb & 13 \\
\text{least plausible} & L\ell b & 14
\end{pmatrix} \tag{16}$$

By Remark 2, two elements of $\doteq$ are $([M], [R]) \doteq ([Mr], [Rr])$ and $([Mm], [Lm]) \doteq ([M], [L])$, which—expressed in terms of the canonical ordinal representation $\rho$—can be written as

$$\begin{aligned}
(0, 1) &\doteq (3, 7) \\
(2, 6) &\doteq (0, 4)
\end{aligned}$$

Then $B_{left} = \{1\} \uplus \{3, 4, 5, 6\} = \{1, 3, 4, 5, 6\}$ and $B_{right} = \{4, 5, 6, 7) \uplus \{1, 2, 3, 4\} = \{1, 2, 3, 4, 4, 5, 6, 7\}$. Thus, since $B_{left} \sqsubset B_{right}$, by Part (C) of Proposition 3 $\precsim$ is not choice measurable.

As a further application of Proposition 3 consider the total pre-order $\precsim$ given in (13) together with the subset of the equivalence relation $\doteq$ given in (14). Then there is no cardinal representation of

$\precsim$ that satisfies the constraints expressed by $\doteq$, because of Part (C) of the above proposition and the following sequence[27]:

$$\langle((0,1) \doteq (3,4)),((1,3) \doteq (4,6)),((3,4) \doteq (1,3)),((4,6) \doteq (0,2))\rangle$$

where $B_{left} = \{1\} \uplus \{2,3\} \uplus \{4\} \uplus \{5,6\} = \{1,2,3,4,5,6\} \sqsubset B_{right} = \{4\} \uplus \{5,6\} \uplus \{2,3\} \uplus \{1,2\} = \{1,2,2,3,4,5,6\}$.

In fact, the above sequence corresponds to the following system of equations:

$$\begin{array}{rll} x_1 = x_4 & \text{corresponding to} & (0,1) \doteq (3,4) \\ x_2 + x_3 = x_5 + x_6 & \text{corresponding to} & (1,3) \doteq (4,6) \\ x_4 = x_2 + x_3 & \text{corresponding to} & (3,4) \doteq (1,3) \\ x_5 + x_6 = x_1 + x_2 & \text{corresponding to} & (4,6) \doteq (0,2) \end{array}$$

Adding the four equations we get $x_1 + x_2 + x_3 + x_4 + x_5 + x_6 = x_1 + 2x_2 + x_3 + x_4 + x_5 + x_6$ which simplifies to $0 = x_2$, which is incompatible with a positive solution.

**Remark 3.** *In [15] an algorithm is provided for determining whether a system of linear equations has a positive solution and for calculating such a solution if one exists. Furthermore, if the coefficients of the equations are integers and a positive solution exists, then the algorithm yields a solution consisting of positive integers.*

## 5. Related Literature

As noted in Section 1, the quest in the literature for a "simple" solution concept intermediate between subgame-perfect equilibrium and sequential equilibrium has produced several attempts to provide a general definition of perfect Bayesian equilibrium.

In [16] a notion of perfect Bayesian equilibrium was provided for a small subset of extensive-form games (namely the class of multi-stage games with observed actions and independent types), but extending that notion to arbitrary games proved to be problematic[28].

In [14] a notion of perfect Bayesian equilibrium is provided that can be applied to general extensive-form games (although it was defined only for games without chance moves); however, the proposed definition is in terms of a more complex object, namely a "tree-extended assessment" $(\nu, \sigma, \mu)$ where $\nu$ is a conditional probability system on the set of terminal nodes. The main idea underlying the notion of perfect Bayesian equilibrium proposed in [14] is what the author calls "strategic independence": when forming beliefs, the strategic choices of different players should be regarded as independent events.

Several more recent contributions [5,17,18] have re-addressed the issue of providing a definition of perfect Bayesian equilibrium that applies to general extensive-form games. Since [5] has been the focus of this paper, here we shall briefly discuss [17,18]. In [17] the notion of "simple perfect Bayesian equilibrium" is introduced and it is shown to lie strictly between subgame-perfect equilibrium and sequential equilibrium. This notion is based on an extension of the definition of sub-tree, called "quasi sub-tree", which consists of an information set $I$ together with all the histories that are successors of histories in $I$ (that is, $\Gamma_I$ is a quasi-subtree that starts at $I$ if $h' \in \Gamma_I$ if and only if there exists an $h \in I$ such that $h$ is a prefix of $h'$). A quasi sub-tree $\Gamma_I$ is called regular if it satisfies the following property: if $h \in \Gamma_I$ and $h' \in I(h)$ then $h' \in \Gamma_I$ (that is, every information set that has a non-empty intersection with $\Gamma_I$ is entirely included in $\Gamma_I$). An information set $I$ is called regular if the quasi-subtree that starts at $I$ is regular. For example, in the game of Figure 4, the singleton information set $\{b\}$ of

[27] By symmetry of $\doteq$, we can express the third and fourth constraints as $(4,6) \doteq (0,2)$ and $(3,4) \doteq (1,3)$ instead of $(0,2) \doteq (4,6)$ and $(1,3) \doteq (3,4)$, respectively.

[28] The main element of the notion of PBE put forward in [16] is the "no signaling what you don't know" condition on beliefs. For example, if Player 2 observes Player 1's action and Player 1 has observed nothing about a particular move of Nature, then Player 2 should not update her beliefs about Nature's choice based on Player 1's action.

Player 2 is *not* regular. An assessment $(\sigma, \mu)$ is defined to be a "simple perfect Bayesian equilibrium" if it is sequentially rational and, for every regular quasi-subtree $\Gamma_I$, Bayes' rule is satisfied at every information set that is reached with positive probability by $\sigma$ in $\Gamma_I$ (in other words, if the restriction of $(\sigma, \mu)$ to every regular quasi-subtree is a weak sequential equilibrium of the quasi-subtree). This notion of perfect Bayesian equilibrium is weaker than the notion considered in this paper (Definition 4). For example, in the game of Figure 4, the pure-strategy profile $\sigma = (c, d, f)$ (highlighted by double edges), together with the system of beliefs $\mu(a) = \mu(bd) = 0$, $\mu(be) = 1$, is a simple perfect Bayesian equilibrium, while (as shown in [5]) there is no system of beliefs $\mu'$ such that $(\sigma, \mu')$ is a perfect Bayesian equilibrium. *A fortiori*, the notion of simple perfect Bayesian equilibrium is weaker than the refinements of PBE discussed in the Section 3.

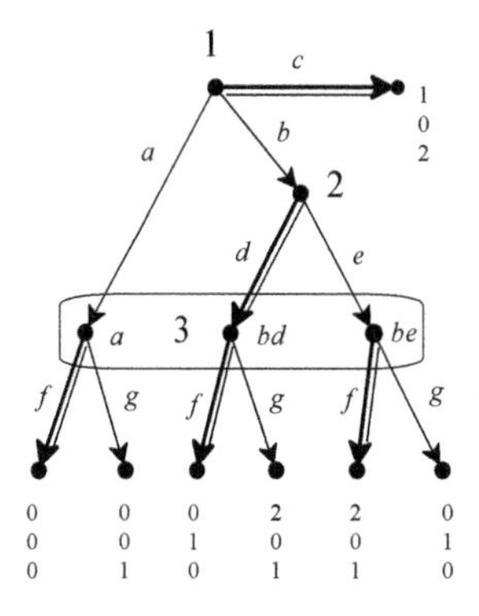


**Figure 4**

In [18], the author proposes a definition of perfect Bayesian equilibrium which is framed not in terms of assessments but in terms of "appraisals". Each player is assumed to have a (possibly artificial) information set representing the beginning of the game and an appraisal for player $i$ is a map that associates with every information set of player $i$ a probability distribution over the set of pure-strategy profiles that reach that information set. Thus an appraisal for player $i$ captures, for every information set of hers, her conjecture about how the information set was reached and what will happen from this point in the game. An appraisal system is defined to be "plainly consistent" if, whenever an information set of player $i$ has a product structure (each information set is identified with the set of pure-strategy profiles that reach that information set), the player's appraisal at that information set satisfies independence[29]. A strategy profile $\sigma$ is defined to be a perfect Bayesian equilibrium if there is a plainly consistent appraisal system $P$ that satisfies sequential rationality and is such that at their "initial" information sets all the players assign probability 1 to $\sigma$; in [18] (p. 15), the author summarizes the notion of PBE as being based on "a simple foundation: sequential rationality and preservation of independence and Bayesian updating where applicable" (that is, on subsets of strategy profiles that have the appropriate product structure and independence property). Despite the fact that the notion of PBE suggested in [18] incorporates a notion of independence, it can be weaker than the notion of PBE discussed in Section 2 (Definition 4) and thus, *a fortiori*, weaker than the notion of weakly independent PBE (Definition 8, Section 3). This can be seen from the game of Figure 5, which essentially reproduces an example given in [18]. The strategy profile $\sigma = (b, d, e)$ (highlighted by double edges), together with any system of beliefs $\mu$ such that $\mu(ac) > 0$ cannot be a

[29] Intuitively, on consecutive information sets, a player does not change her beliefs about the actions of other players, if she has not received information about those actions.

PBE according to Definition 4 (Section 2). In fact, since $\sigma(d) > 0$ while $\sigma(c) = 0$, any plausibility order that rationalizes $(\sigma, \mu)$ must be such that $a \sim ad \prec ac$, which implies that $\mu(ac) = 0$ (because $ac$ cannot be most plausible in the set $\{ac, ad, bc\}$). On the other hand, $\sigma$ can be a PBE according to the definition given in [18] (p. 15), since the information set of Player 3 does not have a product structure so that Player 3 is not able to separate the actions of Players 1 and 2. For example, consider the appraisal system $P$ where, initially, all the players assign probability 1 to $\sigma$ and, at his information set, Player 2 assigns probability 1 to the strategy profile $(b, e)$ of Players 1 and 3 and, at her information set, Player 3 assigns probability $\frac{1}{3}$ to each of the strategy profiles $(a, c)$, $(a, d)$ and $(b, c)$ of Players 1 and 2. Then $P$ is plainly consistent and sequentially rational, so that $\sigma$ is a PBE as defined in [18].

Thus, *a fortiori*, the notion of perfect Bayesian equilibrium given in [18] can be weaker than the notions of weakly/strongly independent PBE introduced in Section 3.

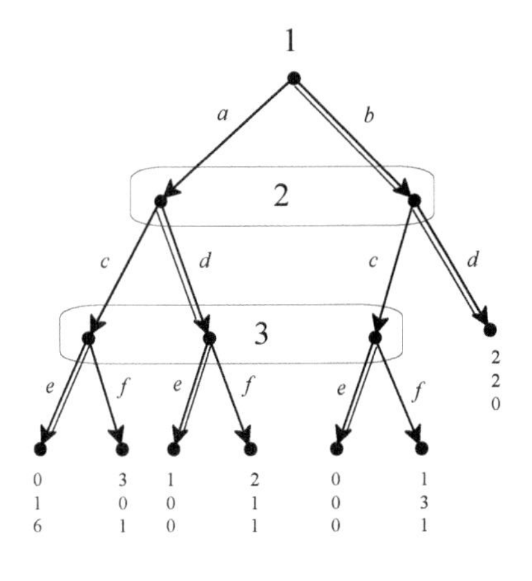


**Figure 5**

## 6. Conclusions

Besides sequential rationality, the notion of perfect Bayesian equilibrium (Definition 4) introduced in [5] is based on two elements: (1) rationalizability of the assessment by a plausibility order (Definition 2); and (2) the notion of Bayesian consistency relative to the plausibility order. The first property identifies the set of decision histories that can be assigned positive conditional probability by the system of beliefs, while the second property imposes constraints on how conditional probabilities can be distributed over that set in order to guarantee "Bayesian updating as long as possible"[30]. In [8] it was shown that by strengthening these two conditions one obtains a "limit free" characterization of sequential equilibrium. The strengthening of the first condition is that the plausibility order that rationalizes the given assessment be choice measurable, that it, that there be a cardinal representation of it (which can be interpreted as measuring the plausibility distance between histories in a way that is preserved by the addition of a common action). The strengthening of the second condition imposes "uniform consistency" on the conditional probability density functions on the equivalence classes of the plausibility order, by requiring that there be a full-support common prior

[30] By "Bayesian updating as long as possible" we mean the following: (1) when information causes no surprises, because the play of the game is consistent with the most plausible play(s) (that is, when information sets are reached that have positive prior probability), then beliefs should be updated using Bayes' rule; and (2) when information is surprising (that is, when an information set is reached that had zero prior probability) then new beliefs can be formed in an arbitrary way, but from then on Bayes' rule should be used to update those new beliefs, whenever further information is received that is consistent with those beliefs.

that preserves the relative probabilities of two decision histories in the same information set when a common action is added. There is a "substantial" gap between the notion of PBE and that of sequential equilibrium. In this paper we identified two solution concepts that lie in this gap. The first notion, weakly independent PBE (Definition 8), is obtained by adding a restriction on the belief revision policy encoded in the plausibility order that rationalizes the given assessment (together with strengthening Bayes consistency to uniform Bayes consistency). This restriction says that observation of a new action at an information set should not alter the relative plausibility of any two histories in that information set (condition $IND_1$); it can be interpreted as an "independence" or "conservative" principle, in the sense that observation of a new action should not lead to a reversal of judgment of plausibility concerning past histories. The second notion, strongly independent PBE (Definition 9), is obtained by adding to the first notion a further restriction, according to which the implicit plausibility ranking of two actions available at the same information set should be independent of the history at which the actions are taken.

A further contribution of this paper has been to provide a method for determining if a plausibility order is choice measurable, which is one of the two conditions that, together, are necessary and sufficient for a PBE to be a sequential equilibrium.

This paper highlights the need to conceptualize refinements of subgame-perfect equilibrium in extensive form games in terms of principles of belief revision. Through the notion of plausibility order and AGM-consistency we have appealed to the principles of belief revision that underlie the so-called AGM theory [6]. However, in a dynamic game, beliefs typically need to be revised several times in a sequence as new information sets are reached. Thus the relevant notion of belief revision is *iterated belief revision*. There is an extensive literature on the topic of iterated belief revision (see, for example, [19–22] and the special issue of the *Journal of Philosophical Logic*, Vol. 40 (2), April 2011). An exploration of different solution concepts in the gap between PBE and sequential equilibrium, based on different principles of iterated belief revision, seems to be a promising area of research.

**Conflicts of Interest:** The author declares no conflict of interest.

## Appendix A. Proofs

**Proof of Lemma 1.** Let $\precsim$ be a plausibility order on the set of histories $H$ and let $F : H \to \mathbb{N}$ be an integer-valued representation of $\precsim$. We want to show that properties $CM$ and $CM'$ below are equivalent (for arbitrary $h, h' \in H$, with $h' \in I(h)$, and $a, b \in A(h)$)

$$F(h') - F(h) = F(h'a) - F(ha). \tag{CM}$$

$$F(hb) - F(ha) = F(h'b) - F(h'a). \tag{CM'}$$

First of all, note that, without loss of generality, we can assume that $F(\emptyset) = 0$[31].

First we show that $CM \Rightarrow CM'$. Let $F$ be an integer-valued representation of $\precsim$ that satisfies $CM$. For every decision history $h$ and action $a \in A(h)$, define

$$\lambda(a) = F(ha) - F(h). \tag{A1}$$

The function $\lambda : A \to \mathbb{N}$ is well defined, since, by $CM$, $h' \in I(h)$ implies that $F(h'a) - F(h') = F(ha) - F(h)$. Then, for every history $h = \langle a_1, a_2, ..., a_m \rangle$, $F(h) = \sum_{i=1}^{m} \lambda(a_i)$. In fact,

$$\begin{aligned}
&\lambda(a_1) + \lambda(a_2) + ... + \lambda(a_m) = \\
&\quad = (F(a_1) - F(\emptyset)) + (F(a_1a_2) - F(a_1)) + ... + (F(a_1a_2...a_m) - F(a_1a_2...a_{m-1})) = \\
&\quad = F(a_1a_2...a_m) = F(h) \text{ (recall the hypothesis that } F(\emptyset) = 0).
\end{aligned}$$

[31] It is straightforward to check that if $F' : H \to \mathbb{N}$ is an integer-valued representation of $\precsim$ then so is $F : H \to \mathbb{N}$ defined by $F(h) = F'(h) - F'(\emptyset)$; furthermore if $F'$ satisfies property $CM$ ($CM'$) then so does $F$.

Thus, for every $h \in D$ and $a \in A(h)$, $F(ha) = F(h) + \lambda(a)$. Hence, $F(hb) - F(ha) = F(h) + \lambda(b) - F(h) - \lambda(a) = \lambda(b) - \lambda(a)$ and $F(h'b) - F(h'a) = F(h') + \lambda(b) - F(h') - \lambda(a) = \lambda(b) - \lambda(a)$ and, therefore, $F(hb) - F(ha) = F(h'b) - F(h'a)$.

Next we show that $CM' \Rightarrow CM$. Let $\precsim$ be a plausibility order and let $F : H \to \mathbb{N}$ be an integer-valued representation that satisfies $CM'$. Select arbitrary $h' \in I(h)$ and $a \in A(h)$. Let $b \in A(h)$ be a plausibility-preserving action at $h$ (there must be at least one such action: see Definition 1); then, $h \sim hb$ and $h' \sim h'b$. Hence, since $F$ is a representation of $\precsim$, $F(hb) = F(h)$ and $F(h'b) = F(h')$ and thus

$$F(h') - F(h) = F(h'b) - F(hb). \tag{A2}$$

By $CM'$, $F(h'b) - F(hb) = F(h'a) - F(ha)$. From this and (A2) it follows that $F(h') - F(h) = F(h'a) - F(ha)$. □

**Proof of Proposition 2.** Let $(\sigma, \mu)$ be a sequential equilibrium. We want to show that $(\sigma, \mu)$ is a strongly independent PBE (Definition 9). By Proposition 1, it is sufficient to show that $(\sigma, \mu)$ is rationalized by a plausibility order $\precsim$ that satisfies Properties $IND_1$ and $IND_2$. By Proposition 1 there is a choice measurable plausibility order $\precsim$ that rationalizes $(\sigma, \mu)$. Let $F$ be an integer-valued representation of $\precsim$ that satisfies Property $CM$. Let $h$ and $h'$ be decision histories that belong to the same information set and let $a \in A(h)$. Then, by $CM$,

$$F(h) - F(h') = F(ha) - F(h'a). \tag{A3}$$

If $h \precsim h'$ then $F(h) \leq F(h')$; by (A3), it follows that $F(ha) \leq F(h'a)$ and thus $ha \precsim h'a$. Conversely, if $ha \precsim h'a$ then $F(ha) \leq F(h'a)$ and thus, by (A3), $F(h) \leq F(h')$ so that $h \precsim h'$. Hence $\precsim$ satisfies $IND_1$.

Let $h$ and $h'$ be decision histories that belong to the same information set and let $a, b \in A(h)$. We want to show that $IND_2$ holds, that is, that $ha \precsim hb$ if and only if $h'a \precsim h'b$. Let $F$ be an integer-valued representation of $\precsim$ that satisfies Property $CM$. By Lemma 1 $F$ satisfies Property $CM'$, that is,

$$F(ha) - F(hb) = F(h'a) - F(h'b). \tag{A4}$$

If $ha \precsim hb$ then $F(ha) \leq F(hb)$ and thus, by (A4), $F(h'a) \leq F(h'b)$, that is, $h'a \precsim h'b$. Conversely, if $h'a \precsim h'b$ then $F(h'a) \leq F(h'b)$ and thus, by (A4), $F(ha) \leq F(hb)$, so that $ha \precsim hb$. □

**Proof of Proposition 3.** $(A) \Rightarrow (B)$. Let $F' : S \to \mathbb{N}$ satisfy the properties of Part $(A)$. Select an arbitrary $s_0 \in S_0 = \{s \in S : s \precsim t, \forall t \in S\}$ and define $F : S \to \mathbb{N}$ by $F(s) = F'(s) - F'(s_0)$. Then $F$ is also a function that satisfies the properties of Part $(A)$ (note that since, for all $s \in S$, $F'(s_0) \leq F'(s)$, $F(s) \in \mathbb{N}$; furthermore, $F(s') = 0$ for all $s' \in S_0$). Let $K = \{k \in \mathbb{N} : k = \rho(s)$ for some $s \in S\}$ (where $\rho$ is the canonical ordinal representation of $\precsim$: see Footnote 23). For every $k \in K$, define

$$\begin{aligned} &\hat{x}_0 = 0 \\ &\text{and, for } k > 0, \\ &\hat{x}_k = F(t) - F(s) \quad \text{for some } s,t \in S \text{ such that } \rho(t) = k \text{ and } \rho(s) = k-1. \end{aligned} \tag{A5}$$

Thus $\hat{x}_k$ is the distance, as measured by $F$, between the equivalence class of some $t$ such that $\rho(t) = k$ and the immediately preceding equivalence class (that is, the equivalence class of some $s$ such that $\rho(s) = k - 1)$[32]. Note that $\hat{x}_k$ is well defined since, if $x, y \in S$ are such that $\rho(y) = k$ and $\rho(x) = k - 1$, then $x \sim s$ and $y \sim t$ and thus, by (1) of Property $(A)$, $F(x) = F(s)$ and $F(y) = F(t)$. Note also that, for all $k \in K \setminus \{0\}$, $\hat{x}_k$ is a positive integer, since $\rho(t) = k$ and $\rho(s) = k - 1$ imply that $s \prec t$

[32] For example, if $S = \{a, b, c, d, e, f\}$ and $\precsim$ is given by $a \sim b \prec c \prec d \sim e \prec f$ then $\rho(a) = \rho(b) = 0, \rho(c) = 1, \rho(d) = \rho(e) = 2$ and $\rho(f) = 3$; if $F$ is given by $F(a) = F(b) = 0, F(c) = 3, F(d) = F(e) = 5$ and $F(f) = 9$ then $\hat{x}_0 = 0, \hat{x}_1 = 3, \hat{x}_2 = 2$ and $\hat{x}_3 = 4$.

and thus, by (1) of Property $(A)$, $F(s) < F(t)$. We want to show that the values $\{\hat{x}_k\}_{k\in K\setminus\{0\}}$ defined in (A5) provide a solution to the system of equations corresponding to $\doteq$ (Definition 10). Select an arbitrary element of $\doteq$, $([s_1],[s_2]) \doteq ([t_1],[t_2])$ (with $s_1 \prec s_2$ and $t_1 \prec t_2$) and express it, using the canonical ordinal representation $\rho$ (see Footnote 23), as $(i_1,i_2) \doteq (j_1,j_2)$ (thus $i_1 = \rho(s_1)$, $i_2 = \rho(s_2)$, $j_1 = \rho(t_1)$, $j_2 = \rho(t_2)$, $i_1 < i_2$ and $j_1 < j_2$). Then the corresponding equation (see Definition 10) is: $x_{i_1+1} + x_{i_1+2} + ... + x_{i_2} = x_{j_1+1} + x_{j_1+2} + ... + x_{j_2}$. By (2) of Property $(A)$,

$$F(s_2) - F(s_1) = F(t_2) - F(t_1) \tag{A6}$$

Using (A5), $F(s_2) - F(s_1) = \hat{x}_{i_1+1} + \hat{x}_{i_1+2} + ... + \hat{x}_{i_2}$. To see this, for every $k \in \{i_1+1, i_1+2, ..., i_2-1\}$, select an arbitrary $r_k \in S$ such that $\rho(r_k) = k$; then, by (A5),

$$F(s_2) - F(s_1) = \underbrace{\hat{x}_{i_1+1}}_{=F(r_{i_1+1})-F(s_1)} + \underbrace{\hat{x}_{i_2+2}}_{=F(r_{i_1+2})-F(r_{i_1+1})} + ... + \underbrace{\hat{x}_{i_2}}_{=F(s_2)-F(r_{i_2-1})}.$$

Similarly, $F(t_2) - F(t_1) = \hat{x}_{j_1+1} + \hat{x}_{j_1+2} + ... + \hat{x}_{j_2}$. Thus, by (A6), $\hat{x}_{i_1+1} + \hat{x}_{i_1+2} + ... + \hat{x}_{i_2} = \hat{x}_{j_1+1} + \hat{x}_{j_1+2} + ... + \hat{x}_{j_2}$.

$(B) \Rightarrow (A)$. Assume that the system of equations corresponding to $\doteq$ has a solution consisting of positive integers $\hat{x}_1, ..., \hat{x}_m$. Define $F : S \to \mathbb{N}$ as follows: if $\rho(s) = 0$ (equivalently, $s \in S_0$) then $F(s) = 0$ and if $\rho(s) = k > 0$ (equivalently, $s \in S_k$ for $k > 0$) then $F(s) = \hat{x}_1 + \hat{x}_2 + ... + \hat{x}_k$ (where $\rho$ and the sets $S_k$ are as defined in Footnote 23). We need to show that $F$ satisfies the properties of Part $(A)$. Select arbitrary $s, t \in S$ with $s \precsim t$. Then $\rho(s) \le \rho(t)$ and thus $F(s) = \hat{x}_1 + \hat{x}_2 + ... + \hat{x}_{\rho(s)} \le F(t) = \hat{x}_1 + \hat{x}_2 + ... + \hat{x}_{\rho(s)} + \hat{x}_{\rho(s)+1} + ... + \hat{x}_{\rho(t)}$. Conversely, suppose that $s, t \in S$ are such that $F(s) \le F(t)$. Then $\hat{x}_1 + \hat{x}_2 + ... + \hat{x}_{\rho(s)} \le \hat{x}_1 + \hat{x}_2 + ... + \hat{x}_{\rho(t)}$ and thus $\rho(s) \le \rho(t)$, so that $s \precsim t$. Thus Property (1) of Part $(A)$ is satisfied. Now let $s, t, x, y \in S$ be such that $s \prec t$, $x \prec y$ and $([s],[t]) \doteq ([x],[y])$. Let $\rho(s) = i$, $\rho(t) = j$, $\rho(x) = k$ and $\rho(y) = \ell$ (thus $i < j$ and $k < \ell$). Then, by (A5), $F(t) - F(s) = \hat{x}_{i+1} + \hat{x}_{i+2} + ... + \hat{x}_j$ and $F(y) - F(x) = \hat{x}_{k+1} + \hat{x}_{k+2} + ... + \hat{x}_\ell$. Since $x_{i+1} + x_{i+2} + ... + x_j = x_{k+1} + x_{k+2} + ... + x_\ell$ is the equation corresponding to $([s],[t]) \doteq ([x],[y])$ (which - using $\rho$ - can be expressed as $(i,j) \doteq (k,\ell)$), by our hypothesis $\hat{x}_{i+1} + \hat{x}_{i+2} + ... + \hat{x}_j = \hat{x}_{k+1} + \hat{x}_{k+2} + ... + \hat{x}_\ell$ and thus $F(t) - F(s) = F(y) - F(x)$, so that (2) of Property $(A)$ is satisfied.

*not* $(B) \Rightarrow$ *not* $(C)$. Suppose that there is a sequence in $\doteq$ (expressed in terms of the canonical representation $\rho$ of $\precsim$) $\langle((i_1,j_1) \doteq (k_1,\ell_1)), ..., ((i_m,j_m) \doteq (k_m,\ell_m))\rangle$ such that

$$B_{left} \sqsubset B_{right} \tag{A7}$$

where $B_{left} = B_{(i_1,j_1)} \uplus ... \uplus B_{(i_m,j_m)}$ and $B_{right} = B_{(k_1,\ell_1)} \uplus ... \uplus B_{(k_m,\ell_m)}$. Let $\mathbb{E} = \{E_1, ..., E_m\}$ be the system of equations corresponding to the above sequence (for example, $E_1$ is the equation $x_{i_1+1} + x_{i_1+2} + ... + x_{j_1} = x_{k_1+1} + x_{k_1+2} + ... + x_{\ell_1}$). Let $L$ be the sum of the left-hand-side and $R$ be the sum of the right-hand-side of the equations $E_1, ..., E_m$. Note that for every integer $i$, $nx_i$ is a summand of $L$ if and only if $i$ appears in $B_{left}$ exactly $n$ times and similarly $nx_i$ is a summand of $R$ if and only if $i$ appears in $B_{right}$ exactly $n$ times. By (A7), if $nx_i$ is a summand of $L$ then $mx_i$ is a summand of $R$ with $m \ge n$ and, furthermore, $L \ne R$. Thus there cannot be a positive solution of $\mathbb{E}$, because it would be incompatible with $L = R$. Since $\mathbb{E}$ is a subset of the system of equations corresponding to $\doteq$, it follows that the latter cannot have a positive solution either.

It only remains to prove that *not* $(C) \Rightarrow$ *not* $(B)$. We will return to this below after providing an additional result. □

First some notation. Given two vectors $x, y \in \mathbb{R}^m$ we write (1) $x \le y$ if $x_i \le y_i$, for every $i = 1, ..., m$; (2) $x < y$ if $x \le y$ and $x \ne y$; and (3) $x \ll y$ if $x_i < y_i$, for every $i = 1, ..., m$.

**Lemma A1.** *Let $A$ be the $m \times n$ matrix such that the system of equations corresponding to $\doteq$ (Definition 10) can be expressed as $Ax = 0$ (note that each entry of $A$ is either $-1$, $0$ or $1$; furthermore, by symmetry of $\doteq$,* *for each row $a_i$ of $A$ there is another row $a_k$ such that $a_k = -a_i$)*[33]. *If the system of equations $Ax = 0$ does not have a positive integer solution then there exist $r$ rows of $A$, $a_{i_1}, ..., a_{i_r}$ with $1 < r \leq \frac{m}{2}$ and $r$ positive integers $\alpha_1, ..., \alpha_r \in \mathbb{N}\backslash\{0\}$ such that if $B$ is the submatrix of $A$ consisting of the $r$ rows $a_{i_1}, ..., a_{i_r}$ (thus for every $k = 1, ..., r$, $b_k = a_{i_k}$, where $b_k$ is the $k^{th}$ row of $B$) then $\sum_{k=1}^{r} \alpha_k b_k < 0$.*

**Proof.** By Stiemke's theorem[34] if the system of equations $Ax = 0$ does not have a positive integer solution then there exists a $y \in \mathbb{Z}^m$ (where $\mathbb{Z}$ denotes the set of integers) such that $yA < 0$ (that is, $\sum_{i=1}^{m} y_i a_i < 0$). Let $K = \{k \in \mathbb{Z} : y_k \neq 0\}$. Let $r$ be the cardinality of $K$; then, without loss of generality, we can assume that $r \leq \frac{m}{2}$[35]. Furthermore, again without loss of generality, we can assume that for every $k \in K$, $y_k > 0$[36]. Let $B$ be the $r \times n$ submatrix of $A$ consisting of those rows $a_k$ of $A$ such that $k \in K$ and for $i = 1, ..., r$ let $\alpha = (\alpha_1, ..., \alpha_r)$ be the vector corresponding to $(y_k)_{k \in K}$[37]. Then $\alpha B = \sum_{j=1}^{r} \alpha_j b_j = yA < 0$ and $\alpha_i \in \mathbb{N}\backslash\{0\}$ for all $i = 1, ..., r$. □

**Completion of Proof of Proposition 3.** It remains to prove that *not* $(C) \Rightarrow$ *not* $(B)$. Let $A$ be the $m \times n$ matrix such that the system of equations corresponding to $\doteq$ can be expressed as $Ax = 0$ and assume that $Ax = 0$ does not have a positive integer solution. Let $B$ be the $r \times n$ submatrix of $A$ and $\alpha = (\alpha_1, ..., \alpha_r)$ the vector of positive integers of Lemma A1 such that $\alpha B = \sum_{j=1}^{r} \alpha_j b_j < 0$. Define two $r \times n$ matrices $C = (c_{ij})_{i=1,...,r;\ j=1,...,n}$ and $D = (d_{ij})_{i=1,...,r;\ j=1,...,n}$ as follows (recall that each entry of $B$ is either $-1$, 0 or 1):

$$c_{ij} = \begin{cases} 1 & \text{if } b_{ij} = 1 \\ 0 & \text{otherwise} \end{cases} \text{ and } d_{ij} = \begin{cases} 1 & \text{if } b_{ij} = -1 \\ 0 & \text{otherwise} \end{cases}.$$

Then, for every $i = 1, ..., r$, $b_i = c_i - d_i$ and thus (since $\sum_{i=1}^{r} \alpha_i b_i < 0$)

$$\sum_{i=1}^{r} \alpha_i c_i < \sum_{i=1}^{r} \alpha_i d_i. \quad \text{(A9)}$$

---

33 For example, the system of Equation (15) can be written as $Ax = 0$, where $x = (x_1, ..., x_5)$ and

$$A = \begin{pmatrix} 1 & 1 & 0 & -1 & 0 \\ -1 & -1 & 0 & 1 & 0 \\ 0 & 0 & 1 & 0 & -1 \\ 0 & 0 & -1 & 0 & 1 \\ 1 & 1 & 1 & -1 & -1 \\ -1 & -1 & -1 & 1 & 1 \end{pmatrix} \quad \text{(A8)}$$

34 See, for example, [23] (p. 216) or [24] (Theorem 1.1, p. 65).

35 Proof. Recall that for each row $a_i$ of $A$ there is a row $a_k$ such that $a_i = -a_k$. If $y_i \neq 0$ and $y_k \neq 0$ for some $i$ and $k$ such that $a_i = -a_k$ then

$$y_i a_i + y_k a_k = \begin{cases} 0 & \text{if } y_i = y_k \\ (y_k - y_i) a_k & \text{if } 0 < y_i < y_k \\ (y_i - y_k) a_i & \text{if } 0 < y_k < y_i \\ (|y_i| + y_k)\, a_k & \text{if } y_i < 0 < y_k \\ (y_i + |y_k|)\, a_i & \text{if } y_k < 0 < y_i \\ (|y_k| - |y_i|)\, a_i & \text{if } y_i < y_k < 0 \\ (|y_i| - |y_k|)\, a_k & \text{if } y_k < y_i < 0 \end{cases}$$

where all the multipliers (of $a_i$ or $a_k$) are positive. Thus one can set one of the two values of $y_i$ and $y_k$ to zero and replace the other value with the relevant of the above values while keeping $yA$ unchanged. For example, if $y_k < y_i < 0$ then one can replace $y_i$ with 0 and $y_k$ with $(|y_i| - |y_k|)$ thereby reducing the cardinality of $K$ by one. This process can be repeated until the multipliers of half of the rows of $A$ have been replaced by zero.

36 Proof. Suppose that $y_k < 0$ for some $k \in K$. Recall that there exists an $i$ such that $a_k = -a_i$. By the argument of the previous footnote, $y_i = 0$. Then replace $y_k$ by 0 and replace $y_i = 0$ by $\tilde{y}_i = -y_k$.

37 For example, if $K = \{3, 6, 7\}$ and $y_3 = 2$, $y_6 = 1$, $y_7 = 3$, then $B$ is the $3 \times n$ matrix where $b_1 = a_3$, $b_2 = a_6$ and $b_3 = a_7$ and $\alpha_1 = 2$, $\alpha_2 = 1$ and $\alpha_3 = 3$.

Let $C'$ be the matrix obtained from $C$ by replacing each row $c_i$ of $C$ with $\alpha_i$ copies of it and let $D'$ be constructed from $D$ similarly. Then, letting $s = \sum_{i=1}^{r} \alpha_i$, $C'$ and $D'$ are $s \times n$ matrices whose entries are either 0 or 1. It follows from (A9) that

$$\sum_{i=1}^{s} c'_i < \sum_{i=1}^{s} d'_i. \tag{A10}$$

Consider the system of equations

$$C'x = D'x. \tag{A11}$$

For every $j = 1, ..., n$, the $j^{th}$ coordinate of $\sum_{i=1}^{s} c'_i$ is the number of times that the variable $x_j$ appears on the left-hand-side of (A11) and the $j^{th}$ coordinate of $\sum_{i=1}^{s} d'_i$ is the number of times that the variable $x_j$ appears on the right-hand-side of (A11). Hence, by (A10), for every $j = 1, ..., n$, the number of times that the variable $x_j$ appears on the left-hand-side of (A11) is less than or equal to the number of times that it appears on the right-hand-side of (A11) and for at least one $j$ it is less. Thus, letting $\langle((i_1, j_1) \doteq (k_1, \ell_1)), ..., ((i_s, j_s) \doteq (k_s, \ell_s))\rangle$ be the sequence of elements of $\doteq$ corresponding to the equations in (A11), we have that $B_{left} \sqsubset B_{right}$ where $B_{left} = B_{(i_1,j_1)} \uplus ... \uplus B_{(i_m,j_m)}$ and $B_{right} = B_{(k_1,\ell_1)} \uplus ... \uplus B_{(k_m,\ell_m)}$. □

## References


1. Kreps, D.; Wilson, R. Sequential equilibrium. *Econometrica* **1982**, *50*, 863–894.
2. Mas-Colell, A.; Whinston, M.D.; Green, J.R. *Microeconomic Theory*; Oxford University Press: Oxford, UK, 1995.
3. Myerson, R. *Game Theory: Analysis of Conflict*; Harvard University Press: Cambridge, MA, USA, 1991.
4. Selten, R. Re-examination of the perfectness concept for equilibrium points in extensive games. *Int. J. Game Theory* **1975**, *4*, 25–55.
5. Bonanno, G. AGM-consistency and perfect Bayesian equilibrium. Part I: Definition and properties. *Int. J. Game Theory* **2013**, *42*, 567–592.
6. Alchourrón, C.; Gärdenfors, P.; Makinson, D. On the logic of theory change: Partial meet contraction and revision functions. *J. Symb. Log.* **1985**, *50*, 510–530.
7. Bonanno, G. AGM belief revision in dynamic games. In Proceedings of the 13th Conference on Theoretical Aspects of Rationality and Knowledge (TARK XIII), Groningen, The Netherlands, 12–14 July 2011; Apt, K.R., Ed.; ACM: New York, NY, USA, 2011; pp. 37–45.
8. Bonanno, G. AGM-consistency and perfect Bayesian equilibrium. Part II: From PBE to sequential equilibrium. *Int. J. Game Theory* **2005**, doi:10.1007/s00182-015-0506-6.
9. Osborne, M.; Rubinstein, A. *A Course In Game Theory*; MIT Press: Cambridge, MA, USA, 1994.
10. Hendon, E.; Jacobsen, J.; Sloth, B. The one-shot-deviation principle for sequential rationality. *Games Econ. Behav.* **1996**, *12*, 274–282.
11. Perea, A. A note on the one-deviation property in extensive form games. *Games Econ. Behav.* **2002**, *40*, 322–338.
12. Kohlberg, E.; Reny, P. Independence on relative probability spaces and consistent assessments in game trees. *J. Econ. Theory* **1997**, *75*, 280–313.
13. Perea, A.; Jansen, M.; Peters, H. Characterization of consistent assessments in extensive-form games. *Games Econ. Behav.* **1997**, *21*, 238–252.
14. Battigalli, P. Strategic independence and perfect Bayesian equilibria. *J. Econ. Theory* **1996**, *70*, 201–234.
15. Dines, L. On positive solutions of a system of linear equations. *Ann. Math.* **1926–1927**, *28*, 386–392.
16. Fudenberg, D.; Tirole, J. Perfect Bayesian equilibrium and sequential equilibrium. *J. Econ. Theory* **1991**, *53*, 236–260.
17. González-Díaz, J.; Meléndez-Jiménez, M.A. On the notion of perfect Bayesian equilibrium. *TOP J. Span. Soc. Stat. Oper. Res.* **2014**, *22*, 128–143.
18. Watson, J. *Perfect Bayesian Equilibrium: General Definitions and Illustrations*; Working Paper; University of California San Diego: San Diego, CA, USA, 2016.
19. Bonanno, G. Belief change in branching time: AGM-consistency and iterated revision. *J. Philos. Log.* **2012**, *41*, 201–236.

20. Boutilier, C. Iterated revision and minimal change of conditional beliefs. *J. Philos. Log.* **1996**, *25*, 262–305.
21. Darwiche, A.; Pearl, J. On the logic of iterated belief revision. *Artif. Intell.* **1997**, *89*, 1–29.
22. Stalnaker, R. Iterated belief revision. *Erkenntnis* **2009**, *70*, 189–209.
23. Schrijver, A. *Theory of Linear and Integer Programming*; John Wiley & Sons: Hoboken, NJ, USA, 1986.
24. Fishburn, P.C. Finite linear qualitative probability. *J. Math. Psychol.* **1996**, *40*, 64–77.

 *games*

*Article*

# Epistemically Robust Strategy Subsets

**Geir B. Asheim [1,*], Mark Voorneveld [2] and Jörgen W. Weibull [2,3,4]**

1 Department of Economics, University of Oslo, P.O. Box 1095 Blindern, NO-0317 Oslo, Norway
2 Department of Economics, Stockholm School of Economics, Box 6501, SE-113 83 Stockholm, Sweden; nemv@hhs.se (M.V.); nejw@hhs.se (J.W.W.)
3 Institute for Advanced Study in Toulouse, 31000 Toulouse, France
4 Department of Mathematics, KTH Royal Institute of Technology, SE-100 44 Stockholm, Sweden
* Correspondence: g.b.asheim@econ.uio.no; Tel.: +47-455-051-36

Academic Editors: Paul Weirich and Ulrich Berger
Received: 31 August 2016; Accepted: 17 November 2016; Published: 25 November 2016

**Abstract:** We define a concept of *epistemic robustness* in the context of an epistemic model of a finite normal-form game where a player type corresponds to a belief over the profiles of opponent strategies and types. A Cartesian product $X$ of pure-strategy subsets is epistemically robust if there is a Cartesian product $Y$ of player type subsets with $X$ as the associated set of best reply profiles such that the set $Y_i$ contains all player types that believe with sufficient probability that the others are of types in $Y_{-i}$ and play best replies. This robustness concept provides epistemic foundations for set-valued generalizations of strict Nash equilibrium, applicable also to games without strict Nash equilibria. We relate our concept to closedness under rational behavior and thus to strategic stability and to the best reply property and thus to rationalizability.

**Keywords:** epistemic game theory; epistemic robustness; rationalizability; closedness under rational behavior; mutual $p$-belief

**JEL Classification:** C72; D83

---

## 1. Introduction

In most applications of noncooperative game theory, Nash equilibrium is used as a tool to predict behavior. Under what conditions, if any, is this approach justified? In his Ph.D. thesis, Nash [1] suggested two interpretations of Nash equilibrium, one rationalistic, in which all players are fully rational, know the game, and play it exactly once. In the other, "mass action" interpretation, there is a large population of actors for each player role of the game, and now and then exactly one actor from each player population is drawn at random to play the game in his or her player role, and this is repeated (i.i.d.) indefinitely over time. Whereas the latter interpretation is studied in the literature on evolutionary game theory and social learning, the former—which is the interpretation we will be concerned with here—is studied in a sizeable literature on epistemic foundations of Nash equilibrium. It is by now well-known from this literature that players' rationality and beliefs or knowledge about the game and each others' rationality in general do not imply that they necessarily play a Nash equilibrium or even that their conjectures about each others' actions form a Nash equilibrium; see Bernheim [2], Pearce [3], Aumann and Brandenburger [4].

The problem is not only a matter of coordination of beliefs (conjectures or expectations), as in a game with multiple equilibria. It also concerns the fact that, in Nash equilibrium interpreted as an equilibrium in belief (see [4], Theorems A and B), beliefs are supposed to correspond to *specific* randomizations over the others' strategies. In particular, a player might have opponents with multiple pure strategies that maximize their expected payoffs, given their equilibrium beliefs. Hence, for these opponents, any randomization over their pure best replies maximizes their expected payoffs. Yet in Nash equilibrium, the player is assumed to have a belief that singles out a randomization over the best replies of her opponents that serves to keep this player indifferent across the support of her equilibrium strategies, and ensures that none of the player's other strategies are better replies. In addition, a player's belief concerning the behavior of others assigns positive probability *only* to best replies; players are not allowed to entertain any doubt about the rationality of their opponents.

Our aim is to formalize a notion of epistemic robustness that relaxes these requirements. In order to achieve this, we have to move away from point-valued to set-valued solution concepts. In line with the terminology of epistemic game theory, let a player's epistemic *type* correspond to a belief over the profiles of opponent strategies and types. Assume that the epistemic model is complete in the sense that all possible types are represented in the model. Let non-empty Cartesian products of (pure-strategy or type) subsets be referred to as (strategy or type) *blocks* [5]. Say that a strategy block $X = X_1 \times \cdots \times X_n$ is epistemically robust if there exists a corresponding type block $Y = Y_1 \times \cdots \times Y_n$ such that: for each player $i$,

(I) the strategy subset $X_i$ coincides with the set of best replies of the types in $Y_i$;
(II) the set $Y_i$ contains all player types that believe with sufficient probability that the others are of types in $Y_{-i}$ and play best replies.

Here, for each player, (II) requires the player's type subset to be robust in the sense of including all possible probability distributions over opponent pure-strategy profiles that consist of best replies to the beliefs of opponent types that are included in the opponents' type subsets, even including player types with a smidgen of doubt that only these strategies are played. In particular, our epistemic model does not allow a player to pinpoint a specific opponent type or a specific best reply for an opponent type that has multiple best replies. The purpose of (I) is, for each player, to map this robust type subset into a robust subset of pure strategies by means of the best reply correspondence.

Consider, in contrast, the case where point (II) above is replaced by:

(II′) the set $Y_i$ contains *only* player types that believe with probability 1 that the others are of types in $Y_{-i}$ and play best replies.

Tan and Werlang [6] show that the strategy block $X$ is a *best reply set* [3] if there exists a corresponding type block $Y$ such that (I) and (II′) hold for all players. This epistemic characterization of a best reply set $X$ explains why, for each player $i$, all strategies in $X_i$ are included. In contrast, the concept of epistemic robustness explains why all strategies outside $X_i$ are excluded, as a rational player will never choose such a strategy, not even if the player with small probability believes that opponents will not stick to their types $Y_{-i}$ or will not choose best replies.

Any strict Nash equilibrium, viewed as a singleton strategy block, is epistemically robust. In this case, each player has opponents with unique pure strategies that maximize their expected payoffs, given their equilibrium beliefs. The player's equilibrium strategy remains her unique best reply, as long as she is *sufficiently sure* that the others stick to their unique best replies. By contrast, non-strict pure-strategy Nash equilibria by definition have 'unused' best replies and are consequently not epistemically robust: a player, even if she is sure that her opponents strive to maximize their expected payoffs given their equilibrium beliefs, might well believe that her opponents play such alternative best replies.

In informal terms, our Proposition 1 establishes that epistemic robustness is sufficient and necessary for the non-existence of such 'unused' best replies. Consequently, epistemic robustness captures, through restrictions on the players' beliefs, a property satisfied by strict Nash equilibria, but not by non-strict pure-strategy Nash equilibria. The restrictions on players' beliefs implied by epistemic robustness can be imposed also on games without strict Nash equilibria. Indeed, our Propositions 2–5 show how epistemic robustness is achieved by variants of CURB sets. A CURB set (mnemonic for 'closed under rational behavior') is a strategy block that contains, for each player, all best replies to all probability distributions over the opponent strategies in the block[1]. Hence, if a player believes that her opponents stick to strategies from their components of a CURB set, then she'd better stick to her strategies as well.

A strategy block is fixed under rational behavior (FURB; or 'tight' CURB in the terminology of Basu and Weibull [7]) if each player's component not only contains, but is identical with the set of best replies to all probability distributions over the opponent strategies in the block. Basu and Weibull [7] show that minimal CURB (MINCURB) sets and the unique largest FURB set are important special cases of FURB sets. The latter equals the strategy block of rationalizable strategies [2,3]. At the other extreme, MINCURB is a natural set-valued generalization of strict Nash equilibrium. The main purpose of this paper is to provide epistemic foundations for set-valued generalizations of strict Nash equilibrium. Our results are not intended to advocate any particular point- or set-valued solution concept, only to propose a definition of epistemic robustness and apply this to some set-valued solution concepts currently in use[2].

In order to illustrate our line of reasoning, consider first the two-player game

| | $l$ | $c$ |
|---|---|---|
| $u$ | 3,1 | 1,2 |
| $m$ | 0,3 | 2,1 |

In its unique Nash equilibrium, player 1's equilibrium strategy assigns probability 2/3 to her first pure strategy and player 2's equilibrium strategy assigns probability 1/4 to his first pure strategy. However, even if player 1's belief about the behavior of player 2 coincides with his equilibrium strategy, $(1/4, 3/4)$, player 1 would be indifferent between her two pure strategies. Hence, any pure or mixed strategy would be optimal for her, under the equilibrium belief about player 2. For all other beliefs about her opponent's behavior, only one of her pure strategies would be optimal, and likewise for player 2. The unique CURB set and unique epistemically robust subset in this game is the full set $S = S_1 \times S_2$ of pure-strategy profiles.

Add a third pure strategy for each player to obtain the two-player game

| | $l$ | $c$ | $r$ |
|---|---|---|---|
| $u$ | 3,1 | 1,2 | 0,0 |
| $m$ | 0,3 | 2,1 | 0,0 |
| $d$ | 5,0 | 0,0 | 6,4 |

(1)

Strategy profile $x^* = (x_1^*, x_2^*) = \left(\left(\frac{2}{3}, \frac{1}{3}, 0\right), \left(\frac{1}{4}, \frac{3}{4}, 0\right)\right)$ is a Nash equilibrium (indeed a perfect and proper equilibrium). However, if player 2's belief concerning the behavior of 1 coincides with $x_1^*$, then 2 is indifferent between his pure strategies $l$ and $c$, and if 1 assigns equal probability to these two pure strategies of player 2, then 1 will play the unique best reply $d$, a pure strategy outside the support of the equilibrium[3]. Moreover, if player 2 expects 1 to reason this way, then 2 will play $r$: the smallest epistemically robust subset containing the support of the mixed equilibrium $x^*$ is the entire pure

1 CURB sets and variants were introduced by Basu and Weibull [7] and have since been used in many applications. Several classes of adaptation processes eventually settle down in a minimal CURB set; see Hurkens [8], Sanchirico [9], Young [10], and Fudenberg and Levine [11]. Moreover, minimal CURB sets give appealing results in communication games [12,13] and network formation games [14]. For closure properties under generalizations of the best reply correspondence, see Ritzberger and Weibull [15].

2 Clearly, if a strategy block is *not* epistemically robust, then our concept does not imply that players should or will avoid strategies in the block.

3 We emphasize that we are concerned with rationalistic analysis of a game that is played once, and where players have beliefs about the rationality and beliefs of their opponents. If the marginal of a player's belief on an opponent's strategy set is non-degenerate—so that the player is uncertain about the behavior of the opponent—then this can be interpreted as the player believing that the opponent is playing a mixed strategy.

strategy space. By contrast, the pure-strategy profile $(d, r)$ is a strict equilibrium. In this equilibrium, no player has any alternative best reply and each equilibrium strategy remains optimal also under some uncertainty as to the other player's action: the set $\{d\} \times \{r\}$ is epistemically robust. In this game, all pure strategies are rationalizable, $S = S_1 \times S_2$ is a FURB set, and the game's unique MINCURB set (thus, the unique minimal FURB set) is $T = \{d\} \times \{r\}$. These are also the epistemically robust subsets; in particular, $\{u, m\} \times \{l, c\}$ is not epistemically robust.

Our results can be described as follows. First, the intuitive link between strict Nash equilibria and our concept of epistemic robustness in terms of ruling out the existence of 'unused' best replies is formalized in Proposition 1: a strategy block $X$ is *not* epistemically robust if and only if for each type block $Y$ raised in its defense—so that $X$ is the set of best reply profiles associated with $Y$—there is a player $i$ and a type $t_i$ with a best reply outside $X_i$, even if $t_i$ believes with high probability that his opponents are of types in $Y_{-i}$ and play best replies. Second, in part (a) of Proposition 2, we establish that epistemically robust strategy blocks are CURB sets. As a consequence (see [15]), every epistemically robust strategy block contains at least one strategically stable set in the sense of Kohlberg and Mertens [16]. In part (b) of Proposition 2, although not every CURB set is epistemically robust (since a CURB set may contain non-best replies), we establish that every CURB set contains an epistemically robust strategy block and we also characterize the largest such subset. As a by-product, we obtain the existence of epistemically robust strategy blocks in all finite games. Third, in Proposition 3, we show that a strategy block is FURB if and only if it satisfies the definition of epistemic robustness with equality, rather than inclusion, in (II). FURB sets thus have a clean epistemic robustness characterization in the present framework. Fourth, in Proposition 4, instead of starting with strategy blocks, we start from a type block and show how an epistemically robust strategy block can be algorithmically obtained; we also show that this is the smallest CURB set that contains all best replies for the initial type block. Fifth, Proposition 5 shows how MINCURB sets (which are necessarily FURB and hence epistemically robust) can be characterized by initiating the above algorithm with a single type profile, while no proper subset has this property. We argue that this latter result shows how MINCURB sets capture characteristics of strict Nash equilibrium.

As our notion of epistemic robustness checks for player types with 'unused' best replies on the basis of their beliefs about the opponents' types and rationality, we follow, for instance, Asheim [17] and Brandenburger, Friedenberg, and Keisler [18], and model players as having beliefs about the opponents without modeling the players' actual behavior. Moreover, we consider epistemic models that are complete in the sense of including all possible beliefs. In these respects, our modeling differs from that of Aumann and Brandenburger [4]'s characterization of Nash equilibrium. In other respects, our modeling resembles that of Aumann and Brandenburger [4]. They assume that players' beliefs about opponent play is commonly known. Here, we require the existence of a type block $Y$ and consider, for each player $i$, types of player $i$ who believe that opponent types are in $Y_{-i}$. In addition, as do Aumann and Brandenburger [4], we consider types of players that believe that their opponents are rational.

The notion of persistent retracts [19] goes part of the way towards epistemic robustness. These are product sets requiring the presence of *at least one* best reply to arbitrary beliefs *close to* the set. In other words, they are robust against small belief perturbations, but admit alternative best replies outside the set, in contrast to our concept of epistemic robustness. Moreover, as pointed out by (van Damme [20] Section 4.5) and Myerson and Weibull [5], persistence is sensitive to certain game details that might be deemed strategically inessential.

The present approach is related to Tercieux [21]'s analysis in its motivation in terms of epistemic robustness of solution concepts and in its use of $p$-belief. His epistemic approach, however, is completely different from ours. Starting from a two-player game, he introduces a Bayesian game where payoff functions are perturbations of the original ones and he investigates which equilibria are robust against this kind of perturbation. Zambrano [22] studies the stability of non-equilibrium concepts in terms of mutual belief and is hence more closely related to our analysis. In fact, our

Proposition 3 overlaps with but is distinct from his main results. Also Hu [23] restricts attention to rationalizability, but allows for $p$-beliefs, where $p < 1$. In the games considered in Hu [23], pure strategy sets are permitted to be infinite. By contrast, our analysis is restricted to finite games, but under the weaker condition of mutual, rather than Hu [23]'s common, $p$-belief of opponent rationality and of opponents' types belonging to given type sets.

The remainder of the paper is organized as follows. Section 2 contains the game theoretic and epistemic definitions used. Section 3 characterizes variants of CURB sets in terms of epistemic robustness. An appendix contains proofs of the propositions.

## 2. The Model

### 2.1. Game Theoretic Definitions

Consider a finite normal-form game $\langle N, (S_i)_{i \in N}, (u_i)_{i \in N} \rangle$, where $N = \{1, \ldots, n\}$ is the non-empty and finite set of players. Each player $i \in N$ has a non-empty, finite set of pure strategies $S_i$ and a payoff function $u_i : S \to \mathbb{R}$ defined on the set $S := S_1 \times \cdots \times S_n$ of pure-strategy profiles. For any player $i$, let $S_{-i} := \times_{j \neq i} S_j$. It is over this set of *other* players' pure-strategy combinations that player $i$ will form his or her probabilistic beliefs. These beliefs may, but need not be, product measures over the other player's pure-strategy sets. We extend the domain of the payoff functions to probability distributions over pure strategies as usual.

For each player $i \in N$, pure strategy $s_i \in S_i$, and probabilistic belief $\sigma_{-i} \in \mathcal{M}(S_{-i})$, where $\mathcal{M}(S_{-i})$ is the set of all probability distributions on the finite set $S_{-i}$, write

$$u_i(s_i, \sigma_{-i}) := \sum_{s_{-i} \in S_{-i}} \sigma_{-i}(s_{-i}) u_i(s_i, s_{-i}).$$

Define $i$'s *best reply correspondence* $\beta_i : \mathcal{M}(S_{-i}) \to 2^{S_i}$ as follows: for all $\sigma_{-i} \in \mathcal{M}(S_{-i})$,

$$\beta_i(\sigma_{-i}) := \{ s_i \in S_i \mid u_i(s_i, \sigma_{-i}) \geq u_i(s_i', \sigma_{-i}) \text{ for all } s_i' \in S_i \}.$$

Let $\mathcal{S} := \{ X \in 2^S \mid \varnothing \neq X = X_1 \times \cdots \times X_n \}$ denote the collection of strategy blocks. For $X \in \mathcal{S}$, we abuse notation slightly by writing, for each $i \in N$, $\beta_i(\mathcal{M}(X_{-i}))$ as $\beta_i(X_{-i})$. Let $\beta(X) := \beta_1(X_{-1}) \times \cdots \times \beta_n(X_{-n})$. Each constituent set $\beta_i(X_{-i}) \subseteq S_i$ in this strategy block is the set of best replies of player $i$ to all probabilistic beliefs over the others' strategy choices $X_{-i} \subseteq S_{-i}$.

Following Basu and Weibull [7], a set $X \in \mathcal{S}$ is:

*closed under rational behavior* (CURB) if $\beta(X) \subseteq X$;

*fixed under rational behavior* (FURB) if $\beta(X) = X$;

*minimal* CURB (MINCURB) if it is CURB and does not properly contain another one: $\beta(X) \subseteq X$ and there is no $X' \in \mathcal{S}$ with $X' \subsetneq X$ and $\beta(X') \subseteq X'$.

Basu and Weibull [7] call a FURB set a 'tight' CURB set. The reversed inclusion, $X \subseteq \beta(X)$, is the *best reply property* ([3] p. 1033). It is shown in (Basu and Weibull [7] Propositions 1 and 2) that a MINCURB set exists, that all MINCURB sets are FURB, and that the block of rationalizable strategies is the game's largest FURB set. While Basu and Weibull [7] require that players believe that others' strategy choices are statistically independent, $\sigma_{-i} \in \times_{j \neq i} \mathcal{M}(S_j)$, we here allow players to believe that others' strategy choices are correlated, $\sigma_{-i} \in \mathcal{M}(S_{-i})$[4]. Our results carry over—with minor modifications in the proofs—to the case of independent strategy choices. Thus, in games with more than two players, the present definition of CURB is somewhat more demanding than that in Basu and Weibull [7], in the

[4] In doing so, we follow (Osborne and Rubinstein [24] Chapter 5).

sense that we require closedness under a wider space of beliefs. Hence, the present definition may, in games with more than two players, lead to different MINCURB sets[5].

*2.2. Epistemic Definitions*

The epistemic analysis builds on the concept of player types, where a type of a player is characterized by a probability distribution over the others' strategies and types.

For each $i \in N$, denote by $T_i$ player $i$'s non-empty type space. The *state space* is defined by $\Omega := S \times T$, where $T := T_1 \times \cdots \times T_n$. For each player $i \in N$, write $\Omega_i := S_i \times T_i$ and $\Omega_{-i} := \times_{j \neq i} \Omega_j$. To each type $t_i \in T_i$ of every player $i$ is associated a probabilistic belief $\mu_i(t_i) \in \mathcal{M}(\Omega_{-i})$, where $\mathcal{M}(\Omega_{-i})$ denotes the set of Borel probability measures on $\Omega_{-i}$ endowed with the topology of weak convergence. For each player $i$, we thus have the player's pure-strategy set $S_i$, type space $T_i$ and a mapping $\mu_i : T_i \to \mathcal{M}(\Omega_{-i})$ that to each of $i$'s types $t_i$ assigns a probabilistic belief, $\mu_i(t_i)$, over the others' strategy choices and types. Assume that, for each $i \in N$, $\mu_i$ is continuous and $T_i$ is compact. The structure $(S_1, \ldots, S_n, T_1, \ldots, T_n, \mu_1, \ldots, \mu_n)$ is called an *S-based (interactive) probability structure*. Assume in addition that, for each $i \in N$, $\mu_i$ is onto: all Borel probability measures on $\Omega_{-i}$ are represented in $T_i$. A probability structure with this additional property is called *complete*.[6] The completeness of the probability structure is essential for our analysis and results. In particular, the assumption of completeness is invoked in all proofs.

For each $i \in N$, denote by $\mathbf{s}_i(\omega)$ and $\mathbf{t}_i(\omega)$ $i$'s strategy and type in state $\omega \in \Omega$. In other words, $\mathbf{s}_i : \Omega \to S_i$ is the projection of the state space to $i$'s strategy set, assigning to each state $\omega \in \Omega$ the strategy $s_i = \mathbf{s}_i(\omega)$ that $i$ uses in that state. Likewise, $\mathbf{t}_i : \Omega \to T_i$ is the projection of the state space to $i$'s type space. For each player $i \in N$ and positive probability $p \in (0, 1]$, the *p-belief operator* $B_i^p$ maps each event (Borel-measurable subset of the state space) $E \subseteq \Omega$ to the set of states where player $i$'s type attaches at least probability $p$ to $E$:

$$B_i^p(E) := \{\omega \in \Omega \mid \mu_i(\mathbf{t}_i(\omega))(E^{\omega_i}) \geq p\},$$

where $E^{\omega_i} := \{\omega_{-i} \in \Omega_{-i} \mid (\omega_i, \omega_{-i}) \in E\}$. This is the same belief operator as in Hu [23][7]. One may interpret $B_i^p(E)$ as the event 'player $i$ believes $E$ with probability at least $p$'. For all $p \in (0, 1]$, $B_i^p$ satisfies $B_i^p(\varnothing) = \varnothing$, $B_i^p(\Omega) = \Omega$, $B_i^p(E') \subseteq B_i^p(E'')$ if $E' \subseteq E''$ (monotonicity), and $B_i^p(E) = E$ if $E = \text{proj}_{\Omega_i} E \times \Omega_{-i}$. The last property means that each player $i$ always $p$-believes his own strategy-type pair, for any positive probability $p$. Since also $B_i^p(E) = \text{proj}_{\Omega_i} B_i^p(E) \times \Omega_{-i}$ for all events $E \subseteq \Omega$, each operator $B_i^p$ satisfies both positive ($B_i^p(E) \subseteq B_i^p(B_i^p(E))$) and negative ($\neg B_i^p(E) \subseteq B_i^p(\neg B_i^p(E))$) introspection. For all $p \in (0, 1]$, $B_i^p$ violates the truth axiom, meaning that $B_i^p(E) \subseteq E$ need not hold for all $E \subseteq \Omega$. In the special case $p = 1$, we have $B_i^p(E') \cap B_i^p(E'') \subseteq B_i^p(E' \cap E'')$ for all $E', E'' \subseteq \Omega$. Finally, note that $B_i^p(E)$ is monotone with respect to $p$ in the sense that, for all $E \subseteq \Omega$, $B_i^{p'}(E) \supseteq B_i^{p''}(E)$ if $p' < p''$.

We connect types with the payoff functions by defining $i$'s *choice correspondence* $C_i : T_i \to 2^{S_i}$ as follows: For each of $i$'s types $t_i \in T_i$,

$$C_i(t_i) := \beta_i(\text{marg}_{S_{-i}} \mu_i(t_i))$$

---

[5] We also note that a pure strategy is a best reply to some belief $\sigma_{-i} \in \mathcal{M}(S_{-i})$ if and only if it is not strictly dominated (by any pure or mixed strategy). This follows from Lemma 3 in Pearce [3], which, in turn, is closely related to (Ferguson [25] p. 86, Theorem 1) and (van Damme [26] Lemma 3.2.1).

[6] An adaptation of the proof of (Brandenburger, Friedenberg, and Keisler [18] Proposition 7.2) establishes the existence of such a complete probability structure under the assumption that, for all $i \in N$, player $i$'s type space $T_i$ is Polish (separable and completely metrizable). The exact result we use is Proposition 6.1 in an earlier working paper version [27]. Existence can also be established by constructing a universal state space [28,29].

[7] See also Monderer and Samet [30].

consists of $i$'s best replies when player $i$ is of type $t_i$. Let $\mathcal{T} := \{Y \in 2^T \mid \emptyset \neq Y = Y_1 \times \cdots \times Y_n\}$ denote the collection of type blocks. For any such set $Y \in \mathcal{T}$ and player $i \in N$, write $C_i(Y_i) := \bigcup_{t_i \in Y_i} C_i(t_i)$ and $C(Y) := C_1(Y_1) \times \cdots \times C_n(Y_n)$. In other words, these are the choices and choice profiles associated with $Y$. If $Y \in \mathcal{T}$ and $i \in N$, write

$$[Y_i] := \{\omega \in \Omega \mid \mathbf{t}_i(\omega) \in Y_i\}.$$

This is the event that player $i$ is of a type in the subset $Y_i$. Likewise, write $[Y] := \bigcap_{i \in N}[Y_i]$ for the event that the type profile is in $Y$. Finally, for each player $i \in N$, write $R_i$ for the event that player $i$ uses a best reply:

$$R_i := \{\omega \in \Omega \mid \mathbf{s}_i(\omega) \in C_i(\mathbf{t}_i(\omega))\}.$$

One may interpret $R_i$ as the event that $i$ is rational: if $\omega \in R_i$, then $\mathbf{s}_i(\omega)$ is a best reply to $\text{marg}_{S_{-i}} \mu_i(\mathbf{t}_i(\omega))$.

## 3. Epistemic Robustness

We define a strategy block $X \in \mathcal{S}$ to be *epistemically robust* if there exists a $\bar{p} < 1$ such that, for each probability $p \in [\bar{p}, 1]$, there is a type block $Y \in \mathcal{T}$ (possibly dependent on $p$) such that

$$C(Y) = X \tag{2}$$

and

$$B_i^p\left(\bigcap_{j \neq i}(R_j \cap [Y_j])\right) \subseteq [Y_i] \quad \text{for all } i \in N. \tag{3}$$

Hence, epistemic robustness requires the existence of a type block $Y$ satisfying, for each player $i$, that $X_i$ is the set of best replies of the types in $Y_i$, and that every type of player $i$ who $p$-believes that opponents are rational and of types in $Y_{-i}$ is included in $Y_i$. Condition (2) is thus not an equilibrium condition as it is not interactive: it relates each player's type subset to the same player's strategy subset. The interactivity enters through condition (3), which relates each player's type subset to the type subsets of the other players. For each $p < 1$, condition (3) allows each player $i$ to attach a positive probability to the event that others do not play best replies and/or are of types outside $Y$. It follows from the monotonicity of $B_i^p(\cdot)$ with respect to $p$ that, for a fixed type block $Y$, if inclusion (3) is satisfied for $p = \bar{p}$, then inclusion (3) is satisfied also for all $p \in (\bar{p}, 1]$.

Note that if condition (2) is combined with a variant of condition (3), with the weak inclusion reversed and $p$ set to 1, then we obtain a characterization of Pearce [3]'s best reply set; see [6].

In line with what we mentioned in the introduction, we can now formally show that if $s \in S$ is a strict Nash equilibrium, then $\{s\}$ is epistemically robust. To see this, define for all $i \in N$, $Y_i := \{t_i \in T_i \mid C_i(t_i) = \{s_i\}\}$. Since the game is finite, there is, for each player $i \in N$, a $\underline{p}_i \in (0, 1)$ such that $\beta_i(\sigma_{-i}) = \{s_i\}$ for all $\sigma_{-i} \in \mathcal{M}(S_{-i})$ with $\sigma_{-i}(\{s_{-i}\}) \geq \underline{p}_i$. Let $\underline{p} = \max\{\underline{p}_1, \ldots, \underline{p}_n\}$. Then it holds for each $p \in [\underline{p}, 1]$:

$$B_i^p\left(\bigcap_{j \neq i}(R_j \cap [Y_j])\right) \subseteq B_i^p(\{\omega \in \Omega \mid \forall j \neq i,\, \mathbf{s}_j(\omega) \in X_j\}) \subseteq [Y_i] \quad \text{for all } i \in N.$$

Thus, by condition (2) and condition (3), $\{s\}$ is epistemically robust.

Also, as discussed in the introduction, non-strict pure-strategy Nash equilibria have 'unused' best replies. Our first result demonstrates that epistemic robustness is sufficient and necessary for the non-existence of such 'unused' best replies.

**Proposition 1.** *The following two statements are equivalent:*

*(a)* $X \in \mathcal{S}$ *is* not *epistemically robust.*

(b) *For all $\bar{p} < 1$, there exists $p \in [\bar{p}, 1]$ such that if $Y \in \mathcal{T}$ satisfies $C(Y) = X$, then there exist $i \in N$ and $t_i \in T_i$ such that $C(t_i) \nsubseteq X_i$ and $[\{t_i\}] \subseteq B_i^p \left(\bigcap_{j \neq i}(R_j \cap [Y_j])\right)$.*

Hence, while an epistemically robust subset is defined by a *set* of profiles of player types, it suffices with one player and one possible type of this player to determine that a strategy block is not epistemically robust.

We now relate epistemically robust subsets to CURB sets. To handle the fact that all strategy profiles in any epistemically robust subset are profiles of best replies, while CURB sets may involve strategies that are not best replies, introduce the following notation: For each $i \in N$ and $X_i \subseteq S_i$, let

$$\beta_i^{-1}(X_i) := \{\sigma_{-i} \in \mathcal{M}(S_{-i}) \mid \beta_i(\sigma_{-i}) \subseteq X_i\}$$

denote the pre-image (upper inverse) of $X_i$ under player $i$'s best reply correspondence[8]. For a given subset $X_i$ of $i$'s pure strategies, $\beta_i^{-1}(X_i)$ consists of the beliefs over others' strategy profiles having the property that all best replies to these beliefs are contained in $X_i$.

**Proposition 2.** *Let $X \in \mathcal{S}$.*

(a) *If $X$ is epistemically robust, then $X$ is a CURB set.*
(b) *If $X$ is a CURB set, then $\times_{i \in N}\beta_i(\beta_i^{-1}(X_i)) \subseteq X$ is epistemically robust. Furthermore, it is the largest epistemically robust subset of $X$.*

Claim (a) implies that every epistemically robust subset contains at least one strategically stable set, both as defined in Kohlberg and Mertens [16] and as defined in Mertens [32], see Ritzberger and Weibull [15] and Demichelis and Ritzberger [33], respectively[9]. Claim (a) also implies that subsets of epistemically robust sets need not be epistemically robust. Concerning claim (b), note that $\times_{i \in N}\beta_i(\beta_i^{-1}(S_i))$ equals the set of profiles of pure strategies that are best replies to some belief. Hence, since for each $i \in N$, both $\beta_i(\cdot)$ and $\beta_i^{-1}(\cdot)$ are monotonic with respect to set inclusion, it follows from Proposition 2(b) that any epistemically robust subset involves only strategies surviving one round of strict elimination. Thus, $\times_{i \in N}\beta_i(\beta_i^{-1}(S_i))$ is the largest epistemically robust subset, while the characterization of the smallest one(s) will be dealt with by Proposition 5.

Our proof shows that Proposition 2(a) can be slightly strengthened, as one only needs the robustness conditions with $p = 1$; as long as there is a $Y \in \mathcal{T}$ such that $C(Y) = X$ and condition (3) holds with $p = 1$, $X$ is CURB.[10] Moreover, although epistemic robustness allows that $Y \in \mathcal{T}$ depends on $p$, the proof of (b) defines $Y$ independently of $p$.

The following result shows that FURB sets are characterized by epistemic robustness when player types that do *not* believe with sufficient probability that the others play best replies are removed:

**Proposition 3.** *The following two statements are equivalent:*

(a) *$X \in \mathcal{S}$ is a FURB set.*
(b) *There exists a $\bar{p} < 1$ such that, for each probability $p \in [\bar{p}, 1]$, there is a type block $Y \in \mathcal{T}$ satisfying condition (2) such that condition (3) holds with equality.*

The block of rationalizable strategies [2,3] is the game's largest FURB set [7]. Thus, it follows from Proposition 3 that epistemic robustness yields a characterization of the block of rationalizable strategies,

---

8 Harsanyi and Selten [31] refer to such pre-images of strategy sets as *stability sets*.

9 In fact, these inclusions hold under the slightly weaker definition of CURB sets in Basu and Weibull [7], in which a player's belief about other players is restricted to be a product measure over the others' pure-strategy sets.

10 In the appendix we also prove that if $p \in (0, 1]$ and $Y \in \mathcal{T}$ are such that $C(Y) = X$ and (3) holds for all $i \in N$, then $X$ is a $p$-best reply set in the sense of Tercieux [21].

without involving any explicit assumption of common belief of rationality. Instead, only mutual $p$-belief of rationality and type sets are assumed. Proposition 3 also applies to MINCURB sets, as these sets are FURB. In particular, it follows from Propositions 2(a) and 3 that a strategy block is MINCURB if and only if it is a minimal epistemically robust subset[11].

As much of the literature on CURB sets (recall footnote 1) focuses on minimal ones, we now turn to how smallest CURB sets can be characterized in terms of epistemic robustness. This characterization is presented through Propositions 4 and 5.

Proposition 4 starts from an arbitrary block $Y$ of types and generates an epistemically robust subset by including all beliefs over the opponents' best replies, and all beliefs over opponents' types that have such beliefs over their opponents, and so on. Formally, define for any $Y \in \mathcal{T}$ the sequence $\langle Y(k) \rangle_k$ by $Y(0) = Y$ and, for each $k \in \mathbb{N}$ and $i \in N$,

$$[Y_i(k)] := [Y_i(k-1)] \cup B_i^1 \left( \bigcap\nolimits_{j \neq i} (R_j \cap [Y_j(k-1)]) \right) . \tag{4}$$

Define the correspondence $E : \mathcal{T} \to 2^S$, for any $Y \in \mathcal{T}$, by

$$E(Y) := C \left( \bigcup\nolimits_{k \in \mathbb{N}} Y(k) \right) .$$

We show that the strategy block $E(Y)$ of best replies is epistemically robust and is the smallest CURB set that includes $C(Y)$.[12]

**Proposition 4.** *Let $Y \in \mathcal{T}$. Then $X = E(Y)$ is the smallest* CURB *set satisfying $C(Y) \subseteq X$. Furthermore, $E(Y)$ is epistemically robust.*

**Remark 1.** *If the strategy block $C(Y)$ contains strategies that are not rationalizable, then $E(Y)$ will not be* FURB. *Therefore, the epistemic robustness of $E(Y)$ does not follow from Proposition 3; its robustness is established by invoking Proposition 2(b).*

Note that if a strategy block $X$ is epistemically robust, then there exists a type block $Y$ satisfying condition (2) such that condition (3) is satisfied for $p = 1$. Thus, $X = C(Y) = E(Y)$, showing that all epistemically robust strategy blocks can be obtained using the algorithm of Proposition 4.

The final Proposition 5 shows how MINCURB sets can be characterized by epistemically robust subsets obtained by initiating the algorithm of Proposition 4 with a single type profile: a strategy block $X$ is a MINCURB set if and only if (a) the algorithm leads to $X$ from a single type profile, and (b) no single type profile leads to a strict subset of $X$.

**Proposition 5.** *$X \in \mathcal{S}$ is a* MINCURB *set if and only if there exists a $t \in T$ such that $E(\{t\}) = X$ and there exists no $t' \in T$ such that $E(\{t'\}) \subsetneq X$.*

Strict Nash equilibria (interpreted as equilibria in beliefs) satisfy 'coordination', in the sense that there is mutual belief about the players' sets of best replies, 'concentration', in the sense that each player has only one best reply, and epistemic robustness (as defined here), implying that each player's set of beliefs about opponent choices contains all probability distributions over opponent strategies that are best replies given their beliefs. In Proposition 5, starting with a single type profile $t$ that corresponds to 'coordination', using the algorithm of Proposition 4 and ending up with $E(\{t\}) = X$ ensures epistemic

---

[11] We thank Peter Wikman for this observation.

[12] For each strategy block $X \in \mathcal{S}$, there exists a unique smallest CURB set $X' \in \mathcal{S}$ with $X \subseteq X'$ (that is, $X'$ is a subset of all CURB sets $X''$ that include $X$). To see that this holds for all finite games, note that the collection of CURB sets including a given block $X \in \mathcal{S}$ is non-empty and finite, and that the intersection of two CURB sets that include $X$ is again a CURB set including $X$.

robustness, while the non-existence of $t' \in T$ such that $E(\{t'\})$ is a proper subset of $X$ corresponds to 'concentration'. Hence, these three characteristics of strict Nash equilibria characterize MINCURB sets in Proposition 5.

In order to illustrate Propositions 4 and 5, consider the Nash equilibrium $x^*$ in game (1) in the introduction. This equilibrium corresponds to a type profile $(t_1, t_2)$ where $t_1$ assigns probability $1/4$ to $(l, t_2)$ and probability $3/4$ to $(c, t_2)$, and where $t_2$ assigns probability $2/3$ to $(u, t_1)$ and probability $1/3$ to $(m, t_1)$. We have that $C(\{t_1, t_2\}) = \{u, m\} \times \{l, c\}$, while the full strategy space $S$ is the smallest CURB set that includes $C(\{t_1, t_2\})$. Proposition 4 shows that $C(\{t_1, t_2\})$ is not epistemically robust, since it does not coincide with the smallest CURB set that includes it. Recalling the discussion from the introduction: if player 2's belief concerning the behavior of 1 coincides with $x_1^*$, then 2 is indifferent between his pure strategies $l$ and $c$, and if 1 assigns equal probability to these two pure strategies of player 2, then 1 will play the unique best reply $d$, a pure strategy outside the support of the equilibrium. Moreover, if player 2 expects 1 to reason this way, then 2 will play $r$. Hence, to assure epistemic robustness, starting from type set $\{t_1, t_2\}$, the repeated inclusion of all beliefs over opponents' best replies eventually leads to the smallest CURB set, here $S$, that includes the Nash equilibrium that was our initial point of departure. By contrast, for the type profile $(t_1', t_2')$ where $t_1'$ assigns probability 1 to $(r, t_2')$ and $t_2'$ assigns probability 1 to $(d, t_1')$ we have that $C(\{t_1', t_2'\}) = \{(d, r)\}$ coincides with the smallest CURB set that includes it. Thus, the strict equilibrium $(d, r)$ to which $(t_1', t_2')$ corresponds is epistemically robust, when viewed as a singleton set. Furthermore, by Proposition 5, $\{(d, r)\}$ is the unique MINCURB set.

**Acknowledgments:** We thank four anonymous referees, Itai Arieli, Stefano Demichelis, Daisuke Oyama, Olivier Tercieux, Peter Wikman, and seminar participants in Montreal, Paris, Singapore and Tsukuba for helpful comments and suggestions. Voorneveld's research was supported by the Wallander-Hedelius Foundation under grant P2010-0094:1. Weibull's research was supported by the Knut and Alice Wallenberg Research Foundation, and by the Agence Nationale de la Recherche, chaire IDEX ANR-11-IDEX-0002-02.

**Author Contributions:** The authors contributed equally to this work.

**Conflicts of Interest:** The authors declare no conflict of interest.

## Appendix

**Proof of Proposition 1.** Let $\mathcal{T}(X) := \{Y \in \mathcal{T} \mid C(Y) = X\}$ denote the collection of type blocks having the property that $X$ is the strategy block of best replies. By the completeness of the probability structure, we have that $\mathcal{T}(X)$ is non-empty if and only if $X \subseteq \times_{i \in N} \beta_i(\beta_i^{-1}(S_i))$. Furthermore, by completeness, if $\mathcal{T}(X)$ is non-empty, then $\mathcal{T}(X)$ has a largest element, $\bar{Y}(X)$, which is constructed by letting $\bar{Y}_i(X) = \{t_i \in T_i \mid C_i(t_i) \subseteq X_i\}$ for all $i \in N$.

By the definition of epistemic robustness, a strategy block $X \in \mathcal{S}$ is *not* epistemically robust if and only if, for all $\bar{p} < 1$, there exists $p \in [\bar{p}, 1]$ such that for all $Y \in \mathcal{T}(X)$, there exists $i \in N$ such that

$$B_i^p \left( \bigcap_{j \neq i} (R_j \cap [Y_j]) \right) \not\subseteq [Y_i] .$$

Hence, $X \in \mathcal{S}$ is *not* epistemic robust if and only if

($*$) $X \not\subseteq \times_{i \in N} \beta_i(\beta_i^{-1}(S_i))$ so that $\mathcal{T}(X) = \emptyset$, or
($**$) $X \subseteq \times_{i \in N} \beta_i(\beta_i^{-1}(S_i))$ so that $\mathcal{T}(X) \neq \emptyset$, and, for all $\bar{p} < 1$, there exists $p \in [\bar{p}, 1]$ such that if $Y \in \mathcal{T}(X)$, then there exist $i \in N$ and $t_i \notin Y_i$ such that

$$[\{t_i\}] \subseteq B_i^p \left( \bigcap_{j \neq i} (R_j \cap [Y_j]) \right) .$$

*(b) implies (a).* Assume that, for all $\bar{p} < 1$, there exists $p \in [\bar{p}, 1]$ such that if $Y \in \mathcal{T}(X)$, then there exist $i \in N$ and $t_i \in T_i$ such that $C(t_i) \not\subseteq X_i$ and $[\{t_i\}] \subseteq B_i^p \left( \bigcap_{j \neq i} (R_j \cap [Y_j]) \right)$. Note that if $Y \in \mathcal{T}(X)$ and $C(t_i) \not\subseteq X_i$, then $t_i \notin Y_i$.

Either $\mathcal{T}(X) = \emptyset$, so that $(*)$ is satisfied, or $\mathcal{T}(X) \neq \emptyset$ and, for all $\bar{p} < 1$, there exists $p \in [\bar{p}, 1]$ such that if $Y \in \mathcal{T}(X)$, then there exist $i \in N$ and $t_i \notin Y_i$ such that $[\{t_i\}] \subseteq B_i^p\left(\bigcap_{j\neq i}(R_j \cap [Y_j])\right)$, so that $(**)$ is satisfied.

*(a) implies (b).* Assume that $(*)$ or $(**)$ is satisfied.

Assume that $(*)$ is satisfied, and fix $p < 1$. Then, it holds trivially that if $Y \in \mathcal{T}(X)$, then there exist $i \in N$ and $t_i \in T_i$ such that $C(t_i) \nsubseteq X_i$ and $[\{t_i\}] \subseteq B_i^p\left(\bigcap_{j\neq i}(R_j \cap [Y_j])\right)$.

Assume that $(**)$ is satisfied. Then, since $\bar{Y}(X) \in \mathcal{T}(X)$, it must also hold that for all $\bar{p} < 1$, there exist $p(\bar{p}) \in [\bar{p}, 1]$, $i(\bar{p}) \in N$ and $t_{i(\bar{p})}(\bar{p}) \notin \bar{Y}_{i(\bar{p})}(X)$ such that $[\{t_{i(\bar{p})}(\bar{p})\}] \subseteq B_{i(\bar{p})}^{p(\bar{p})}\left(\bigcap_{j\neq i(\bar{p})}(R_j \cap [\bar{Y}_j(X)])\right)$. By the definition of $\bar{Y}(X)$, $C(t_{i(\bar{p})}(\bar{p})) \nsubseteq X_{i(\bar{p})}$. It is sufficient to construct, for all $\bar{p} < 1$ and $Y \in \mathcal{T}(X)$, a type $t_{i(\bar{p})} \in T_i$ such that $C(t_{i(\bar{p})}) = C(t_{i(\bar{p})}(\bar{p}))$ and $[\{t_{i(\bar{p})}\}] \subseteq B_{i(\bar{p})}^{p(\bar{p})}\left(\bigcap_{j\neq i(\bar{p})}(R_j \cap [Y_j])\right)$.

For all $s_{-i(\bar{p})} \in X_{-i(\bar{p})}$ with $\text{marg}_{S_{-i(\bar{p})}}\mu_{-i(\bar{p})}(t_{i(\bar{p})}(\bar{p}))(s_{-i(\bar{p})}) > 0$, select $t_{-i(\bar{p})} \in Y_{-i(\bar{p})}$ such that $s_j \in C_j(t_j)$ for all $j \neq i(\bar{p})$ (which exists since $C(Y) = X$) and let

$$\mu_{i(\bar{p})}(t_{i(\bar{p})})(s_{-i(\bar{p})}, t_{-i(\bar{p})}) = \text{marg}_{S_{-i(\bar{p})}}\mu_{-i(\bar{p})}(t_{i(\bar{p})}(\bar{p}))(s_{-i(\bar{p})}).$$

For all $s_{-i(\bar{p})} \notin X_{-i(\bar{p})}$ with $\text{marg}_{S_{-i(\bar{p})}}\mu_{-i(\bar{p})}(t_{i(\bar{p})}(\bar{p}))(s_{-i(\bar{p})}) > 0$, select $t_{-i(\bar{p})} \in Y_{-i(\bar{p})}$ arbitrary and let again

$$\mu_{i(\bar{p})}(t_{i(\bar{p})})(s_{-i(\bar{p})}, t_{-i(\bar{p})}) = \text{marg}_{S_{-i(\bar{p})}}\mu_{-i(\bar{p})}(t_{i(\bar{p})}(\bar{p}))(s_{-i(\bar{p})}).$$

Then $\text{marg}_{S_{-i(\bar{p})}}\mu_{-i(\bar{p})}(t_{i(\bar{p})})(s_{-i(\bar{p})}) = \text{marg}_{S_{-i(\bar{p})}}\mu_{-i(\bar{p})}(t_{i(\bar{p})}(\bar{p}))(s_{-i(\bar{p})})$, implying that $C(t_{i(\bar{p})}) = C(t_{i(\bar{p})}(\bar{p}))$. Furthermore, by the construction of $t_{i(\bar{p})}$:

$$\begin{aligned}
&\mu_{i(\bar{p})}(t_{i(\bar{p})})\left(\{(s_{-i(\bar{p})}, t_{-i(\bar{p})}) \in S_{-i(\bar{p})} \times Y_{-i(\bar{p})} \mid s_j \in C_j(t_j) \text{ for all } j \neq i(\bar{p})\}\right) \\
&= \mu_{i(\bar{p})}(t_{i(\bar{p})})\left(X_{-i(\bar{p})} \times T_{-i(\bar{p})}\right) \\
&= \mu_{i(\bar{p})}(t_{i(\bar{p})}(\bar{p}))\left(X_{-i(\bar{p})} \times T_{-i(\bar{p})}\right) \\
&\geq \mu_{i(\bar{p})}(t_{i(\bar{p})}(\bar{p}))\left(\{(s_{-i(\bar{p})}, t_{-i(\bar{p})}) \in S_{-i(\bar{p})} \times \bar{Y}_{-i(\bar{p})}(X) \mid s_j \in C_j(t_j) \text{ for all } j \neq i(\bar{p})\}\right) \\
&\geq p(\bar{p}),
\end{aligned}$$

since $C(Y) = X = C(\bar{Y}(X))$.[13] Thus, $[\{t_{i(\bar{p})}\}] \subseteq B_{i(\bar{p})}^{p(\bar{p})}\left(\bigcap_{j\neq i(\bar{p})}(R_j \cap [Y_j])\right)$. □

**Proof of Proposition 2.** *Part (a).* By assumption, there is a $Y \in \mathcal{T}$ with $C(Y) = X$ such that for each $i \in N$, $B_i^1\left(\bigcap_{j\neq i}(R_j \cap [Y_j])\right) \subseteq [Y_i]$.

Fix $i \in N$, and consider any $\sigma_{-i} \in \mathcal{M}(X_{-i})$. Since $C(Y) = X$, it follows that, for each $s_{-i} \in S_{-i}$ with $\sigma_{-i}(s_{-i}) > 0$, there exists $t_{-i} \in Y_{-i}$ such that, for all $j \neq i$, $s_j \in C_j(t_j)$. Hence, since the probability structure is complete, there exists a

$$\omega \in B_i^1\left(\bigcap\nolimits_{j\neq i}(R_j \cap [Y_j])\right) \subseteq [Y_i]$$

[13] To see that the first equality in the expression above holds, note first that, since $C(Y) = X$,

$$\{(s_{-i(\bar{p})}, t_{-i(\bar{p})}) \in S_{-i(\bar{p})} \times Y_{-i(\bar{p})} \mid s_j \in C_j(t_j) \text{ for all } j \neq i(\bar{p})\} \subseteq X_{-i(\bar{p})} \times T_{-i(\bar{p})}.$$

However, by construction, for any $(s_{-i(\bar{p})}, t_{-i(\bar{p})}) \in X_{-i(\bar{p})} \times T_{-i(\bar{p})}$ assigned positive probability by $\mu_{i(\bar{p})}(t_{i(\bar{p})})$, it is the case that $t_{-i(\bar{p})} \in Y_{-i(\bar{p})}$ and $s_j \in C_j(t_j)$ for all $j \neq i(\bar{p})$. Hence, the two sets are given the same probability by $\mu_{i(\bar{p})}(t_{i(\bar{p})})$.

with $\text{marg}_{S_{-i}} \mu_i(\mathbf{t}_i(\omega)) = \sigma_{-i}$. So

$$\beta_i(X_{-i}) := \beta_i(\mathcal{M}(X_{-i})) \subseteq \bigcup_{t_i \in Y_i} \beta_i(\text{marg}_{S_{-i}} \mu_i(t_i)) := C_i(Y_i) = X_i\,.$$

Since this holds for all $i \in N$, $X$ is a CURB set.

*Part (b).* Assume that $X \in \mathcal{S}$ is a CURB set, i.e., $X$ satisfies $\beta(X) \subseteq X$. It suffices to prove that $\times_{i \in N} \beta_i(\beta_i^{-1}(X_i)) \subseteq X$ is epistemically robust. That it is the largest epistemically robust subset of $X$ then follows immediately from the fact that, for each $i \in N$, both $\beta_i(\cdot)$ and $\beta_i^{-1}(\cdot)$ are monotonic with respect to set inclusion.

Define $Y \in \mathcal{T}$ by taking, for each $i \in N$, $Y_i := \{t_i \in T_i \mid C_i(t_i) \subseteq X_i\}$. Since the probability structure is complete, it follows that $C_i(Y_i) = \beta_i(\beta_i^{-1}(X_i))$. For notational convenience, write $X_i' = \beta_i(\beta_i^{-1}(X_i))$ and $X' = \times_{i \in N} X_i'$. Since the game is finite, there is, for each player $i \in N$, a $\underline{p}_i \in (0,1)$ such that $\beta_i(\sigma_{-i}) \subseteq \beta_i(X_{-i}')$ for all $\sigma_{-i} \in \mathcal{M}(S_{-i})$ with $\sigma_{-i}(X_{-i}') \geq \underline{p}_i$. Let $\underline{p} = \max\{\underline{p}_1, \ldots, \underline{p}_n\}$.

We first show that $\beta(X') \subseteq X'$. By definition, $X' \subseteq X$, so for each $i \in N$: $\mathcal{M}(X_{-i}') \subseteq \mathcal{M}(X_{-i})$. Moreover, as $\beta(X) \subseteq X$ and, for each $i \in N$, $\beta_i(X_i) := \beta_i(\mathcal{M}(X_{-i}))$, it follows that $\mathcal{M}(X_{-i}) \subseteq \beta_i^{-1}(X_i)$. Hence, for each $i \in N$,

$$\beta_i(X_i') := \beta_i(\mathcal{M}(X_{-i}')) \subseteq \beta_i(\mathcal{M}(X_{-i})) \subseteq \beta_i(\beta_i^{-1}(X_i)) = X_i'\,.$$

For all $p \in [\underline{p}, 1]$ and $i \in N$, we have that

$$\begin{aligned}
& B_i^p\left(\bigcap\nolimits_{j \neq i}(R_j \cap [Y_j])\right) \\
= \; & B_i^p\left(\bigcap\nolimits_{j \neq i}\{\omega \in \Omega \mid \mathbf{s}_j(\omega) \in C_j(\mathbf{t}_j(\omega)) \subseteq X_j'\}\right) \\
\subseteq \; & \left\{\omega \in \Omega \mid \mu_i(\mathbf{t}_i(\omega))\{\omega_{-i} \in \Omega_{-i} \mid \text{for all } j \neq i,\ \mathbf{s}_j(\omega) \in X_j'\} \geq p\right\} \\
\subseteq \; & \{\omega \in \Omega \mid \text{marg}_{S_{-i}} \mu_i(\mathbf{t}_i(\omega))(X_{-i}') \geq p\} \\
\subseteq \; & \{\omega \in \Omega \mid C_i(\mathbf{t}_i(\omega)) \subseteq \beta_i(X_{-i}')\} \\
\subseteq \; & \{\omega \in \Omega \mid C_i(\mathbf{t}_i(\omega)) \subseteq X_{-i}'\} \; = \; [Y_i],
\end{aligned}$$

using $\beta(X') \subseteq X'$. □

For $X \in \mathcal{S}$ and $p \in (0,1]$, write, for each $i \in N$,

$$\begin{aligned}
\beta_i^p(X_{-i}) := \{s_i \in S_i \mid \exists \sigma_{-i} \in \mathcal{M}(S_{-i}) \text{ with } \sigma_{-i}(X_{-i}) \geq p \\
\text{such that } u_i(s_i, \sigma_{-i}) \geq u_i(s_i', \sigma_{-i})\ \forall s_i' \in S_i\}\,.
\end{aligned}$$

Let $\beta^p(X) := \beta_1^p(X_{-1}) \times \cdots \times \beta_n^p(X_{-n})$. Following Tercieux [21], a set $X \in \mathcal{S}$ is a *p-best reply set* if $\beta^p(X) \subseteq X$.

**Claim:** *Let $X \in \mathcal{S}$ and $p \in (0,1]$. If $Y \in \mathcal{T}$ is such that $C(Y) = X$ and condition* (3) *holds for each $i \in N$, then $X$ is a p-best reply set.*

**Proof.** By assumption, there is a $Y \in \mathcal{T}$ with $C(Y) = X$ such that for each $i \in N$, $B_i^p\left(\bigcap_{j \neq i}(R_j \cap [Y_j])\right) \subseteq [Y_i]$.

Fix $i \in N$ and consider any $\sigma_{-i} \in \mathcal{M}(S_{-i})$ with $\sigma_{-i}(X_{-i}) \geq p$. Since $C(Y) = X$, it follows that, for each $s_{-i} \in X_{-i}$, there exists $t_{-i} \in Y_{-i}$ such that $s_j \in C_j(t_j)$ for all $j \neq i$. Hence, since the probability structure is complete, there exists a

$$\omega \in B_i^p \left( \bigcap_{j \neq i} (R_j \cap [Y_j]) \right) \subseteq [Y_i]$$

with $\text{marg}_{S_{-i}} \mu_i(\mathbf{t}_i(\omega)) = \sigma_{-i}$. Thus, by definition of $\beta_i^p(X_{-i})$:

$$\beta_i^p(X_{-i}) \subseteq \bigcup_{t_i \in Y_i} \beta_i(\text{marg}_{S_{-i}} \mu_i(t_i)) := C_i(Y_i) = X_i .$$

Since this holds for all $i \in N$, $X$ is a $p$-best reply set. □

**Proof of Proposition 3.** *(b) implies (a).* By assumption, there is a $Y \in \mathcal{T}$ with $C(Y) = X$ such that for all $i \in N$, $B_i^1 \left( \bigcap_{j \neq i} (R_j \cap [Y_j]) \right) = [Y_i]$.

Fix $i \in N$. Since $C(Y) = X$, and the probability structure is complete, there exists, for any $\sigma_{-i} \in \mathcal{M}(S_{-i})$, an

$$\omega \in B_i^1 \left( \bigcap_{j \neq i} (R_j \cap [Y_j]) \right) = [Y_i]$$

with $\text{marg}_{S_{-i}} \mu_i(\mathbf{t}_i(\omega)) = \sigma_{-i}$ if and only if $\sigma_{-i} \in \mathcal{M}(X_{-i})$. Thus,

$$\beta_i(X_{-i}) := \beta_i(\mathcal{M}(X_{-i})) = \bigcup_{t_i \in Y_i} \beta_i(\text{marg}_{S_{-i}} \mu_i(t_i)) := C_i(Y_i) = X_i .$$

Since this holds for all $i \in N$, $X$ is a FURB set.

*(a) implies (b).* Assume that $X \in \mathcal{S}$ satisfies $X = \beta(X)$. Since the game is finite, there exists, for each player $i \in N$, a $\underline{p}_i \in (0,1)$ such that $\beta_i(\sigma_{-i}) \subseteq \beta_i(X_{-i})$ if $\sigma_{-i}(X_{-i}) \geq \underline{p}_i$. Let $\underline{p} = \max\{\underline{p}_1, \ldots, \underline{p}_n\}$.

For each $p \in [\underline{p}, 1]$, construct the sequence of type blocks $\langle Y^p(k) \rangle_k$ as follows: for each $i \in N$, let $Y_i^p(0) = \{t_i \in T_i \mid C_i(t_i) \subseteq X_i\}$. Using continuity of $\mu_i$, the correspondence $C_i : T_i \rightrightarrows S_i$ is upper hemi-continuous. Thus $Y_i^p(0) \subseteq T_i$ is closed, and, since $T_i$ is compact, so is $Y_i^p(0)$. There exists a closed set $Y_i^p(1) \subseteq T_i$ such that

$$[Y_i^p(1)] = B_i^p \left( \bigcap_{j \neq i} (R_j \cap [Y_j^p(0)]) \right) .$$

It follows that $Y_i^p(1) \subseteq Y_i^p(0)$. Since $Y_i^p(0)$ is compact, so is $Y_i^p(1)$. By induction,

$$[Y_i^p(k)] = B_i^p \left( \bigcap_{j \neq i} (R_j \cap [Y_j^p(k-1)]) \right) \tag{A1}$$

defines, for each player $i$, a decreasing chain $\left\langle Y_i^p(k) \right\rangle_k$ of compact and non-empty subsets: $Y_i^p(k+1) \subseteq Y_i^p(k)$ for all $k$. By the finite-intersection property, $Y_i^p := \bigcap_{k \in \mathbb{N}} Y_i^p(k)$ is a non-empty and compact subset of $T_i$. For each $k$, let $Y^p(k) = \times_{i \in N} Y_i^p(k)$ and let $Y^p := \bigcap_{k \in \mathbb{N}} Y^p(k)$. Again, these are non-empty and compact sets.

Next, $C(Y^p(0)) = \beta(X)$, since the probability structure is complete. Since $X$ is FURB, we thus have $C(Y^p(0)) = X$. For each $i \in N$,

$$[Y_i^p(1)] \subseteq \{\omega \in \Omega \mid \text{marg}_{S_{-i}} \mu_i(\mathbf{t}_i(\omega))(X_{-i}) \geq p\} ,$$

implying that $C_i(Y_i^p(1)) \subseteq \beta_i(X_{-i}) = X_i$ by the construction of $\underline{p}$. Moreover, since the probability structure is complete, for each $i \in N$ and $\sigma_{-i} \in \mathcal{M}(X_{-i})$, there exists $\omega \in [Y_i^p(1)] = B_i^p(\bigcap_{j \neq i}(R_j \cap [Y_j^p(0)]))$ with $\text{marg}_{S_{-i}} \mu_i(\mathbf{t}_i(\omega)) = \sigma_{-i}$, implying that $C_i(Y_i^p(1)) \supseteq \beta_i(X_{-i}) = X_i$. Hence, $C_i(Y_i^p(1)) = \beta_i(X_{-i}) = X_i$. By induction, it holds for all $k \in N$ that $C(Y^p(k)) = \beta(X) = X$ . Since $\left\langle Y_i^p(k) \right\rangle_k$ is a decreasing chain, we also have that $C(Y^p) \subseteq X$. The converse inclusion follows by upper hemi-continuity of the correspondence $C$. To see this, suppose that $x^o \in X$ but $x^o \notin C(Y^p)$. Since $x^o \in X$,

$x^o \in C(Y^p(k))$ for all $k$. By the Axiom of Choice: for each $k$, there exists a $y_k \in Y^p(k)$ such that $(y_k, x^o) \in graph(C)$. By the Bolzano–Weierstrass Theorem, we can extract a convergent subsequence for which $y_k \to y^o$, where $y^o \in Y^p$, since $Y^p$ is closed. Moreover, since the correspondence $C$ is closed-valued and upper hemi-continuous, with $S$ compact (it is in fact finite), $graph(C) \subseteq T \times S$ is closed, and thus $(y^o, x^o) \in graph(C)$, contradicting the hypothesis that $x^o \notin C(Y^p)$. This establishes the claim that $C(Y^p) \subseteq X$.

It remains to prove that, for each $i \in N$, condition (3) holds with equation for $Y^p$. Fix $i \in N$, and let

$$E_k = \bigcap_{j \neq i}(R_j \cap [Y_j^p(k)]) \qquad \text{and} \qquad E = \bigcap_{j \neq i}(R_j \cap [Y_j^p]).$$

Since, for each $j \in N$, $\langle Y_j^p(k)\rangle_k$ is a decreasing chain with limit $Y_j^p$, it follows that $\langle E_k\rangle_k$ is a decreasing chain with limit $E$.

To show $B_i^p(E) \subseteq [Y_i^p]$, note that by (A1) and monotonicity of $B_i^p$, we have, for each $k \in \mathbb{N}$, that

$$B_i^p(E) \subseteq B_i^p(E_{k-1}) = [Y_i^p(k)].$$

As the inclusion holds for all $k \in \mathbb{N}$:

$$B_i^p(E) \subseteq \bigcap_{k \in \mathbb{N}}[Y_i^p(k)] = [Y_i^p].$$

To show $B_i^p(E) \supseteq [Y_i^p]$, assume that $\omega \in [Y_i^p]$.[14] This implies that $\omega \in [Y_i^p(k)]$ for all $k$, and, using (A1): $\omega \in B_i^p(E_k)$ for all $k$. Since $E_k = \Omega_i \times \text{proj}_{\Omega_{-i}} E_k$, we have that $E_k^{\omega_i} = \text{proj}_{\Omega_{-i}} E_k$. It follows that

$$\mu_i(\mathbf{t}_i(\omega))(\text{proj}_{\Omega_{-i}} E_k) \geq p \quad \text{for all } k.$$

Thus, since $\langle E_k\rangle_k$ is a decreasing chain with limit $E$,

$$\mu_i(\mathbf{t}_i(\omega))(\text{proj}_{\Omega_{-i}} E) \geq p.$$

Since $E = \Omega_i \times \text{proj}_{\Omega_{-i}} E$, we have that $E^{\omega_i} = \text{proj}_{\Omega_{-i}} E$. Hence, the inequality implies that $\omega \in B_i^p(E)$. □

**Proof of Proposition 4.** Let $X \in \mathcal{S}$ be the smallest CURB set containing $C(Y)$: (i) $C(Y) \subseteq X$ and $\beta(X) \subseteq X$ and (ii) there exists no $X' \in \mathcal{S}$ with $C(Y) \subseteq X'$ and $\beta(X') \subseteq X' \subsetneq X$. We must show that $X = E(Y)$.

Consider the sequence $\langle Y(k)\rangle_k$ defined by $Y(0) = Y$ and condition (4) for each $k \in \mathbb{N}$ and $i \in N$. We show, by induction, that $C(Y(k)) \subseteq X$ for all $k \in \mathbb{N}$. By assumption, $Y(0) = Y \in \mathcal{T}$ satisfies this condition. Assume that $C(Y(k-1)) \subseteq X$ for some $k \in \mathbb{N}$, and fix $i \in N$. Then, $\forall j \neq i$, $\beta_j(\text{marg}_{S_{-j}} \mu_j(\mathbf{t}_j(\omega))) \subseteq X_j$ if $\omega \in [Y_j(k-1)]$ and $\mathbf{s}_j(\omega) \in X_j$ if, in addition, $\omega \in R_j$. Hence, if $\omega \in B_i^1(\bigcap_{j \neq i}(R_j \cap [Y_j(k-1)]))$, then $\text{marg}_{S_{-i}} \mu_i(\mathbf{t}_i(\omega)) \in \mathcal{M}(X_{-i})$ and $C_i(\mathbf{t}_i(\omega)) \subseteq \beta_i(X_{-i}) \subseteq X_i$. Since this holds for all $i \in N$, we have $C(Y(k)) \subseteq X$. This completes the induction.

Secondly, since the sequence $\langle Y(k)\rangle_k$ is non-decreasing and $C(\cdot)$ is monotonic with respect to set inclusion, and the game is finite, there exist a $k' \in \mathbb{N}$ and some $X' \subseteq X$ such that $C(Y(k)) = X'$ for all $k \geq k'$. Let $k > k'$ and consider any player $i \in N$. Since the probability structure is complete, there exists, for each $\sigma_{-i} \in \mathcal{M}(X'_{-i})$ a state $\omega \in [Y_i(k)]$ with $\text{marg}_{S_{-i}} \mu_i(\mathbf{t}_i(\omega)) = \sigma_{-i}$, implying that $\beta_i(X'_{-i}) \subseteq C_i(Y_i(k)) = X'_i$. Since this holds for all $i \in N$, $\beta(X') \subseteq X'$. Therefore, if $X' \subsetneq X$ would hold, then this would contradict that there exists no $X' \in \mathcal{S}$ with $C(Y) \subseteq X'$ such that $\beta(X') \subseteq X' \subsetneq X$. Hence, $X = C(\bigcup_{k \in \mathbb{N}} Y(k)) = E(Y)$.

---

[14] We thank Itai Arieli for suggesting this proof of the reversed inclusion, shorter than our original proof. A proof of both inclusions can also be based on property (8) of Monderer and Samet [30].

Write $X = E(Y)$. To establish that $X$ is epistemically robust, by Proposition 2(b), it is sufficient to show that

$$X \subseteq \times_{i \in N} \beta_i(\beta_i^{-1}(X_i)),$$

keeping in mind that, for all $X' \in \mathcal{S}$, $X' \supseteq \times_{i \in N} \beta_i(\beta_i^{-1}(X_i'))$.

Fix $i \in \mathbb{N}$. Define $Y_i' \in \mathcal{T}$ by taking $Y_i' := \{t_i \in T_i \mid C_i(t_i) \subseteq X_i\}$. Since the probability structure is complete, it follows that $C_i(Y_i') = \beta_i(\beta_i^{-1}(X_i))$. Furthermore, for all $k \in \mathbb{N}$, $Y(k) \subseteq Y'$ and, hence, $\bigcup_{k \in \mathbb{N}} Y(k) \subseteq Y'$. This implies that

$$X_i = C\left(\bigcup\nolimits_{k \in \mathbb{N}} Y(k)\right) \subseteq C_i(Y_i') = \beta_i(\beta_i^{-1}(X_i)),$$

since $C_i(\cdot)$ is monotonic with respect to set inclusion. □

**Proof of Proposition 5.** *(Only if)* Let $X \in \mathcal{S}$ be a MINCURB set. Let $t \in T$ satisfy $\text{marg}_{S_{-i}} \mu_i(t_i)(X_{-i}) = 1$ for all $i \in N$. By construction, $C(\{t\}) \subseteq X$, as $X$ is a CURB set. By Proposition 4, $E(\{t\})$ is the *smallest* CURB set with $C(\{t\}) \subseteq E(\{t\})$. But then $E(\{t\}) \subseteq X$. The inclusion cannot be strict, as $X$ is a MINCURB set. Hence, there exists a $t \in T$ such that $E(\{t\}) = X$. Moreover, as $E(\{t'\})$ is a CURB set for all $t' \in T$ and $X$ is a MINCURB set, there exists no $t' \in T$ such that $E(\{t'\}) \subsetneq X$.

*(If)* Assume that there exists a $t \in T$ such that $E(\{t\}) = X$ and there exists no $t' \in T$ such that $E(\{t'\}) \subsetneq X$. Since $E(\{t\}) = X$, it follows from Proposition 4 that $X$ is a CURB set. To show that $X$ is a *minimal* CURB set, suppose—to the contrary—that there is a CURB set $X' \subsetneq X$. Let $t' \in T$ be such that $\text{marg}_{S_{-i}} \mu_i(t_i')(X_{-i}') = 1$ for each $i \in N$. By construction, $C(\{t'\}) \subseteq X'$, so $X'$ is a CURB set containing $C(\{t'\})$. By Proposition 4, $E(\{t'\})$ is the smallest CURB set containing $C(\{t'\})$. However, by assumption, there exists no $t' \in T$ such that $E(\{t'\}) \subsetneq X$, so it must be that $E(\{t'\}) \supseteq X$. This contradicts $X' \subsetneq X$. □

## References


1. Nash, J.F. Non-Cooperative Games. Ph.D. Thesis, Princeton University, Princeton, NJ, USA, 1950.
2. Bernheim, D. Rationalizable strategic behavior. *Econometrica* **1984**, *52*, 1007–1028.
3. Pearce, D.G. Rationalizable strategic behavior and the problem of perfection. *Econometrica* **1984**, *52*, 1029–1050.
4. Aumann, R.J.; Brandenburger, A. Epistemic conditions for Nash equilibrium. *Econometrica* **1995**, *63*, 1161–1180.
5. Myerson, R.; Weibull, J.W. Tenable strategy blocks and settled equilibria. *Econometrica* **2015**, *83*, 943–976.
6. Tan, T.; Werlang, S.R.C. The Bayesian foundations of solution concepts of games. *J. Econ. Theory* **1988**, *45*, 370–391.
7. Basu, K.; Weibull, J.W. Strategy subsets closed under rational behavior. *Econ. Lett.* **1991**, *36*, 141–146.
8. Hurkens, S. Learning by forgetful players. *Games Econ. Behav.* **1995**, *11*, 304–329.
9. Sanchirico, C. A probabilistic model of learning in games. *Econometrica* **1996**, *64*, 1375–1394.
10. Young, P.H. *Individual Strategy and Social Structure*; Princeton University Press: Princeton, NJ, USA, 1998.
11. Fudenberg, D.; Levine, D.K. *The Theory of Learning in Games*; MIT Press: Cambridge, MA, USA, 1998.
12. Blume, A. Communication, risk, and efficiency in games. *Games Econ. Behav.* **1998**, *22*, 171–202.
13. Hurkens, S. Multi-sided pre-play communication by burning money. *J. Econ. Theory* **1996**, *69*, 186–197.
14. Galeotti, A.; Goyal, S.; Kamphorst, J.J.A. Network formation with heterogeneous players. *Games Econ. Behav.* **2006**, *54*, 353–372.
15. Ritzberger, K.; Weibull, J.W. Evolutionary selection in normal-form games. *Econometrica* **1995**, *63*, 1371–1399.
16. Kohlberg, E.; Mertens J.-F. On the strategic stability of equilibria. *Econometrica* **1986**, *54*, 1003–1037.
17. Asheim, G.B. *The Consistent Preferences Approach to Deductive Reasoning in Games*; Springer: Dordrecht, The Netherlands, 2006.
18. Brandenburger, A.; Friedenberg, A.; Keisler, H.J. Admissibility in games. *Econometrica* **2008**, *76*, 307–352.
19. Kalai, E.; Samet, D. Persistent equilibria in strategic games. *Int. J. Game Theory* **1984**, *14*, 41–50.

20. Van Damme, E. Strategic Equilibrium. In *Handbook of Game Theory*; Aumann, R.J., Hart, S., Eds.; Chapter 41; Elsevier: Amsterdam, The Netherlands, 2002; Volume 3.
21. Tercieux, O. *p*-Best response set. *J. Econ. Theory* **2006**, *131*, 45–70.
22. Zambrano, E. Epistemic conditions for rationalizability. *Games Econ. Behav.* **2008**, *63*, 395–405.
23. Hu, T.-W. On *p*-rationalizability and approximate common certainty of rationality. *J. Econ. Theory* **2007**, *136*, 379–391.
24. Osborne, M.J.; Rubinstein, A. *A Course in Game Theory*; The MIT Press: Cambridge, MA, USA, 1994.
25. Ferguson, T.S. *Mathematical Statistics, A Decision Theoretic Approach*; Academic Press: New York, NY, USA, 1967.
26. Van Damme, E. *Refinements of the Nash Equilibrium Concept*; Springer: Berlin, Germany, 1983.
27. Brandenburger, A.; Friedenberg, A.; Keisler, H.J. *Admissibility in Games*; New York University: New York, NY, USA; Washington University: St. Louis, MO, USA; University of Wisconsin-Madison: Madison, WI, USA, 12 December 2004.
28. Brandenburger, A.; Dekel, E. Hierarchies of beliefs and common knowledge. *J. Econ. Theory* **1993**, *59*, 189–198.
29. Mertens, J.-F.; Zamir, S. Formulation of Bayesian analysis for games with incomplete information. *Int. J. Game Theory* **1985**, *14*, 1–29.
30. Monderer, D.; Samet, D. Approximating common knowledge with common beliefs. *Games Econ. Behav.* **1989**, *1*, 170–190.
31. Harsanyi, J.C.; Selten, R. *A General Theory of Equilibrium Selection in Games*; The MIT Press: Cambridge, MA, USA, 1988.
32. Mertens, J.-F. Stable equilibria—A reformulation, part I: Definition and basic properties. *Math. Oper. Res.* **1989**, *14*, 575–625.
33. Demichelis, S.; Ritzberger, K. From Evolutionary to Strategic Stability. *J. Econ. Theory* **2003**, *113*, 51–75.

 *games*

MDPI

*Article*

# The Welfare Cost of Signaling

**Fan Yang [1,2] and Ronald M. Harstad [2,*]**

[1] Economics area, New York University Shanghai, 1555 Century Ave, Pudong, Shanghai 200122, China; fyang@nyu.edu
[2] Economics Department, University of Missouri, 901 University Avenue, Columbia, MO 65211-1140, USA
* Correspondence: ron.harstad@gmx.us

Academic Editor: Paul Weirich
Received: 17 August 2016; Accepted: 18 January 2017; Published: 9 February 2017

**Abstract:** Might the resource costliness of making signals credible be low or negligible? Using a job market as an example, we build a signaling model to determine the extent to which a transfer from an applicant might replace a resource cost as an equilibrium method of achieving signal credibility. Should a firm's announcement of hiring for an open position be believed, the firm has an incentive to use a properly-calibrated fee to implement a separating equilibrium. The result is robust to institutional changes, outside options, many firms or many applicants and applicant risk aversion, though a sufficiently risk-averse applicant who is sufficiently likely to be a high type may lead to a preference for a pooling equilibrium.

**Keywords:** costly signaling; social cost of signaling; asymmetric information; separating equilibrium

---

Adverse selection becomes a concern when a party A faces a decision based on information possessed by a party B, whose utility is also affected by A's decision. That is, under what circumstances can party A rely on information communicated by party B?

Spence's 1973 paper [1] introduced a model in which employers may use education as a screening device. In a related context, Akerlof (1970) [2] provided perhaps the most widely-taught adverse-selection example. Stiglitz (1975) [3] discussed the concept of screening in the context of employment and education. All of the above mechanisms are costly ways of solving an adverse-selection problem by creating an incentive to self-select. By contrast, in cheap-talk games (Crawford and Sobel, 1982 [4], Chakraborty and Harbaugh, 2010 [5] and the references in the latter), where communication is privately and (probably) socially costless, information that can credibly be transmitted is limited, usually severely. This paper asks: Since a sender must incur a cost of transmitting if the message is to be credible (for present purposes, the cost of obtaining, say, an MBA degree, is here labeled a cost of transmitting), to what extent can the cost be reduced for society by using a transfer instead of a pure resource cost?

We address this question not to explain common occurrences in markets, but to better understand the foundations of the economics of transacting under asymmetric information. To explore these foundations, suppose that a firm can credibly commit to considering only those applicants who pay an application fee that might be substantial. [1] A test that might distinguish between applicants in some aspect of their suitability could still be conducted, but only if the firm's resource costs of administering the test and evaluating the effectiveness shown are quite small and an applicant's resource costs of preparing for and taking the test are negligible compared both to the resource costs of a usual signal and to the size of the application fee (the firm comparing a privately known threshold to the applicant's credit score, or her/his driving record, or her/his number of semesters on the Dean's list

---

[1] Also credibly commit to not collecting application fees as a profitable activity without an appropriately compensated job waiting for the chosen applicant.

or quickly searching Facebook or Instagram or YouTube for incidents). With this setup, the question becomes whether a suitably calibrated application fee can achieve the same types of signaling equilibria that are accomplished by calibrating the resource cost of the usual sort of signal, such as obtaining a particular level of education. That substantial application fees may not be common in labor markets of acquaintance does not bear on the relevance of this research. [2]

Two papers touch on this question, both less directly and subject to objections. Wang (1997) [6] introduces an employment model in which only if the firm commits to a wage schedule before the applicants pay the fee might an application-fee equilibrium be possible. Though set-of-wages, positive-application-fee equilibria may be possible below, this possibility need not require the firm to commit to a wage schedule before an applicant decides whether to pay the fee (cf. Section 2). As to using the necessity of commitment to explain why no application fee is observed in reality as Wang (1997) [6] does, no practical reason can be given why a firm could not commit to a schedule of multiple wages corresponding to multiple estimated productivities (indeed, this is a feature of nearly every job posting seen in the market for Ph.D. economists). Furthermore, the pre-commitment argument is based on the assumption that firms have full control over wages. If the wage is instead determined through, say, alternating-offer bargaining, [3] it is obvious that applicants can still expect to attain some surplus, making a positive application-fee possible.

Guasch and Weiss (1981) [7] suggest that applicants' risk aversion and an assumption that applicants do not have perfect information about themselves may prevent a positive-application-fee equilibrium. As shown below, risk aversion alone is insufficient to prevent a positive-application-fee equilibrium. The Guasch/Weiss model [7] requires the assumption that the labor-supply constraint is not binding, which is problematic: If there are more than enough high types applying, why test all of them? Where do the "extra" high-type applicants go? Firm profit maximization implies that they are not hired, while applicants' expected returns show that they get paid and hired. [4]

By assumption, the firm genuinely wishes to hire someone, and this is believed by the applicant (the credibility of many firms, especially prestigious ones, is in fact too valuable to risk fraudulently collecting application fees for nonexistent jobs).

Although the models use job-application settings, to varying degrees, they can apply to other contexts, as well. For example, the job vacancy can easily be interpreted as a promotional opportunity within a firm. The model may extend to payment below productivity during a required internship period. Another possibility is that a firm may attempt to signal credibly the quality of a product or product line or a service by a donation to a charity that it knows will be publicized at no cost to the firm. [5]

In the following model and variations, a separating equilibrium always exhibits a positive application fee. Whether there is a separating equilibrium depends entirely on the firm's incentives (or the firm's and headhunter's incentives in Section 2). Should no separating equilibrium exist, the firm's only options are not to hire or to charge no application fee and hire any applicant without testing. With testing cost sufficiently low, separating equilibrium almost always exists.

## 1. The Base Model

Consider a game between a profit-seeking monopsony employer and a potential applicant, both risk neutral. The applicant is either Type 1 or Type 2, and knows her/his own type. The firm does not know the type, but correctly knows the distribution of types (probability $p \in (0, 1)$ of being

[2] Among many markets with adverse-selection issues exhibiting application fees are college admissions, scholarly journals and nightclubs with live music.

[3] Cf. van Damme, Selten and Winter (1990) [8] and references there.

[4] The models used in this paper are similar to those of Guasch and Weiss (1981) [7]. In fact, the one-firm, multiple-applicant case (Section 5) can be regarded as a simpler version of their model, while avoiding the "labor-supply constraint not binding" problem.

[5] For a context yielding several more examples, see Spence (2002) [9].

Type 1; $1-p$ of being Type 2). Type $t$ worker has productivity $t$ if working for the firm ($t=1,2$). Both types can produce $r \in (1,2)$ at home if not hired. [6] At cost $c \geq 0$, the firm can conduct a small test, with probability $q \in (0.5,1)$ of correctly revealing the applicant's type and probability $1-q$ of being misleading, thus possibly of very little reliability. The hiring game is played as follows:

Step 1 The firm chooses its strategy $s=(w_L, w_H, f)$, where $w_L$ is the wage offered to an applicant with Test Result 1, $w_H$ is the wage offered to an applicant with Test Result 2 and $f$ is the application fee.
Step 2 The potential applicant sees the wage/fee schedule $s$ and decides whether or not to apply for the position. If she/he applies, she/he must pay the application fee.
Step 3 If the applicant has applied in Step 2, she/he takes the test and the result is revealed for both the firm and the applicant. The applicant then decides whether to accept the wage offer fitting the test result.

For the above defined game, the firm's strategy space is $\mathbb{R}^3_+$. The applicant chooses $(App(f, w_L, w_H, t), Acc(f, w_L, w_H, t, x))$, in which $t$ is her/his type and $x$ is the realized test result. "App" can be either "apply" or "not"; "Acc" can be either "accept" or "not".

To avoid trivialities, assume:

$$c < 2 - r. \tag{1}$$

That is, the test cost cannot be so large that the firm would not make an offer to a known high type. For simplicity, also assume that the applicant accepts the offer if she/he is indifferent in Step 3 and that she/he applies if she/he is indifferent in Step 2. Assume, of course, that both the firm and the applicant play to maximize their expected payoff. The above specifications yield the following theorem, proven in Appendix A.

**Theorem 1** (Main theorem). *A strategy profile is a subgame-perfect equilibrium of the above-defined game if and only if: In Step 1, the firm implements a separating equilibrium in which the potential applicant applies if and only if she/he is a high type and hires anyone that applies while setting $w_L$, $w_H$ and $f$, such that (i) the firm makes an acceptable offer:*

$$w_H > w_L \geq r, \tag{2}$$

*(ii) the firm maximizes profit:*

$$qw_H + (1-q)w_L - r = f. \tag{3}$$

*In Step 2, a type t (t= 1 or 2) potential applicant applies if and only if (iii) the wage structure is incentive compatible:*

$$\begin{aligned} &q * max\{w_L, r\} + (1-q) * max\{w_H, r\} - r \geq f, \text{ (if t=1)}, \\ &(1-q) * max\{w_L, r\} + q * max\{w_H, r\} - r \geq f, \text{ (if t=2)}. \end{aligned} \tag{4}$$

*In Step 3, the applicant accepts the offer if and only if the wage is no less than r. That is, for an applicant with test result x, accept if and only if $w_x \geq r$.*

Equation (4) has the potential applicant apply if the expected value added by applying is no less than the application fee $f$.

[6] This assumption is relaxed in Appendix D. A common default productivity accords with Spence's 1973 [1] assumptions and fits reasonably a case in which the differential productivity the firm seeks to uncover is firm specific, or perhaps industry specific, rather than yielding a similar productivity difference to most potential employers.

From (3), $w_L \geq r$ and $w_H \geq r$ ensure that if the applicant applies, she/he is hired, at a test-dependent wage level. $w_H > w_L$ separates the value of applying for different types, in favor of the high type, given $q > 0.5$.

The fee determined by Equation (3) leads the high type to apply, though indifferent. Any higher fee prevents the high type from applying. The low type does not apply because, compared to the high type, she/he has a lower chance of receiving the high wage, but faces the same application fee. For given $w_L$, $w_H$ satisfying (2), $f^{'} = (1-q)w_H + qw_L - r$ is the highest fee that induces the low type to apply; fees in the interval $(f^{'}, f)$ reduce fee income and lead the high type to strictly prefer applying, without otherwise affecting the outcome.

For example, setting $w_L = r$, $w_H > r$ and letting $f$ be determined by Equation (3) yield a subgame-perfect equilibrium. In such an equilibrium, both types are in their most productive positions (firm for high types and home for low types), and perfect separation is achieved without testing the low type, thus saving on testing cost. The application fee serves to make an imperfect sorting device (the test) perfect, even though the fee is purely a private cost rather than a social cost.

This welfare reassessment, that separating equilibria can be achieved by having the signal's cost be a transfer, rather than a resource cost, is robust to institutional changes. The online supplement extends the analysis to institutions in which: (i) the applicant is freely considered, but only if she/he has paid a fee will the firm observe the test result; (ii) the fee must be paid for a positive (but possibly quite low) probability that the firm will observe the test results (and otherwise, the firm does not incur a testing fee); and (iii) the fee must be paid for a positive probability that the firm will observe the test results, but with the fee refunded in the random event that the firm does not observe the results. [7]

Note that our base model is a screening game rather than a signaling game. [8] That is, the firm makes all of the decisions first and then lets nature and the applicant do all of the separating, rather than observing some signals sent by the applicant and then making decisions based on updated information about the applicant type. Note also that, while the fee is an effective screening device, it cannot be made into a signaling device simply by moving the wage decision to the last step. Subgame perfection would require that the firm pay no more than $r$ in the last stage, and as a result, no applicant would pay a positive fee to apply. A variation in which the firm's wage decisions occur after a potential applicant has decided whether to pay an application fee is analyzed in the next section.

**Comparison with Spence (1973)** [1]: Spence's model differs from ours in two ways. His costly signal is an effort that per se serves no economic or social purpose, but is be assumed to create significantly less disutility for a high type than a low type (the same differential capabilities that make a high type a more productive employee are assumed to yield the lower disutility of the communicative effort). In this model, the signal cost is a transfer, and it would be untenable to assume that the high type had a significantly lower marginal utility of income. The monetary nature of the signal carries the social-welfare advantage that the money may be put to an equally productive purpose. It carries the disadvantage that paying the application fee generates the same disutility for the high type and the low type, so paying the application fee cannot by itself credibly signal type. Thus, we add to Spence's model an inexpensive test with an informational content that may be nearly meaningless (e.g., the fraction of the courses taken during her/his senior year that are numbered by her/his university as senior-level courses), but simply is more likely to be passed by a high type than a low type. Now, setting test-result-dependent wages so that the high type is nearly indifferent over paying the application fee separates: a lower likelihood of a passing test result leads the low type to see an insufficient expected advantage to paying the same application fee.

---

[7] The latter two institutions require the firm in Step 1 to set two additional wage levels: that offered to an applicant who paid the fee, but whose test result was not observed, and that offered to an applicant who declined to pay the fee.

[8] For a discussion of the difference between signaling and screening games, see Sections 13.C and 13.D in MasColell, Green and Whinston (1995) [10].

Spence's signal is often described as obtaining a university degree at a job-appropriate level (e.g., MBA), which if not justified as a human-capital investment, might be seen as very costly to a high type and, thus, to society. Spence does not himself so limit his model; the signal in the right circumstances might be exceeding a carefully-chosen threshold on a standardized test, if attaining the threshold yields sufficiently high disutility for a low type. Thus, in some circumstances, Spence's separating equilibrium might have a high type incurring such a small cost transmitting a signal (that would have been highly costly to a low type) as to approximate "first-best": very low social costliness of the signals for types that actually transmit them. In this paper, the approximate first-best simply comes from the lack of social cost of the signal no matter how high the private cost to the signaler.

In the theorem above, the possibility of approaching a low private cost can be attained similarly to Spence's model. The $f$ in the theorem must be positive, but can be made arbitrarily small, adjusting to $w_L = r$, lowering $w_H$ to satisfy (4).

## 2. Application Fee in a Signaling Game: Adding a Third Party

As discussed in Section 1, the base model's application fee cannot be shifted directly to a signaling device. Suppose there is perfect competition by firms hiring in this labor market, but an applicant can only be considered by a firm after she/he pays a fee to that particular firm. Once an applicant has paid the fee to a particular firm, that firm no longer faces any competition in hiring that worker and so offers at most wage $r$. Any positive fee is then impossible.

This issue may be resolved by involving a third party. The applicant must pay a fee to this third party to enter the market; upon entry, all firms in this market can compete for her/his employment.

Consider a job market with multiple firms competing with each other, while still only having one applicant, with type assumptions as before. Now, assume there is a headhunter, who holds some monopoly power in the market: firms can only hire a job applicant through the headhunter, who may demand a fee for the applicant to be available for hire. [9]

The hiring game is played in sequence as follows:

1. The headhunter sets a fee $f$.
2. The applicant decides whether to pay the fee to enter the market.
3. The firms quote wages.
4. If she/he has entered the market, the applicant chooses a firm and applies.
5. The applicant is tested, costing the firm $c$; she/he signs a waiver ceding the right to apply to or negotiate with any other firm. [10]
6. The applicant and firm learn the test results; previously-set wages are offered to the applicant; she/he decides whether to accept or not; if she/he accepts, she/he is hired. If not, she/he returns home and produces $r$.

The waiver is a convenient way to: (i) keep both $w_L$ and $w_H$ wage quotations relevant to applicant decisions; and (ii) prevent the applicant from applying to another firm if she/he tests low at the current firm. It yields the most straightforward comparison to the base model.

For a natural choice of tiebreakers, the main result extends sensibly:

**Corollary 1** (Third-party corollary). *For the headhunter to set $f = 2 - r - c$, firms to set $(w_L, w_H)$ so that $w_H > w_L$ and* (3) *and* (4) *are satisfied, high types to apply and the low types not to apply constitute a separating equilibrium.*

---

[9] A frequent example is a government agency that has to specify that an applicant meets certain criteria before she/he can be hired into a particular field or for a particular job. This model considers the agency setting the fee in excess of its cost of certification.

[10] Having the headhunter cover the testing cost out of fee revenue only yields obvious adjustments to the equilibria.

The setup and proof are in Appendix B.

Note that this result does not require a competitive industry seeking to hire a high type: suppose there is but a single firm who nonetheless can only hire an applicant who paid the fee to a headhunter; if the headhunter sets a fee as above, then the firm optimally sets wages $(w_L, w_H)$ satisfying (2), (3) and (4), generating a separating equilibrium.

## 3. Risk-Averse Applicants

This section returns to a single (risk-neutral) firm, to consider applicant risk aversion. Guasch and Weiss (1981) [7] noted the obvious: a risk-averse applicant requires an expected return greater than $r$ to accept the risk of an uncertain test result implying an uncertain wage. Intuitively, if the risk premium required is high enough and the cost of hiring a low type is low enough, the firm may be unwilling to pay the risk premium as the cost of separating equilibrium and hire everyone without testing instead. Assume both types of applicant have the same pattern of risk tolerance.

Under what conditions can a separating equilibrium be preserved? Instead of positing a particular risk-averse utility function, consider a wage/fee schedule and ask how high a risk premium is needed for a high type to accept. Specifically, in the base model, the firm may set the two wages arbitrarily close; the focal issue is the risk premium required if $w_L$ is close to $w_H$.

Above, as is usual, the applicant is indifferent between a wage of 12 and fee of two and a wage of 13 and fee of three. However, this section's analysis of risk aversion is clarified by generalizing to a utility function $U(w-f,f)$, for either type of applicant, with the usual concavity maintained via assuming $w \geq f \implies \frac{\partial U}{\partial(w-f)}(w-f,f)$ is decreasing in $w-f$ for any $f$.

Let $w > r$; there exists an $\epsilon > 0$ small enough such that $w - \epsilon \geq r$. Then, the wage/fee schedule $w-\epsilon$, $w+\frac{1-q}{q}\epsilon$ and $f = w - r$ is a viable schedule to implement separating equilibrium in the risk-neutral case. As above, $q$ is the probability the test correctly identifies the applicant's type. Since this involves risk, a risk-averse high type would demand a risk premium to accept such an offer; for clarity, treat the risk premium as being subtracted from $f$. [11]

Naturally, assume the risk premium increases with $\epsilon$. Therefore, the firm would prefer to offer wages as close to each other as possible.

A schedule $s$ that makes the high type indifferent over accepting would not be accepted by the low type, who would end up receiving the low wage with a greater probability than the high type. Therefore, the firm only has to make sure that high types are indifferent in order to implement a separating equilibrium.

Let $RP$ be a function that maps $f$ and $\epsilon$ into the amount of risk premium that makes the high type indifferent. Then, we have the following corollary:

**Corollary 2** (Third-party corollary). *$s^*(\epsilon) = [w-\epsilon, w+\frac{1-q}{q}\epsilon, f - RP(f,\epsilon)]$ is a potential schedule in a separating equilibrium. The applicant types separate under this wage/fee schedule, provided the firm is willing.*

Note that all possible sets of wages satisfying Equation (2) can be represented by the above wage schedule, via changing $\epsilon$. Since any separating equilibrium must satisfy Equation (2), the wage schedule can be represented as above. Given the above wage schedule, the fee must be $f - RP(f,\epsilon)$, as the high type would not accept any higher fee, and the firm's profit is suboptimal for any lower fee. Therefore:

**Proposition 1.** *Any separating equilibrium takes the form described in Corollary 2.*

---

[11] The risk could be addressed by increasing $w_H$, but as a high type cannot ensure the high wage, subtracting from $f$ is more straightforward.

Appendix B delves into risk aversion in more detail, finding: a minimal assumption for separation to occur, separation impossible with truly catatonic risk aversion and how unusually risk aversion must be modeled for pooling possibly to be preferred.

## 4. Multiple Firms Competing for the Applicant

This section examines whether multiple firms competing for one risk-neutral hire can affect the realization of separating equilibrium. If separating equilibrium is still achievable, it must be allowing the high type to get all of the surplus, since otherwise, another firm would offer a higher wage and attract the high-type worker away. On the other hand, the low type must be expected to get less than $r$ if she/he were to apply. This immediately means that any wage/fee schedule that implements separating equilibrium with multiple firms needs to separate the two types sufficiently far. Opportunities to separate with $w_L$ closely below $w_H$ are more restricted, perhaps preventing separating equilibrium were this section blending firm competition and applicant risk aversion.

For separation, the wage/fee schedule must satisfy:

$$w_H q + w_L(1-q) - f = 2 - c, \tag{5}$$

$$w_H(1-q) + w_L q - f < r. \tag{6}$$

Equation (5) ensures the high type's expected wage minus fee equals social surplus; Equation (6) discourages the low type from applying. Subtracting (6) from (5) yields:

$$(2q-1)(w_H - w_L) > 2 - c - r. \tag{7}$$

Equation (2) is still needed to ensure hiring all that applied.

**Corollary 3.** *Any set of $w_H$, $w_L$ and $f$ that satisfies Equations* (2), (5) *and* (7) *yields a wage/fee schedule for a separating equilibrium.*

Details and a proof are in Appendix B.

## 5. One Firm, Finitely Many Potential Applicants

Instead of one potential applicant, suppose there are $n$; each is independently a low type with probability $p$; $n$ and $p$ are assumed common knowledge. The firm can only use one worker productively. [12]

Seeking a separating equilibrium, the firm has neither the desire, nor the need to test all applicants. Let it adopt the strategy of testing one randomly-selected applicant, hiring her/him if her/his test result is high, and otherwise hiring a second randomly-selected applicant (possibly the same applicant as the first) without conducting even a second test.

This testing strategy can support a separating equilibrium. If there is no high type in the pool, no one applies, and the firm receives no profit. As long as there is at least one high type, there are fee-paying applicants; the firm tests and hires someone. Therefore, the firm maximizes expected payoff conditioning on at least one high type in the pool.

[12] Note that the number of workers being finite is important because with an infinite number of applicants, no matter how the firm sets up the hiring scheme, all applicants face a zero chance of being hired; therefore, no separation can occur.

Let $m_n$ be the realization of the number of high types in a pool of $n$, $w_u$ be the wage offered to the applicant getting high test results and $w_d$ be the wage offered to the applicant selected through the second random draw. [13] The firm maximizes the payoff:

$$E[2 - c + m_n f - q w_u - (1-q) w_d \mid n, m \geq 1]. \quad (8)$$

This can be simplified to (ignoring the constant $2 - c$):

$$E[m_n \mid n, m \geq 1] f - q w_u - (1-q) w_d. \quad (9)$$

A high type, competing with $n-1$ rival potential applicants who are each a low type with probability $p$, though indifferent, will apply if facing a wage/fee schedule satisfying:

$$E[\frac{w_d}{m_{n-1}+1} + \frac{q(w_u - w_d)}{m_{n-1}+1} - f \mid n] = r, \quad (10)$$

or:

$$f = E[\frac{1}{m_{n-1}+1} \mid n] * (q w_u + (1-q) w_d) - r. \quad (11)$$

Facing the same wage/fee schedule, a low type has the same expected payoff if randomly selected second, but a lower payoff if randomly selected first, as the probability of testing high is less. Therefore:

**Corollary 4.** *The firm can implement a separating equilibrium with any wage/fee schedule that satisfies* (2) *and* (11).

Appendix B details how the base model is a special case of this model.

Adding an option to hire without testing, in a pooling equilibrium, the firm would simply hire the first of the applicants at wage $r$ without testing. Unlike the introduction of more firms, introducing more applicants does not seem to qualitatively change the feasibility of using $f$ and a test to create separating equilibria.

## 6. Discussion

The models presented yield the following conclusions:

- It is possible to use a transfer to implement a separating equilibrium; in that sense, the private cost of signaling need not be a social cost.
- Commitment to a wage by the firm is not necessary to use a transfer as a signaling device. [14]
- Applicant risk aversion alone is normally insufficient to prevent the existence of a separating equilibrium. Considerable differences in home productivity across types may increase the likelihood that equilibrium requires pooling.

An assumption of the base model is that the test cost $c$ is nonnegative. [15] A negative $c$ may make it optimal for the firm to test everyone (the last inequality in Case 3 of the proof may not hold if $c < 0$). For some situations, applying this model would naturally suggest a negative $c$. If the test represents some form of internship or other productive activity and the fee as the reduced pay in this activity, there is a legitimate reason to claim that $c$ can be negative, meaning the interns are producing more than the funds it took the firm to set up such a program.

Similar to the discussion about the internship, if an employee's type may be (imperfectly) revealed only after some periods of employment, the employment period before such a revelation can be

---

[13] When $n = 1$, this model reduces to the model in Appendix E(i), with $w_u$ and $w_d$ playing the role of $w_H$ and $w_L$, respectively.

[14] The firms' credibility concerns can be avoided by having a centralized third party collect the application fee.

[15] Interestingly, the actual cost of the test $c$ does not enter Equation (3) in determining the fee.

considered a test to determine the wage afterwards. Aside from the possibility that $c$ is negative, there are two differences to the base model: the "test" cost $c$ now is less for a high than for a low type, and the "test" is possibly perfect. There will still be a set of separating equilibria if $c$ is still positive.

Are results affected if the firm has to spend money advertising jobs in order to attract applicants? Add to the base model an assumption that the firm needs to incur a fixed cost in order to let the applicant be aware of the opportunity, i.e., to apply for the job. However, once paid, it becomes a sunk cost, so it should not affect the firm's choice of wage and fee. It can affect the firm's choice of whether to enter the market.

Appendix C shows that if instead of having two potential applicant types, types are continuously distributed along an interval on the real line, the separating equilibria are not affected if the reservation wage remains constant across types.

Two main differences exist between Guasch and Weiss (1981) [7] and the models in this paper so far: the models so far allow hiring of an applicant with a low test score and a universal reservation wage across types. What will be the impact of relaxing the latter restriction by introducing a variable reservation-wage based on type to these models? In Appendix D, variable reservation wages are introduced to the base model. Appendix F provides proofs and extends to multiple firms or multiple applicants. It turns out that while variable reservation wages may yield a smaller set of wage/fee schedules implementing separating equilibrium (by requiring a minimum difference between $w_H$ and $w_L$), firms still maintain the option to separate. For models in which a pooling equilibrium is possibly optimal, the firm makes the same choice between separating and pooling as if the reservation wage is a constant.

**Acknowledgments:** We thank, without implicating, Chao Gu, Rick Harbaugh, Oksana Loginova, Christopher Otrok, Joel Sobel, Paul Weirich and two unusually diligent and helpful referees.

**Author Contributions:** F.Y. created the initial models and developed the initial results. R.M.H. recasted the presentation, tightened statements and proofs, and wrote most of the final version.

**Conflicts of Interest:** The authors declare no conflict of interest.

## Appendix A. Proof of the Main Theorem

Step 3 is trivial. For Step 2, the left-hand side of (4) is the expected value of applying for $t = 1, 2$, while the right-hand side is the cost of applying. It remains to show that in Step 1, the firm prefers the wage/fee schedules defined by Equations (2) and (3) to all other schedules. In effect, the firm can decide who gets hired in Step 3 by changing $w_L$ and $w_H$. Given $w_L$ and $w_H$, the firm can decide who applies by changing $f$. Let $s^* = (w_L^*, w_H^*, f^*)$ be an arbitrary schedule satisfying (2) and (3). Facing $s^*$, a high type by assumption applies though indifferent, while a low type's expected payoff is $qw_L^* + (1-q)w_H^* - f^* < r$, preventing applying. An applicant under $s^*$ is thus a high type. With the wage determined by the test result, the firm's expected profit is:

$$(f^* + 2 - c)(1-p) - w_H^* q(1-p) - w_L^*(1-q)(1-p) = (1-p)(2-r-c). \quad \text{(A1)}$$

With probability $1-p$, the potential applicant is a high type, who applies, pays the fee $f^*$, costs the firm $c$ to be tested, is hired, has productivity 2, and is, in expectation, paid $w_L^*(1-q) + w_H^* q = f^* + r$ (from (3)), attaining the strictly positive (from (1)) right-hand side of (A1).

It is trivial to dismiss as suboptimal any schedule $s$ that (a) leads to only low types applying; (b) leads to neither type applying; or (c) leads to hiring only those who test low. Nontrivial alternatives fall into the following three cases.

*Case 1: Only High Types Apply, Only the High-Result Applicant Is Hired*

All such possibilities can be dealt with as if $w_H \geq r > w_L$. Then, adjusting (3), the highest fee acceptable for a high type to apply becomes:

$$f^{**} = q(w_H - r) > (1 - q)(w_H - r). \tag{A2}$$

The equality yields high-types applying though indifferent, the inequality low types not applying. Compared to $s^*$, the firm's profit has fallen by $(1-p)(1-q)(2-r)$, as high types who tested low were profitably hired in $s^*$. Reducing the fee to $f < f^{**}$ at best allows increasing $w_H$ by $\frac{(f^{**}-f)}{q}$, which cannot yield an increase in expected profit, so offering no advantage. Same can be argued for increasing the fee to $f > f^{**}$.

*Case 2: All Types Apply, All Are Hired*

An applicant is always tested (as required in the base model). Initially assuming $w_H \geq w_L \geq r$, sets the highest acceptable fee to $f^{**} = (1-q)(w_H - r) + q(w_L - r)$. With both types hired, expected productivity is $2 - p$, so expected profit is at most $2 - p - r - (1-p)(w_H - w_L)(2q-1) - c \leq 2 - r - p - c = (1-p)(2-r-c) + p(2-r-c) - p < (1-p)(2-r-c)$, which is expected profit for $s^*$, as $r > 1$, $c \geq 0$. Next, reverse the initial assumption: $w_L \geq w_H \geq r$, the analysis corresponds: $2 - p - r - p(w_L - w_H)(2q-1) - c \leq 2 - r - c - p = (1-p)(2-r-c) + p(2-r-c) - p < (1-p)(2-r-c)$.

Again yielding lower expected profit than $s^*$.

*Case 3: All Types Apply, Only a High-Result Applicant Is Hired*

Hiring only an applicant who tests high, as in case 1, it suffices to consider $w = w_H \geq r > w_L$. However, to get the low types to apply, the highest fee becomes $f^{**} = (1-q)(w_H - r)$, which has the low type apply though indifferent (and the high type strictly prefer applying). As no type offered wage $w_L$ accepts, expected profit at $f^{**}$ is

$$\begin{aligned} & f^{**} - c + [(1-p)q(2-w_H)] + \{p(1-q)(1-w_H)\} \\ &= (1-q)(w_H - r) - c + [(1-p)q(2-w_H)] + \{p(1-q)(1-w_H)\} \\ &= (1-2q)(1-p)w_H - c + 2(1-p)q + p(1-q) - (1-q)r. \end{aligned} \tag{A3}$$

where the term in [] is productivity less wage for a high type who tests high, that in {} is the same difference for a low type who tests high, the first equality substitutes for $f^{**}$, the second collects terms in $w_H$. As $q > \frac{1}{2}$, the coefficient of $w_H$ is negative, so expected profit is maximized at $w_H = r$, which sets $f^{**} = 0$. Substituting these values of $w_H$ and $f^{**}$ into the left-hand side of (A3) yields $[(1-p)q(2-r)] + \{p(1-q)(1-r)\} - c \leq [(1-p)q(2-r)] - c < (1-p)(2-r) - (1-p)c = (1-p)(2-c-r)$, where dropping the nonpositive term in provides the weak inequality, and substituting the larger 1 for $q$ and the smaller $(1-p)$ for 1 provides the strict inequality, again yielding lower expected profit than $s^*$.

Thus, an arbitrary wage/fee schedule $s^*$ satisfying (2) and (3) attains a positive expected profit that exceeds all alternative schedules.

## Appendix B. Details and Extensions for Sections 2–5

### *Appendix B.1. The Headhunter Model*

Lexicographic tiebreakers: as before, (i) the applicant is assumed to apply and work if indifferent. Furthermore, (ii) if two firms quote wage offers with the same expected wage, it is convenient to break the tie by assuming the high type applies to the one quoting a higher $w_H$ (thus a lower $w_L$), and the low type applies to the one quoting a higher $w_L$ (lower $w_H$). This tie-breaker follows trembling-hand considerations (Selten, 1975 [11]). (iii) Across firms quoting identical wages, she/he randomizes equiprobably.

Of course, it does not matter to the applicant to whom she/he pays the fee. Therefore, if Equation (4) from section I holds, the high-type applicant continues to apply though indifferent, and the low type continues to strictly prefer not applying.

In a separating equilibrium, each firm for whom the probability of receiving an application is positive must set $(w_L, w_H)$ so that $w_H > w_L$ and (3) is satisfied. Despite introducing the headhunter, (3) still makes the high type indifferent over entering the market and applying, and yields a strict preference for the low type to stay home.

Suppose the headhunter sets $f \leq 2 - r - c$, the high type applies, and the low type may or may not apply. Were a given firm to face a pattern of wage offers in which every other firm offered the high type an expected wage below $2 - c$, its best response would be to offer slightly higher wages.

Thus, for the headhunter to set $f = 2 - r - c$, firms to set $(w_L, w_H)$ so that $w_H > w_L$ and (3) and (4) are satisfied, high type to apply, and the low type not to apply, constitutes a separating equilibrium. In such an equilibrium, the headhunter expropriates all the social surplus, and the high-type expected wage is $2 - c$, leaving firms with zero profit in this labor market. The high type applies to the firm whose wage offer has the largest difference $w_H - w_L$, but the expected wage being driven up to productivity removes any incentive for other firms to deviate to attract the high type.

Are these the only equilibria? The headhunter can be shown, without doing algebra, to set too high a fee to allow a pooling equilibrium in which both types enter. In separating equilibrium, the headhunter extracts all surplus of an efficient labor market. The low type only enters if the chance of being hired justifies paying the fee, and hiring the low type reduces surplus. Wages yielding a high enough expected wage to yield low-type entry must pay some surplus to the high type, who has a greater probability of being offered $w_H$. Therefore, the headhunter would receive only a portion of the smaller surplus. Details are provided in Appendix G.

If application-fee revenue is used by the headhunter for some social purpose with social marginal valuation approximately dollar-for-dollar (or better), then the applicant's private cost of signaling is nearly a transfer, at most a negligible social cost. [16]

*Appendix B.2. Risk-Averse Applicant*

First consider, for later comparison, catatonic risk-aversion: the applicant would always value a lottery at the lowest possible payoff. For this case, $RP(f,\epsilon) = r - (w - \epsilon) + f = \epsilon$, [17] the upper bound of the $RP$ function. $RP = \epsilon$ guarantees a payoff of at least $r$, even if the test result suggests a low type. Therefore, this case yields the low and high types evaluating the wage/fee schedule identically, preventing a separating equilibrium.

In less extreme cases, the firm can choose $f$ to minimize $RP$. Suppose:

$$\inf_{f \in (0,\infty)} \lim_{\epsilon \to 0} RP(f,\epsilon) = 0; \tag{B1}$$

this is equivalent to a continuous utility function at $r$. A separating equilibrium can be implemented by choosing the right $f$ and setting $\epsilon$ as close to zero as possible to pay almost no risk premium. This is different from Guasch and Weiss (1981) [7], because in their model, an applicant with low test result is not considered for hire (equivalently, $w_L$ is forced to be less than the reservation wage $r$), thus selecting $w_L$ and $w_H$ arbitrarily close is not an option. In addition, their model also assumes that the reservation

[16] For a model using a similar setup to discuss the status-seeking motive of charitable donations, see Glazer and Konrad (1996) [12].

[17] The applicant is only willing to apply if the difference between the low wage and the fee is at least $r$, therefore $(w - \epsilon) - (f - RP(f,\epsilon)) = r$, rearrange to get the first equation. Replace $f$ with $w - r$ to get the second equation.

wage varies with type, which prevents $w_L$ and $w_H$ from being arbitrarily close even if the restriction on $w_L$ was relaxed. [18]

How must one model risk aversion so that the firm might prefer a pooling equilibrium? Assume a strictly positive risk premium for any distance between $w_H$ and $w_L$ (less extreme than catatonic risk aversion). Let:

$$0 < z = \inf_{f \in (0,\infty)} \lim_{\epsilon \to 0} RP(f, \epsilon), \tag{B2}$$

i.e., for any $f$, applicant utility function is discontinuous at net income $r$. [19]

Instead of production of 1 and 2, consider production level $l$ for low type and $h$ for high type. Compare firm's surplus in the separating equilibrium and in the pooling equilibrium in which the firm hires both types without testing.

The firm's surplus in separating equilibrium is:

$$(1-p)(h-c-r-z), \tag{B3}$$

and in pooling equilibrium is

$$p(l-r)+(1-p)(h-r). \tag{B4}$$

Subtracting (B4) from (B3):

$$(1-p)(-c-z)-p(l-r), \tag{B5}$$

or,

$$-c-z+p(c+z-l+r). \tag{B6}$$

The firm only seeks a separating equilibrium if expression (B6) is positive. The base model assumes $1 < r$ to give a welfare motivation to not hire the low types. With a similar assumption that $l < r$, separating equilibrium becomes more likely as $p$ goes up (the low type becomes more likely, so hiring without testing becomes more harmful), as $c$ or $z$ goes down (cost of separating equilibrium becomes lower), as $l$ goes down (the cost of hiring the low type becomes more harmful), as $r$ goes up (hiring becomes more costly). Note that $h$ is not in expression (B5), since in both separating and pooling equilibria, high types are hired.

This analysis applies anytime a strictly positive risk premium is required. For example, if Equation (B1) holds, but for any reason $\epsilon$ cannot approach zero, that is, $w_L$ and $w_H$ cannot be arbitrarily close, a risk premium bounded above zero may be needed. A possible reason for $\epsilon$ not to approach zero is at-home productivity $r$ differing with type. See Appendix D.

*Appendix B.3. Multiple Firms*

A separating equilibrium can be achieved unless there is an arrangement that can provide an expected wage minus fee for high types higher than $2-c$, while keeping the firm's return non-negative. Since the firms and the low type are already getting their reservation level, allowing the high type to get even more requires higher social surplus than separating equilibrium can attain. Since the only deviation from the full-information optimum in the separating equilibrium comes from testing the high types, if randomizing tests are disallowed, the only possible way to achieve higher surplus is by hiring everyone without testing, which can be checked by:

$$2-c \geq 2(1-p)+p = 2-p. \tag{B7}$$

[18] The Guasch and Weiss assumption, though the reverse of Spence's common default productivity assumption, could be apt for a situation in which, without applying, a high type would be able to obtain a significantly greater wage in other industries than would a low type.

[19] Since all concave functions on real open intervals are continuous, there does not exist a real-valued utility function yielding (B2).

Separating equilibrium is not possible if $p$ is so low that the firm can hire without testing while offering a sizable wage (close to two), or if $c$ is so high that testing is too costly to be justified.

As in Sections 1 and 2, even when $c$ is zero (costless test administration), separating equilibrium may still be achieved simply by having a positive application fee.

*Appendix B.4. Multiple Applicants*

Substituting (11) into the firm's expected profit (9) yields:

$$(E[m_n \mid n, m \geq 1] * E[\frac{1}{m_{n-1}+1} \mid n] - 1) * (qw_u + (1-q)w_d) - rE[m_n \mid n, m \geq 1]. \tag{B8}$$

The firm does not separately care about $w_u$ and $w_d$, so long as $w_d \geq r$ so that the wage offer is accepted, and $w_u > w_d$ to give high types greater incentive to apply than low types. Only their weighted sum, with test-reliability weights, enters (B8).

For all positive integers, the first term in parentheses in (B8), $E[m_n \mid n, m \geq 1] * E[\frac{1}{m_{n-1}+1} \mid n] - 1$, equals 0 for any $p$. [20]

## Appendix C. Continuous Types

Is separating equilibrium robust to the applicant having continuous types?

Let the applicant's type be any real number in $[1, 2]$, and type $t$ generates output worth $t$ if hired by the firm. The test still only produces two possible results: a high test result and a low test result. Let the test accuracy be $q$, $1 > q > 0.5$ as before, with $(2q-1)t + 2 - 3q$ the probability type $t$ attains a high test result. Thus, the probability of a high test result increases linearly with $t$, from $1-q$ for $t = 1$ to $q$ for $t = 2$. Initially, assume $r$, the applicant's home production level, is the same for all types, and that the firm can only hire an applicant after she/he is tested. For the first part of this section, also assume there is a smallest monetary unit $0 < \delta < (2q-1)^{-1}$.

Hiring necessarily costs at least $r$ in salary plus $c$ for the test, so the firm has no interest in types below $r + c$, but wishes to hire types above $r + c$ if cheap enough. Observing only a high or a low test result, but not observing $t$, limits what is attainable.

For any wage/fee schedule $s$, a type $t$ applicant's expected net wage is:

$$W(t \mid s) = [(2q-1)t + 2 - 3q]w_H + [1 - ((2q-1)t - 2 + 3q]w_L - f, \tag{C1}$$

which is linear in $t$ with slope $(2q-1)(w_H - w_L)$.

Consider a separating equilibrium where only types no less than a certain threshold apply. As (by assumption) both productivity and the chance of testing high increase linearly with type, so will the expected wage. The firm prefers to hire a higher type if and only if the slope of productivity, which is one, exceeds the expected wage slope:

$$1 > (2q-1)(w_H - w_L). \tag{C2}$$

Consider the case in which the firm chooses $s$ that satisfies (2) and:

$$W(r + c \mid s) = f, \tag{C3}$$

$$w_H - w_L = \delta. \tag{C4}$$

Equation (2) ensures that the gain from applying increases with type, and the offer for testing low is accepted. Equation (C3) simply mirrors the firm behavior specification of (3): it has type $r + c$ apply

[20] An intuitive argument is provided in Appendix H.

though indifferent. Equation (C4) (which implies (C2)) minimizes the surplus paid to types above $r + c$, allowing the firm the maximum attainable surplus. Note that the distribution of types does not enter the equations characterizing separating equilibrium.

To consider pooling equilibrium, enable the firm to hire an applicant not taking the test. A pooling equilibrium is implemented if the firm hires everyone without testing at wage $r$. Equilibrium profit depends on the distribution of types.

A direct comparison of a separating equilibrium (where the hiring of an applicant not taking the test is disallowed) and a pooling equilibrium (where such hiring is allowed) would be unconvincing. Therefore, consider a separating equilibrium assuming any type applicant can decline to pay the fee in order to take the test, and may still be hired. Interestingly, an equilibrium schedule $s$ leads to an interval of types applying and choosing to take the test, and the firm does not make an offer to non-test-taking types. To see this, suppose the firm hires the applicant even if test-taking is declined. In equilibrium, there must exist a type $t^* < 2$ such that: (i) types $t \geq t^*$ apply, take the test, and are hired at wage $w_H$ if testing high, $w_L$ if testing low; and (ii) types $t < t^*$ apply, decline to be tested, and are hired at wage $r$. If $t^* \leq r$, the firm is better off not hiring non-test-takers, so an equilibrium requires $t^* > r$. However, in this case, all types are hired, all are paid at least $r$, and the firm incurs test-administration cost $(2 - t^*)c > 0$, so it is strictly better off pooling than separating and hiring non-test-takers.

Therefore, the separating equilibrium specified by Equations (2), (C3) and (C4) still stands. The firm compares the loss of hiring types below $r + c$ at wage $r$ (could be negative depending on distribution of types) with the cost of testing types above $r + c$, if the former is greater the firm implements a separating equilibrium, otherwise it implements the pooling equilibrium. [21]

Now discard the assumption about the smallest monetary unit $\delta$, and the possibility of hiring a non-test-taking applicant. Suppose types are uniformly distributed between one and two, and let $r(t)$ be a continuous increasing function distributed on $[1,2]$, with $r(1) > 1$ and $r(2) < 2$, so that the firm still might profitably hire type 2, but not type 1. Furthermore, assume the firm is again not allowed to hire an applicant not taking the test.

In Figures C1–C3 below, the thin line, $y(t) = t$, represents the gross productivity for each type. The dashed line shows net productivity, productivity reduced by test-administration cost $c$, a downward shift (0.5 in the graph). The thick line represents $r(t)$. By changing $s = (w_L, w_H, f)$, the firm can generate as $W(t \mid s)$, net expected wage, any line with nonnegative slope, obtaining the employ of any types for which $W$ exceeds $r$, profiting by the height (net productivity - $W$).

These figures offer some interesting cases. Figure C1 illustrates the action the firm takes if $r(t)$ is a linear function of $t$, with $r(1) > 1 - c$ and $r(2) < 2 - c$. In this case, matching $W(t \mid s) = r(t)$ is the equilibrium, with the firm obtaining all the surplus (the shaded area). Figure C2 illustrates how a possible optimum of the firm might not involve hiring the highest types. This time $r(t)$ is flat until $t = 1.8$, and then steep. A possible $W$ [22] is the dotted line in the figure, which yields the shaded area as surplus for the firm, while the triangle below the shaded region is applicant surplus. An interval of intermediate types is hired, while lower types are insufficiently productive, higher types overly expensive. Figure C3 shows a case in which $c$ is so high that testing guarantees a loss, thus neither testing nor hiring occurs. Were hiring without testing possible, the firm can separate potential applicants without using the test or wage differentials. A flat wage can separate due to differences in reservation wage across types. In the case shown however, the firm does not wish to do so and prefers not to hire at all instead.

[21] A zero-cost test would make no qualitative difference.

[22] Figure C2 is for illustration purpose only, and the $W$ presented may not be optimal. For a case in which the firm optimally chooses not to hire if the potential applicant is of the highest type, consider Figure C1, but let $r(t)$ be discontinuous at $t = 2$ and that $r(2) = 1.9$. The firm still chooses to match $W$ with the rest of $r$ but a Type 2 potential applicant no longer applies.

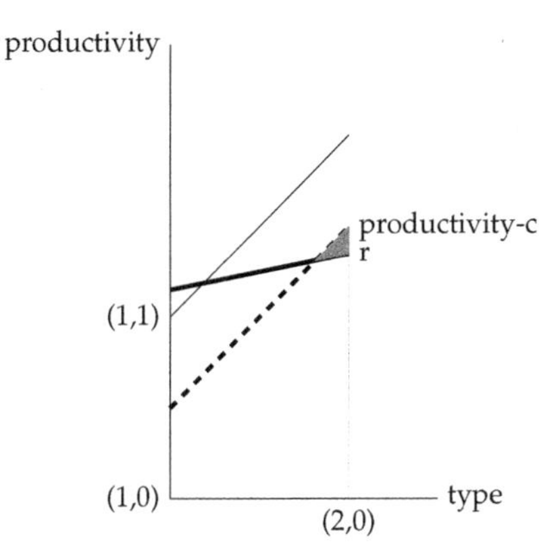


**Figure C1.** Under linear $r(t)$, the firm maximizes profit by matching $W(t \mid s) = r(t)$.

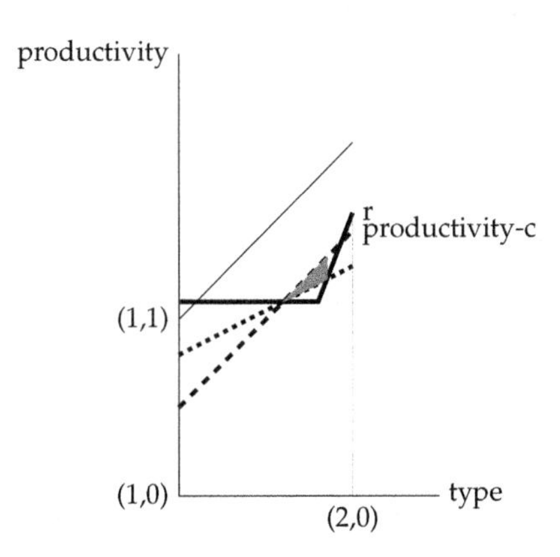


**Figure C2.** If types are continuous and the reservation wage is variable, the firm may only hire intermediate types.

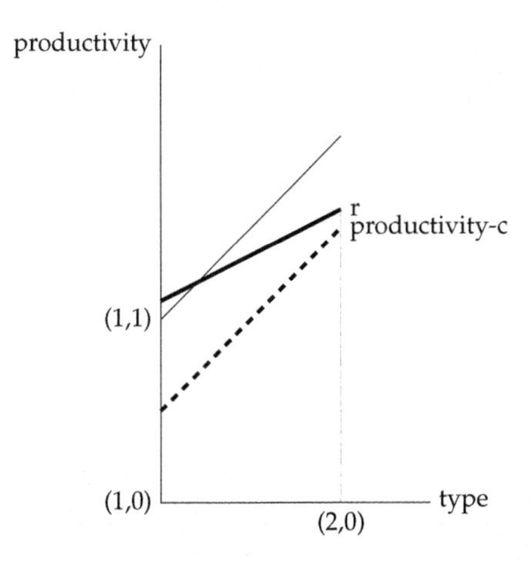


**Figure C3.** A case in which there is no hiring, nor production.

## Appendix D. On Variable Reservation Wages

In the base model, instead of a universal $r$ for both types, assume that the reservation wages of a high type potential applicant and that of a low type potential applicant are $r_H$ and $r_L$, respectively. Naturally assume that $2 > r_H > r_L > 1$. Similar to the assumption given in Equation (1), assume:

$$c < 2 - r_H. \tag{D1}$$

Then, Equation (3) is no longer sufficient to discourage a low type potential applicant from applying. In order to achieve separation, the wage/fee schedule must not only satisfy Equations (2) and (3), but also:

$$(1 - q) * w_H + q * w_L - r_L < f. \tag{D2}$$

Subtracting (D2) from (3) yields:

$$(2q - 1)(w_H - w_L) > r_H - r_L. \tag{D3}$$

**Corollary A1.** *A wage/fee schedule satisfying Equations* (2)*,* (3) *and* (D3) *will implement a separating equilibrium.*

For details and a proof, see Appendix F.

## Appendix E. Modifications to the Institution

In the base model's separating equilibrium, the firm seeks to exclude the low type, has no option but to test an applicant, and the cost $c$ of conducting a test of limited reliability is unavoidable. That a separating equilibrium results may be no surprise. This section considers three institutions via which the firm might hire without necessarily testing.

(i) Suppose, instead of only being considered if the applicant pays the fee, she/he is considered automatically, but to be tested requires paying the fee. In Step 1, the firm, in addition to $(w_L, w_H, f)$, now chooses $w_N$, the wage for an applicant not paying the fee.

For this first modification, there can be pooling equilibria, if (a) the cost of testing is high; (b) the social loss over employing a low type is low; or (c) the probability of a low type is quite small. Specifically, if $\frac{c(1-p)}{p(r-1)} > 1$, the adverse selection problem does not justify spending resources identifying the high type. [23] It offers a fee so high or $w_H$ and $w_L$ so low that the applicant does not take the test and then hires both types at wage $w_N = r$ without testing. Therefore, if the above inequality is met, there is a set of pooling equilibria but no separating equilibria.

(ii) For the above modification or the base model, instead of testing everyone who paid the fee, suppose the firm gives fee payers a random chance $m$ of actually being tested. In addition to $(w_L, w_H, f)$, the firm also selects $w_M$, the wage offered if the applicant paid to take the test but was not randomly selected to take it, and $w_N$, the wage offered if the applicant applied but did not pay for a chance to take the test. The firm can set $m$ as close to zero as possible and can still implement a separating equilibrium. It accomplishes this by setting wage/fee schedule with $w_H > w_L \geq r > w_N, w_M \geq r$ and so that the high type pays the fee though indifferent, and the low type does not pay. This produces a result approaching the full-information labor allocation, while the firm extracts all the surplus. Therefore there cannot be a pooling equilibrium.

---

[23] In a pooling equilibrium where the firm hires without testing, its profit is $p + 2(1-p) - r = 2 - p - r$, in a separating equilibrium, its profit is $(1-p)(2-c-r)$. Comparing the two: $2 - p - r > (1-p)(2-c-r) \iff c(1-p) - p(r-1) > 0 \iff \frac{c(1-p)}{p(r-1)} > 1$.

(iii) As just above, the firm always considers the applicant, and an applicant can decide to pay the fee and request to be tested, with the firm randomly administering the test with chosen probability $m$. Now, however, suppose the application fee is refunded unless the test is actually administered. The firm can again approach the full-information optimum, as in (ii) by setting $w_H > w_L \geq r > w_N$, $w_M \geq r$ and so that high type pays the fee though indifferent, and let $m$ approach 0. [24]

*Appendix E.1. Details of Separating Equilibria*

Here, $w_M$ is used to denote the wage for someone who signs up for the test but not receiving a test, $w_N$ is for someone not signing up for the test.

For (ii), to achieve separation, the firm makes the high type indifferent over paying the application fee:

$$\begin{aligned} &-f + mq(w_H) + m(1-q)(w_L) + (1-m)w_M = r \\ &> -f + m(1-q)(w_H) + mq(w_L) + (1-m)w_M. \end{aligned} \tag{E1}$$

The inequality ensures that the low type does not pay the fee, and is achieved given (2). Conditioning on separation, the firm wants to hire the fee-payer for sure, so $w_L$, $w_H$ and $w_M$ are all no less than $r$. Even if the firm is allowed to hire a non-fee payer, it does not wish to do so, which yields $w_N < r$. The fee achieves separation. Then firm's profit is:

$$(1-p)(2 + f - mq(w_H) - m(1-q)(w_L) - (1-m)w_M) = (1-p)(2 - mc - r), \tag{E2}$$

with equality due to the expected wage being $r + f$ [from (E1)]. As $m$ goes to zero this approaches the full-information optimum, so a pooling equilibrium can never be more profitable (even allowing $w_N$ as in (i)).

For (iii), the separating condition becomes:

$$\begin{aligned} &-fm + mq(w_H) + m(1-q)(w_L) + (1-m)(w_M) = r \\ &> -fm + m(1-q)(w_H) + mq(w_L) + (1-m)w_M, \end{aligned} \tag{E3}$$

and the firm's profit becomes:

$$(1-p)[2 + fm - mc - (r + fm)] = (1-p)(2 - mc - r). \tag{E4}$$

This, again, approaches the full-information optimum.

## Appendix F. Details for Variable Reservation Wage

**Base model:** A firm can always find such a wage/fee schedule by first choosing any pair of $w_H$ and $w_L$ satisfying Equation (2) and with a wide enough difference to satisfy equation (D3). Then calculate the fee using Equation (3). Therefore, a separating equilibrium is achievable by the firm.

In order to check the optimality of such a separating equilibrium for the firm, compare the variable reservation wage model with the one if the reservation wage for both types is $r_H$. The separating equilibria in both cases yield the same return for the firm: the potential applicant is only tested and hired if she/he is a high type, and the firm claims all the surplus. Appendix A shows that in the later case, a separating equilibrium is optimal for the firm. Therefore a separating equilibrium is also optimal in the former case if there is no way for the firm to take advantage of a reduced cost to hire a low type. Indeed, in a separating equilibrium, since the firm hires only if the potential applicant is a

[24] A similar argument to the ones in (ii) and (iii) is made by Stiglitz (1975) [3] and mentioned in Guasch and Weiss (1981) [7].

high type and receives all the surplus, the only possibility to improve comes from hiring the potential applicant if she/he is a low type as well and receive surplus from (a) the production of the low type or (b) savings on the cost of conducting the test. Since the low type reservation wage is still greater than her/his productivity, the surplus from (a) is negative. As for (b), the assumptions in the base model require any hired worker to take the test, preventing the firm from saving on test cost. Therefore, for this model, it is optimal for the firm to implement a separating equilibrium.

**(i):** This adds the possibility of a pooling equilibrium in which the firm hires regardless of the type without testing. However, compared to the case of a single reservation wage $r_H$, the firm does not change its behavior under variable reservation wage. Earlier in this section, separating equilibria yield the firm the same payoff in both cases. In order to implement a pooling equilibrium, the firm needs to pay a wage of $r_H$ regardless of the type, same in both cases. Therefore the variable types model has the same solution (either separating equilibria or pooling) as the one in which both types having a reservation wage of $r_H$.

**Multiple firms:** If the variable reservation wage assumption discussed earlier is added to the model with multiple firms, Equation (6) needs to be adjusted accordingly:

$$w_H(1-q) + w_L q - f < r_L. \tag{F1}$$

Subtracting this from Equation (5) yields:

$$(2q-1)(w_H - w_L) > 2 - c - r_L. \tag{F2}$$

This is comparable to Equation (7). The rest of the analysis remains the same as in Section 4. Therefore, the solution to the multiple firms competing for an applicant model with variable reservation wage is the same as if the reservation wage for both types is $r_L$.

**Multiple applicants:** Apply the same variable reservation wage assumption to Section 5. Equation (10) changes to

$$E[\frac{w_d}{m_{n-1}+1} + \frac{q(w_u - w_d)}{m_{n-1}+1} - f \mid n] = r_H. \tag{F3}$$

Additionally, Equation (11) changes to:

$$f = E[\frac{1}{m_{n-1}+1} \mid n] * (qw_u + (1-q)w_d) - r_H. \tag{F4}$$

In addition, to discourage low types from applying, the following must be satisfied:

$$f > E[\frac{1}{m_{n-1}+1} \mid n] * ((1-q)w_u + qw_d) - r_L. \tag{F5}$$

Subtracting this from Equation (F4) yields:

$$E[\frac{1}{m_{n-1}+1} \mid n] * (qw_u + (1-q)w_d) - r_H > E[\frac{1}{m_{n-1}+1} \mid n] * ((1-q)w_u + qw_d) - r_L, \tag{F6}$$

which can be rearranged into:

$$(2q-1)E[\frac{1}{m_{n-1}+1} \mid n] * (w_u - w_d) > r_H - r_L. \tag{F7}$$

Equation (F7) specifies how much of a wage difference is required in order to implement a separating equilibrium, similar to Equation (D3) in the base model. The firm implements a separating equilibrium by choosing a wage/fee schedule satisfying $w_u > w_d \geq r_H$, (F4) and (F7).

In the pooling equilibrium, the first of the applicants is hired at wage $r_H$. The firm chooses a separating equilibrium or the pooling equilibrium based on its expected payoff.

## Appendix G. Other Equilibria With a Third Party

First, there is no equilibrium in which only the high type enters, but is hired only with a high test result. Were that the situation, the headhunter would get no revenue with a fee higher than $q(2-r-c)$. Competition forces the firms to zero profit, but a firm deviating to hire a low-test-result applicant at wage $r$ attains a positive profit.

Suppose a situation in which both types enter with positive probability. Initially suppose Firms 1 and 2 set wages $(w_{L1}, w_{H1})$ and $(w_{L2}, w_{H2})$, with (by labeling choice) $w_{H1} > w_{H2}$. For Firm 2 to attract the high type requires $w_{L2} > \frac{q(w_{H1}-w_{H2})}{(1-q)} + w_{L1}$. For firm 2 to avoid hiring the low type requires $w_{L2} < \frac{(1-q)(w_{H1}-w_{H2})}{q} + w_{L1}$.[25] Recalling that $q > \frac{1}{2}$ is the probability the test correctly reveals type, there is no value of $w_{L2}$ at which Firm 2 is best responding to Firm 1.

If both types enter, and all firms except firm 1 offer the same wage schedule $(w_L, w_H)$ with $w_L > r$, then firm 1 gains by deviating to $w_{L1}$ in $(r, w_L)$ and $w_{H1} = w_H + (1-q)\frac{(w_L - w_{L1})}{q}$ (by tiebreaker (ii), Firm 1 attracts the high but not the low type).

This leaves two types of candidate equilibria as follows. First, where both types enter and low scores are hired: (a) each firm offers the same $(w_L, w_H)$ with $w_H > w_L = r$ (so that both types accept an offer); (b) $f = (1-q)w_H + qr - r = (1-q)(w_H - r)$ (so that the low type applies); and (c) $2(1-p) + p - [q(1-p) - p(1-q)]w_H - [pq + (1-p)(1-q)]r - c = 0$ (which sets $w_H$ to compete away firm profits). In the second type, both types enter and low scores are not hired: (a') each firm offers the same $(w_L, w_H)$ with $w_H >= r > w_L$; (b') $f = (1-q)w_H - r$; (c') $2q(1-p) + p(1-q) - [p(1-q) + q(1-p)]w_H - c = 0$ (for the same reasons).

Solving (c) for $w_H$: $w_H = \frac{2-p-c-[pq+(1-p)(1-q)]r}{p(1-q)+(1-p)q}$. Substituting into (b):

$$
\begin{aligned}
f &= (1-q)\left(\frac{2-c-p-[pq+(1-p)(1-q)]r}{q(1-p)+p(1-q)} - r\right) \\
\implies [q(1-p)+p(1-q)]f &= (1-q)(\{2-c-p-[pq+(1-p)(1-q)]r\} - r[q(1-p)+p(1-q)]) \\
&= (1-q)\{(2-c-p) - r[pq+(1-p)(1-q)+q(1-p)+p(1-q)]\} \\
&= (1-q)[2-c-p-r(pq+1+pq-p-q+q-pq+p-pq)] \\
&= (1-q)(2-c-p-r)
\end{aligned}
\tag{G1}
$$

Therefore, the maximum application fee if low types apply and low scores are hired: $f = \frac{(1-q)(2-c-p-r)}{q(1-p)+p(1-q)}$.

As mentioned in text, surplus is reduced. To see this, notice that the $(2-c-p-r)$ term is multiplied by a coefficient less than one (subtract the $(1-q)$ term in the numerator from the denominator: $q(1-p)+p(1-q)-(1-q) = q(1-p)-(1-p)(1-q) = (2q-1)(1-p) > 0$, so the ratio $< 1$). Then, conditioning on $2-c-p-r > 0$ yields:

$$
\frac{(1-q)(2-c-p-r)}{q(1-p)+p(1-q)} < 2-c-p-r < 2-c-r-(2-c-r)p = (1-p)(2-c-r), \tag{G2}
$$

which is the headhunter's surplus in a separating equilibrium.

Solving (c') for $w_H$: $w_H = \frac{2q(1-p)+p(1-q)-c}{p(1-q)+q(1-p)}$ Substituting into (b'):

$$
f = (1-q)\frac{2q(1-p)+p(1-q)-c}{p(1-q)+q(1-p)} - r, \tag{G3}
$$

[25] Tiebreaker (ii) above generates the strict inequalities.

which is the maximum application fee if low types apply and low scores are not hired. Then, the headhunter's surplus in this case:

$$\begin{aligned}(1-q)\frac{2q(1-p)+p(1-q)-c}{p(1-q)+q(1-p)}-r &< 2q(1-p)+p(1-q)-c-r \\ &< 2-p-c-r \\ &< (1-p)(2-c-r)\end{aligned} \tag{G4}$$

The first inequality is based on the already established $q(1-p)+p(1-q) > 1-q$; the second inequality is based on $2-p > 2q(1-p)+p(1-q)$. [26] Therefore, the equilibria described in the text are the only equilibria for this model.

## Appendix H. Intuitive Argument for the Coefficient in Equation (B8) to Be Zero

Since $E[m_n \mid n] = n(1-p)$, and the probability that $m_n = 0$ is $p^n$, $E[m_n \mid n, m \geq 1]$ can be calculated:

$$E[m_n \mid n, m \geq 1] = \frac{n(1-p)}{1-p^n} = \frac{n}{1+p+...+p^{n-1}} \tag{H1}$$

To show that $E[m_n \mid n, m \geq 1] * E[\frac{1}{m_{n-1}+1} \mid n] - 1 = 0$, having already known that $E[m_n \mid n, m \geq 1] = \frac{n}{1+p+...+p^{n-1}}$, it only remains to show that $E[\frac{1}{m_{n-1}+1} \mid n] = \frac{1+p+...+p^{n-1}}{n}$.

Suppose there are $n$ balls lining up from left to right. One of the balls is randomly chosen and replaced by a purple ball. Then the rest are randomly painted, each with probability $p$ being white and $1-p$ being red. The white balls are then taken away, leaving only red and purple balls in the line. One of these balls is then randomly chosen. $E[\frac{1}{m_{n-1}+1} \mid n]$ is the probability that the chosen ball is the purple one.

If instead of randomly choosing a ball in the final step, we always choose the first one on the left. The probability of the chosen ball being purple is the same as above (conditioning on there being m+1 red and purple balls, both processes yield $\frac{1}{m+1}$ probability the chosen ball being purple, for all $m \in 0, 1, ..n-1$). However, the probability in the new process is equivalent to the probability that all the balls to the left of the purple ball being white. With probability $\frac{1}{n}$, the purple ball is the leftmost ball in the original line of $n$ balls, then with probability one there will be no white balls to its left. With probability $\frac{1}{n}$, there will be one ball to its left, then the probability of that ball being white is $p$... Therefore, the probability of the chosen ball being purple is $\frac{1+p+...p^{n-1}}{n} = E[\frac{1}{m_{n-1}+1} \mid n]$.

## References


1. Spence, M. Job market signaling. *Q. J. Econ.* **1973**, *87*, 355–374.
2. Akerlof, G.A. The market for 'lemons': Quality uncertainty and the market mechanism. *Q. J. Econ.* **1970**, *84*, 488–500.
3. Stiglitz, J.E. The theory of 'screening,' education, and the distribution of income. *Am. Econ. Rev.* **1975**, *65*, 283–300.
4. Crawford, V.P.; Sobel, J. Strategic Information Transmission. *Econometrica* **1982**, *50*, 1431–1451.
5. Chakraborty, A.; Harbaugh, R. Persuasion by Cheap Talk. *Am. Econ. Rev.* **2010**, *100*, 2361–2382.
6. Wang, R. Competition, wage commitments, and application fees. *J. Labor Econ.* **1997**, *15*, 124–142.
7. Guasch, J.L.; Weiss, A. Self-selection in the labor market. *Am. Econ. Rev.* **1981**, *71*, 275–284.
8. Van Damme, E.; Selten, R.; Winter, E. Alternating Bargaining with Smallest Money Unit. *Games Econ. Behav.* **1990**, *2*, 188–201.


[26] $[2-p]-[2q(1-p)-p(1-q)] = 2(1-p)+p-2q(1-p)-p(1-q) = 2(1-p)(1-q)+pq > 0$

9. Spence, M. Signaling in Retrospect and the Information Structure of Markets. *Am. Econ. Rev.* **2002**, *92*, 434–459.
10. Mas-Colell, A.; Whinston, M.D.; Green, J.R. *Microeconomic Theory*; Oxford University Press: New York, NY, USA, 1995.
11. Selten, R. Reexamination of the perfectness concept for equilibrium points in extensive games. *Int. J. Game Theory* **1975**, *4*, 25–55.
12. Glazer, A.; Konrad, K.A. A Signaling Explanation for Charity. *Am. Econ. Rev.* **1996**, *86*, 1019–1028.

# Section 2:
# Computer Science

 *games*

*Article*

# Leveraging Possibilistic Beliefs in Unrestricted Combinatorial Auctions

**Jing Chen [1,*] and Silvio Micali [2]**

[1] Department of Computer Science, Stony Brook University, Stony Brook, NY 11794, USA
[2] Computer Science and Artificial Intelligence Laboratory, MIT, Cambridge, MA 02139, USA; silvio@csail.mit.edu
* Correspondence: jingchen@cs.stonybrook.edu; Tel.: +1-631-632-1827

Academic Editor: Paul Weirich
Received: 9 August 2016; Accepted: 17 October 2016; Published: 26 October 2016

**Abstract:** In unrestricted combinatorial auctions, we put forward a mechanism that guarantees a meaningful revenue benchmark based on the possibilistic beliefs that the players have about each other's valuations. In essence, the mechanism guarantees, within a factor of two, the maximum revenue that the "best informed player" would be sure to obtain if he/she were to sell the goods to his/her opponents via take-it-or-leave-it offers. Our mechanism is probabilistic and of an extensive form. It relies on a new solution concept, for analyzing extensive-form games of incomplete information, which assumes only mutual belief of rationality. Moreover, our mechanism enjoys several novel properties with respect to privacy, computation and collusion.

**Keywords:** possibilistic beliefs; unrestricted combinatorial auctions; mutual belief of rationality; incomplete information; extensive-form games; distinguishable dominance

## 1. Introduction

In this paper, we study the problem of generating revenue in unrestricted combinatorial auctions, solely relying on the players' possibilistic beliefs about each others' valuations. Let us explain.

In a combinatorial auction, there are multiple indivisible goods for sale and multiple players who are interested in buying. A valuation of a player is a function specifying a non-negative value for each subset of the goods. Many constraints on the players' valuations have been considered in the literature for combinatorial auctions[1]. We instead focus on combinatorial auctions that are unrestricted. That is, in our auctions, a player's value for one subset of the goods may be totally unrelated to his/her value for another subset and to another player's value for any subset. This is the most general class of auctions. It is well known that, for such auctions, the famous Vickrey-Clarke-Groves (VCG) mechanism [4–6] maximizes social welfare in dominant strategies, but offers no guarantee about the amount of revenue it generates. In fact, for unrestricted combinatorial auctions, no known mechanism guarantees any significant revenue benchmark in settings of incomplete information[2].

In our setting, the seller has no information about the players' valuations, and each player knows his/her own valuation, but not necessarily the valuations of his/her opponents. Our players, however, have beliefs about the valuations of their opponents. Typically, beliefs are modeled as probability distributions: for instance, it is often assumed that the valuation profile, $\theta$, is drawn from

[1] Such as monotonicity, single-mindedness and additivity [1–3].
[2] In complete information settings (where the players have common knowledge about their valuations), assuming common knowledge of rationality, [7–9] have designed mechanisms that guarantee revenue arbitrarily close to the maximum social welfare.

a common prior. Our setting is instead non-Bayesian: the players' beliefs are possibilistic and can be arbitrary. That is, a player $i$'s belief consists of a set of valuation profiles, $\mathcal{B}_i$, to which he/she believes $\theta$ belongs. We impose no restriction on $\mathcal{B}_i$ except that, since $i$ knows his/her own valuation, for every profile $v \in \mathcal{B}_i$, we have $v_i = \theta_i$. In a sense, therefore, such possibilistic beliefs are not assumed to exist, but always exist. For instance, if a player $i$ has no information about his/her opponents, then $\mathcal{B}_i$ consists of the set of all valuation profiles $v$, such that $v_i = \theta_i$; if $i$ has complete information about his/her opponents, then $\mathcal{B}_i = \{\theta\}$; and if $\theta$ is indeed drawn from a common prior $D$, then $\mathcal{B}_i$ consists of the support of $D$ conditioned on $\theta_i$.

Possibilistic beliefs are much less structured than Bayesian ones. Therefore, it should be harder for an auction mechanism to generate revenue solely based on the players' possibilistic beliefs. Yet, in single-good auctions, the authors of [10] have constructed a mechanism that guarantees revenue at least as high as the second-highest valuation and, sometimes, much higher. In this paper, for unrestricted combinatorial auctions, we construct a mechanism that guarantees, within a factor of two, another interesting revenue benchmark, $BB$, solely based on the players' possibilistic beliefs.

The benchmark $BB$ is formally defined in Section 3, following the framework put forward by Harsanyi [11] and Aumann [12]. However, it can be intuitively described as follows. Let $BB_i$ (for "best belief") be the maximum social welfare player $i$ can guarantee, based on his/her beliefs, by assigning the goods to his/her opponents. Then, $BB = \max_i BB_i$, and the revenue guaranteed by our main mechanism is virtually $BB/2$. Notice that each $BB_i$ does not depend on $\theta_i$ at all, a property that, as we shall see, gives our mechanism some advantage in protecting the players' privacy.

To ease the discussion of our main mechanism, in Section 4, we construct a first mechanism, of normal form, that guarantees revenue virtually equal to $BB/2$ under two-step elimination of weakly-dominated strategies. The analysis of our first mechanism is very intuitive. However, elimination of weakly-dominated strategies is order-dependent and does not yet have a well-understood epistemic characterization. Moreover, our first mechanism suffers from two problems shared by most normal-form mechanisms. Namely, (1) it reveals all players' true valuations, and (2) it requires an amount of communication that is exponential in the number of goods. Both problems may not be an issue from a pure game-theoretic point of view[3], but are quite serious in several realistic applications, where privacy and communication are, together with collusion and computational complexity, legitimate concerns [14].

Our main mechanism, the best-belief mechanism, significantly decreases the magnitude of the above problems. This second mechanism is designed and analyzed in Section 6 and is of extensive form. In order to analyze it in settings where the players have possibilistic beliefs, we propose a new and compelling solution concept that only assumes mutual belief of rationality, where the notion of rationality is the one considered by Aumann [15].

The Resiliency of the Best-Belief Mechanism.

Besides guaranteeing revenue virtually equal to $BB/2$ under a strong solution concept, the best-belief mechanism enjoys several novel properties with respect to privacy, computation, communication and collusion.

1. Privacy: People value privacy. Thus, "by definition", a privacy-valuing player $i$ de facto receives some "negative utility" if, in an auction, he/she reveals his/her true valuation $\theta_i$ in its entirety, but does not win any goods. Even if he/she wins some goods, his/her traditional utility (namely, his/her value for the goods he/she receives minus the price he/she pays) should be discounted by the loss he/she suffers from having revealed $\theta_i$. One advantage of our best-belief mechanism

[3] Indeed, the revelation principle [13] explicitly asks the players to directly reveal all of their private information.

is that it elicits little information from a player, which presumably diminishes the likelihood that privacy may substantially distort a player's incentives.

2. Computation: Typically, unrestricted combinatorial auctions require the evaluation of complex functions, such as the maximum social welfare. In principle, one may resort to approximating such functions, but approximation may distort incentives[4]. By contrast, our mechanism delegates all difficult computation to the players and ensures that the use of approximation is properly aligned with their incentives.
3. Communication: By eliciting little information from the players, our mechanism also has low communication complexity; quadratic in the number of players and the number of goods.
4. Collusion: Collusion can totally disrupt many mechanisms. In particular, the efficiency of the VCG mechanism can be destroyed by just two collusive players [14]. By contrast, collusion can somewhat degrade the performance of our mechanism, but not totally disrupt it, unless all players are collusive. As long as collusive players are also rational, at least in a very mild sense, the revenue guaranteed by our mechanism is at least half of that obtainable by "the best informed independent player".

For a detailed discussion about these properties of our mechanism, see Section 6.2.

## 2. Related Work

Generating revenue is one of the most important objectives in auction design; see [16,17] for thorough introductions about this area. Following the seminal result of [13], there has been a huge literature on Bayesian auctions [18]. Since we do not assume the existence of a common prior and we focus on the players' possibilistic rather than probabilistic beliefs, our study is different from Bayesian auctions. Spectrum auctions have been widely studied both in theory and in practice, and several interesting auction forms have been proposed recently; see, e.g., [19–23]. Most existing works consider auctions of restricted forms, such as auctions with multiple identical goods and single-demand valuations [24], valuations with free disposal [25], auctions with additive valuations [26], auctions with unlimited supply [27], etc. Revenue in unrestricted combinatorial auctions has been considered by [28], which generalizes the second-price revenue benchmark to such auctions and provides a mechanism guaranteeing a logarithmic fraction of their benchmark in dominant strategies.

The solution concept developed in this paper refines the notion of implementation in undominated strategies [29] and considers a two-round elimination of dominated strategies. In particular, we extend the notion of distinguishably-dominated strategies [30] from extensive-form games of complete information to extensive-form games of incomplete information and possibilistic beliefs. As shown in [30], iterated elimination of distinguishably-dominated strategies is order independent with respect to histories and characterizes extensive-form rationalizability [31,32]. In [10,33], elimination of strictly-dominated strategies has been extended to deal with possibilistic beliefs, but only for normal-form games. Moreover, [34] leverages the players' beliefs for increasing the sum of social welfare and revenue in unrestricted combinatorial auctions.

Preserving the privacy of the players' valuations, or types in general, in the execution of a mechanism has been studied by [35]. The authors present a general method using some elementary physical equipment (i.e., envelopes and an envelope randomizer) so as to execute any given normal-form mechanism, without trusting any one and without revealing any information about the players' true types, beyond what is unavoidably revealed in the final outcome. An alternative way to protect the privacy of the players that has often been considered for auctions is to use encryption and zero-knowledge proofs. In particular, the authors of [36] make efficient use of cryptography to

[4] For instance, the outcome function of the VCG mechanism is NP-hard to compute even when each player only values a single subset of the goods for \$1 and all other subsets for \$0. Moreover, if one replaces this outcome function with an approximation, then VCG would no longer be dominant-strategy truthful.

implement single-good auctions so that, after learning all bids, an untrusted auctioneer can prove who won the good and what price he/she should pay, without having any player learn any information about the bid of another player. Moreover, in differential privacy [37], the mechanisms or databases inject noise to the final outcome to preserve the participants' privacy. By contrast, our main mechanism does not rely on envelopes or any other form of physical equipment, nor on cryptography or noise. It preserves the players' privacy because, despite the fact that all actions are public, a player is asked to reveal little information about herself/himself.

Strong notions of collusion-resilient implementation have been studied in the literature, such as coalition incentive compatibility [38] and bribe proofness [39]. However, the authors prove that many social choice functions cannot be implemented under these solution concepts. The collusive dominant-strategy truthful implementation is defined in [40], together with a mechanism maximizing social welfare in multi-unit auctions under this notion. Other forms of collusion resiliency have also been investigated, in particular by [41–46]. Their mechanisms, however, are not applicable to unrestricted combinatorial auctions in non-Bayesian settings. Moreover, the collusion models there assume various restrictions (e.g., collusive players cannot make side-payments to one another or enter binding agreements, there is a single coalition, no coalition can have more than a given number of players, etc.). By contrast, in unrestricted combinatorial auctions, our main mechanism does not assume any such restrictions. The resiliency of our mechanism is similar to that of [28], where the guaranteed revenue benchmark is defined only on independent players' valuations when collusion exists.

## 3. Preliminaries and the Best-Belief Revenue Benchmark

A combinatorial auction context is specified by a triple $(n, m, \theta)$: the set of players is $\{1, \ldots, n\}$; the set of goods is $\{1, \ldots, m\}$; and the true valuation profile is $\theta$. Adopting a discrete perspective, we assume that a player's value for a set of goods is always an integer. Thus, each $\theta_i$, the true valuation of $i$, is a function from the powerset $2^{\{1,\ldots,m\}}$ to the set of non-negative integers $\mathbb{Z}^+$, with $\theta_i(\emptyset) = 0$. The set of possible valuations of $i$, $\Theta_i$, consists of all such functions, and $\Theta = \Theta_1 \times \cdots \times \Theta_n$. After constructing and analyzing our mechanisms, we will discuss the scenarios where values are real numbers.

An outcome of a combinatorial auction is a pair of profiles $(A, P)$. Here, $A$ is the allocation, with $A_i \subseteq \{1, \ldots, m\}$ being the set of goods each player $i$ gets, and $A_i \cap A_j = \emptyset$ for each player $j \neq i$; and $P$ is the price profile, with $P_i \in \mathbb{R}$ denoting how much each player $i$ pays; if $P_i < 0$, then $i$ receives $-P_i$ from the seller. The set of all possible outcomes is denoted by $\Omega$.

The utility function of $i$, $u_i$, maps each valuation $t_i \in \Theta_i$ and each outcome $\omega = (A, P)$ to a real: $u_i(t_i, \omega) = t_i(A_i) - P_i$. The social welfare of $\omega$ is $SW(\omega) \triangleq \sum_i \theta_i(A_i)$, and the revenue of $\omega$ is $REV(\omega) \triangleq \sum_i P_i$. If $\omega$ is a probability distribution over outcomes, then $u_i(t_i, \omega)$, $SW(\omega)$ and $REV(\omega)$ denote the corresponding expectations.

**Definition 1.** *An augmented combinatorial auction context is a four-tuple $(n, m, \theta, \mathcal{B})$, where $(n, m, \theta)$ is a combinatorial auction context and $\mathcal{B}$ is the belief profile: for each player $i$, $\mathcal{B}_i$, the belief of $i$, is a set of valuation profiles, such that $t_i = \theta_i$ for all $t \in \mathcal{B}_i$.*

In an augmented combinatorial auction context, $\mathcal{B}_i$ is the set of candidate valuation profiles in $i$'s mind for $\theta$. The restriction that $t_i = \theta_i$ for all $t \in \mathcal{B}_i$ corresponds to the fact that player $i$ knows his/her own valuation. Player $i$'s belief is correct if $\theta \in \mathcal{B}_i$ and incorrect otherwise. As we shall see, our result holds whether or not the players' beliefs are correct. From now on, since we do not consider any other type of auctions, we use the terms "augmented" and "combinatorial" for emphasis only.

A revenue benchmark $f$ is a function that maps each auction context $C$ to a real number $f(C)$, denoting the amount of revenue that is desired under this context.

**Definition 2.** *The best-belief revenue benchmark, BB, is defined as follows. For each auction context* $C = (n, m, \theta, \mathcal{B})$,

$$BB(C) \triangleq \max_i BB_i,$$

*where for each player i,*

$$BB_i \triangleq \max_{(A,P)\in\Omega:\ A_i=\emptyset \text{ and } P_j \leq t_j(A_j)\ \forall j\neq i,\ t\in\mathcal{B}_i} REV(A,P).$$

Note that $BB_i$ represents the maximum revenue that player $i$ would be sure to obtain if he/she were to sell the goods to his/her opponents via take-it-or-leave-it offers, which is also the maximum social welfare player $i$ can guarantee, based on his/her beliefs, by assigning the goods to his/her opponents. As an example, consider a combinatorial auction with two items and three players. Player 1 only wants Item 1, and $\theta_1(\{1\}) = 100$; Player 2 only wants Item 2, and $\theta_2(\{2\}) = 100$; and Player 3 only wants the two items together, and $\theta_3(\{1,2\}) = 50$. All the unspecified values are zero. Moreover, Player 1 believes that Player 2's value for Item 2 is at least 25, and Player 3's value for the two items together is at least 10: that is, $\mathcal{B}_1 = \{v \mid v_1 = \theta_1, v_2(\{2\}) \geq 25, v_3(\{1,2\}) \geq 10\}$. Accordingly, $BB_1 = 25$: the best Player 1 can do in selling to others is to offer Item 2 to Player 2 at price 25. Furthermore, $\mathcal{B}_2 = \{v \mid v_2 = \theta_2, v_1(\{1\}) \geq 80, v_3(\{2\}) \geq 20\}$, which implies $BB_2 = 100$, achieved by offering Item 1 to Player 1 at price 80 and Item 2 to Player 3 at price 20. Finally, $\mathcal{B}_3 = \{v \mid v_3 = \theta_3, v_1(\{1\}) \geq 80, v_2(\{2\}) \geq 70\}$, which implies $BB_3 = 150$, achieved by offering Item 1 to Player 1 at price 80 and Item 2 to Player 2 at price 70. Therefore, $BB = 150$ in this example. Note that Player 1's and Player 3's beliefs are correct, but Player 2's beliefs are incorrect because $\theta_3(\{2\}) = 0$.

Furthermore, note that, if there is really a common prior from which the players' valuations are drawn, then the players' possibilistic beliefs consist of the support of the distribution. In this case, it is expected that the optimal Bayesian mechanism generates more revenue than the best-belief benchmark. However, this is a totally different ball game, because Bayesian mechanisms assume that the seller has much knowledge about the players. Besides, little is known in the literature about the structure of the optimal Bayesian mechanism for unrestricted combinatorial auctions or even a good approximation to it.

Finally, the best-belief benchmark is measured based on the players' beliefs about each other, not on their true valuations. If the players all know nothing about each other and believe that the others' values can be anything from close to zero to close to infinity (or a huge finite number), then the benchmark is low. The power of the benchmark comes from the class of contexts where the players know each other well (e.g., as long-time competitors in the same market) and can effectively narrow down the range of the others' values. In this case, our mechanism generates good revenue without assuming a common prior.

## 4. A Normal-Form Mechanism

As a warm up, in this section, we construct a normal-form mechanism that implements the best-belief revenue benchmark within a factor of two, under two-step elimination of weakly-dominated strategies. Indeed, weakly-dominant/dominated strategies have been widely used in analyzing combinatorial auctions where the players only report their valuations: that is, it is weakly dominant for each player to report his/her true valuation. When each player reports both his/her own valuation and his/her beliefs about the other players, it is intuitive that a player $i$ first reasons about what the other players report for their valuations and then reasons about what to report for his/her beliefs about them given their reported valuations: that is, an iterated elimination of dominated strategies. However, in our mechanism, there is no need to go all the way to the end of the iterated procedure, and two steps are sufficient.

Roughly speaking, all players first simultaneously remove all of their weakly-dominated strategies; and then, each player further removes all of his/her strategies that now become weakly dominated, based on all players' surviving strategies. However, care must be taken when defining this solution concept in our setting. Indeed, since a player does not know the other players' true valuations, he/she cannot compute their strategies surviving the first round of elimination, which are needed for him to carry out his/her second round of elimination. To be "on the safe side", we require that the players eliminate their strategies conservatively: that is, a player eliminates a strategy in the second round only if it is dominated by the same alternative strategy with respect to all valuation profiles that he/she believes to be possible. This notion of elimination is the same as the one used by Aumann in [15], except that in the latter, it is strict instead of weak domination. In [33], the authors provide an epistemic characterization for iterated elimination based on the notion of [15].

More precisely, given a normal-form auction mechanism $M$, let $S_i$ be the set of strategies of each player $i$ and $S = S_1 \times \cdots \times S_n$. For any strategy profile $s$, $M(s)$ is the outcome when each player $i$ uses strategy $s_i$. If $T = T_i \times T_{-i}$ is a subset of strategy profiles, $t_i \in \Theta_i$, $s_i \in T_i$, and $\sigma_i \in \Delta(T_i)$[5], then we say that $s_i$ is weakly dominated by $\sigma_i$ with respect to $t_i$ and $T$, in symbols $s_i \preceq_T^{t_i} \sigma_i$, if:

- $u_i(t_i, M(s_i, s_{-i})) \leq u_i(t_i, M(\sigma_i, s_{-i}))$ for all $s_{-i} \in T_{-i}$ and
- $u_i(t_i, M(s_i, s_{-i})) < u_i(t_i, M(\sigma_i, s_{-i}))$ for some $s_{-i} \in T_{-i}$.

That is, $s_i$ is weakly dominated by $\sigma_i$ when the valuation of player $i$ is $t_i$ and the set of strategy sub-profiles of the other players is $T_{-i}$. The set of strategies in $T_i$ that are not weakly dominated with respect to $t_i$ and $T$ is denoted by $U_i(t_i, T)$. For simplicity, we use $U_i$ to denote $U_i(\theta_i, S)$, the set of undominated strategies of player $i$.

**Definition 3.** *Given an auction context $C = (n, m, \theta, \mathcal{B})$ and a mechanism $M$, the set of conservatively weakly-rational strategies of player $i$ is:*

$$C_i \triangleq U_i \setminus \{s_i : \exists \sigma_i \in \Delta(U_i) \text{ s.t. } \forall t \in \mathcal{B}_i,\ s_i \preceq_{U(t)}^{\theta_i} \sigma_i\},$$

*where $U(t) \triangleq \times_j U_j(t_j, S)$ for any $t \in \Theta$. The set of conservatively weakly-rational strategy profiles is $C = C_1 \times \cdots \times C_n$.*

*Mechanism $M$ conservatively weakly implements a revenue benchmark $f$ if, for any auction context $C$ and any strategy profile $s \in C$,*

$$REV(M(s)) \geq f(C).$$

Now, we provide and analyze our normal-form mechanism $M_{Normal}$. Intuitively, the players compete for the right to sell to others, and the mechanism generates revenue by delegating this right to the player who offers the most revenue. Besides the number of players $n$ and the number of goods $m$, the mechanism takes as input a constant $\epsilon \in (0,1]$. The players act only in Step 1, and Steps a through f are "steps taken by the mechanism". The expression "$X := x$" sets or resets variable $X$ to value $x$. Moreover, $[m] = \{1, 2, \ldots, m\}$.

Mechanism $M_{Normal}$:

1: Each player $i$, publicly and simultaneously with the other players, announces:

- a valuation $v_i$ and
- an outcome $\omega^i = (\alpha^i, \pi^i)$, such that: $\alpha_i^i = \emptyset$, and for each player $j$, $\pi_j^i$ is zero whenever $\alpha_j^i = \emptyset$; and is a positive integer otherwise.

[5] As usual, for a set $T$, $\Delta(T)$ is the set of probability distributions over $T$.

After the players simultaneously execute Step 1, the mechanism chooses the outcome $(A, P)$ by means of the following six steps.

a: Set $A_i := \emptyset$, and $P_i := 0$ for each player $i$.

b: Set $R_i := REV(\omega^i)$ for each player $i$, and $w := \text{argmax}_i R_i$ with ties broken lexicographically.

c: Publicly flip a fair coin and denote the result by $r$.

d: If $r = Heads$, then $A_w := \text{argmax}_{a \subseteq [m]} v_w(a)$, with ties broken lexicographically, and halt.

e: (Note that $r = Tails$ when this step is reached.)

For each player $i$, such that $\alpha_i^w \neq \emptyset$:

- If $v_i(\alpha_i^w) < \pi_i^w$, then $P_w := P_w + \pi_i^w$.
- Otherwise, $A_i := \alpha_i^w$ and $P_i := \pi_i^w - \frac{\epsilon}{n}$.

f: For each player $i$, $P_i := P_i - \delta_i$ with $\delta_i = \frac{\epsilon}{n} \cdot \frac{R_i}{1+R_i}$.

The final outcome is $(A, P)$.

In the analysis, we refer to player $w$ as the winner and each $\delta_i$ as player $i$'s reward. Furthermore, given a context $(n, m, \theta, \mathcal{B})$ and an outcome $\omega$, for succinctness, we use $u_i(\omega)$ instead of $u_i(\theta_i, \omega)$ for player $i$'s utility under $\omega$. We have the following.

**Theorem 1.** *For any context $(n, m, \theta, \mathcal{B})$ and constant $\epsilon \in (0, 1]$, mechanism $M_{Normal}$ conservatively weakly implements the revenue benchmark $\frac{BB}{2} - \epsilon$.*

As we will see in the proof of Theorem 1, the mechanism incentivizes each player $i$ to report his/her true valuation and an outcome whose revenue is at least $BB_i$. In particular, the latter is achieved by the fair coin toss: when $r = Heads$, the winner is given his/her favorite subset of goods for free, which is better than any offer he/she can possibly get if somebody else becomes the winner. Moreover, the rewards are strictly increasing with the revenue of the reported outcomes. Accordingly, the players do not have incentives to underbid; that is, to report an outcome whose revenue is lower than the corresponding $BB_i$. Thus, the winner's reported outcome has a revenue of at least $\max_i BB_i$. When $r = Tails$, the mechanism tries to sell the goods as suggested by the winner to the other players, as a take-it-or-leave-it offer. If a player accepts the offer, then he/she pays the suggested price; otherwise, this price is charged to the winner as a fine. Accordingly, with probability 1/2 (that is, when $r = Tails$), the mechanism generates revenue $\max_i BB_i$. Formally, we show the following two lemmas.

**Lemma 1.** *For any context $(n, m, \theta, \mathcal{B})$, constant $\epsilon$, player $i$ and strategy $s_i = (v_i, \omega^i)$, if $s_i \in U_i$, then $v_i = \theta_i$.*

**Proof.** Notice that $v_i$ is used in two places in the mechanism: to select player $i$'s "favorite subset" in Step d when he/she is the winner and to decide whether he/she gets the set $\alpha_i^w$ in Step e when he/she is not the winner. Intuitively, it is $i$'s best strategy to announce his/her true valuation so as to select his/her "truly favorite subset" and to take the allocated set if and only of its price is less than or equal to his/her true value for it.

More precisely, arbitrarily fix a strategy $s_i = (v_i, \omega^i)$ with $v_i \neq \theta_i$, and let $s'_i = (\theta_i, \omega^i)$. We show that $s_i \preceq_S^{\theta_i} s'_i$, where $S$ is the set of all strategy profiles of $M_{Normal}$. To do so, arbitrarily fix a strategy sub-profile $s_{-i}$ of the other players; let $(A, P)$ be the outcome of $s = (s_i, s_{-i})$, and let $(A', P')$ be the outcome of $s' = (s'_i, s_{-i})$. Since $s_i$ and $s'_i$ announce the same outcome $\omega^i$, $i$ is the winner under $s$ if and only if he/she is the winner under $s'$. We discuss these two cases separately.

Case 1: $i$ is the winner under both $s$ and $s'$.

In this case, conditioned on $r = Heads$, we have $A_i = \text{argmax}_{a \subseteq [m]} v_i(a)$, $A'_i = \text{argmax}_{a \subseteq [m]} \theta_i(a)$ and $P_i = P'_i = 0$. Accordingly, $\theta_i(A'_i | r = Heads) \geq \theta_i(A_i | r = Heads)$ and $u_i(A', P' | r = Heads) \geq u_i(A, P | r = Heads)$.

Conditioned on $r = Tails$, we have $A_i = A'_i = \emptyset$ and:

$$P_i = P'_i = \sum_{j: \alpha^i_j \neq \emptyset \text{ and } v_j(\alpha^i_j) < \pi^i_j} \pi^i_j - \delta_i,$$

where $\delta_i = \frac{\epsilon}{n} \cdot \frac{REV(\omega^i)}{1+REV(\omega^i)}$ is player $i$'s reward under both strategy profiles. Accordingly, $u_i(A', P' | r = Tails) = u_i(A, P | r = Tails)$.

In sum, $u_i(A, P) \leq u_i(A', P')$ in Case 1.

Case 2: $i$ is the winner under neither $s$ nor $s'$.

In this case, the winner $w$ is the same under both $s$ and $s'$. Conditioned on $r = Heads$, we have $A_i = A'_i = \emptyset$ and $P_i = P'_i = 0$; thus, $u_i(A, P | r = Heads) = u_i(A', P' | r = Heads)$.

Conditioned on $r = Tails$, if $v_i(\alpha^w_i) < \pi^w_i$ and $\theta_i(\alpha^w_i) < \pi^w_i$, or if both inequalities are reversed, then $(A_i, P_i) = (A'_i, P'_i)$ and $u_i(A, P | r = Tails) = u_i(A', P' | r = Tails)$. Otherwise, if $v_i(\alpha^w_i) < \pi^w_i$ and $\theta_i(\alpha^w_i) \geq \pi^w_i$, then:

$$u_i(A', P' | r = Tails) = \theta_i(\alpha^w_i) - \pi^w_i + \frac{\epsilon}{n} + \delta_i > \delta_i = u_i(A, P | r = Tails), \tag{1}$$

where again $\delta_i$ is $i$'s reward under both strategy profiles. Otherwise, we have $v_i(\alpha^w_i) \geq \pi^w_i$ and $\theta_i(\alpha^w_i) < \pi^w_i$; thus:

$$u_i(A, P | r = Tails) = \theta_i(\alpha^w_i) - \pi^w_i + \frac{\epsilon}{n} + \delta_i \leq -1 + \frac{\epsilon}{n} + \delta_i < \delta_i = u_i(A', P' | r = Tails). \tag{2}$$

In sum, $u_i(A, P) \leq u_i(A', P')$ in Case 2, as well.

It remains to show there exists a strategy sub-profile $s_{-i}$, such that $u_i(A, P) < u_i(A', P')$, and such an $s_{-i}$ has actually appeared in Case 2 above. Indeed, since $v_i \neq \theta_i$, there exists $a \subseteq [m]$, such that $v_i(a) \neq \theta_i(a)$. When $v_i(a) < \theta_i(a)$, arbitrarily fix a player $j \neq i$, and choose strategy $s_j$, such that:

$$\alpha^j_i = a, \pi^j_i = \theta_i(a), \text{ and } REV(\omega^j) > \max\{\pi^j_i, REV(\omega^i)\}.$$

Notice that such a strategy exists in $S_j$: player $j$ can set $\pi^j_k$ to be arbitrarily high for any player $k \notin \{i, j\}$. Moreover, for any player $k \notin \{i, j\}$, choose $s_k$ to be such that $REV(\omega^k) = 0$. By construction, $w = j$ under both $s$ and $s'$, $v_i(\alpha^w_i) < \pi^w_i$ and $\theta_i(\alpha^w_i) \geq \pi^w_i$. Following Case 2 above, $u_i(A, P | r = Heads) = u_i(A', P' | r = Heads)$ and $u_i(A, P | r = Tails) < u_i(A', P' | r = Tails)$ by Inequality 1. Thus, $u_i(A, P) < u_i(A', P')$.

When $v_i(a) > \theta_i(a)$, similarly, choose strategy $s_j$, such that:

$$\alpha^j_i = a, \pi^j_i = v_i(a), \text{ and } REV(\omega^j) > \max\{\pi^j_i, REV(\omega^i)\},$$

and choose strategy $s_k$ the same as above for any $k \notin \{i, j\}$. The analysis again follows from Case 2 above (in particular, Inequality 2); thus, $u_i(A, P) < u_i(A', P')$.

Combining everything together, $s_i \leq^{\theta_i}_S s'_i$, and Lemma 1 holds. □

**Lemma 2.** *For any context $(n, m, \theta, \mathcal{B})$, constant $\epsilon$, player $i$ and strategy $s_i = (v_i, \omega^i)$, if $s_i \in C_i$, then $REV(\omega^i) \geq BB_i$.*

**Proof.** By Lemma 1, we only need to consider strategies, such that $v_i = \theta_i$. Arbitrarily fix a strategy $s_i = (\theta_i, \omega^i) \in U_i$ with $REV(\omega^i) < BB_i$. Consider a strategy $\hat{s}_i = (\theta_i, \hat{\omega}^i)$, such that $\hat{\omega}^i = (\hat{\alpha}^i, \hat{\pi}^i)$ satisfies the following conditions:

$$\hat{\omega}^i \in \underset{(A,P)\in\Omega:\ A_i=\emptyset \text{ and } P_j \leq t_j(A_j)\ \forall j\neq i,\ t\in\mathcal{B}_i}{\operatorname{argmax}} REV(A, P)$$

and

$$\hat{\pi}^i_j > 0 \text{ whenever } \hat{\alpha}^i_j \neq \emptyset.$$

Notice that $REV(\hat{\omega}^i) = BB_i > REV(\omega^i)$. We show that for all $t \in \mathcal{B}_i$, $s_i \leq^{\theta_i}_{U(t)} \hat{s}_i$.

To do so, arbitrarily fix a valuation profile $t \in \mathcal{B}_i$ and a strategy sub-profile $s_{-i}$, such that $s_j \in U_j(t_j, S)$ for each player $j$. Note that $t_i = \theta_i$ by the definition of $\mathcal{B}_i$. Moreover, by Lemma 1, each $s_j$ is of the form $(t_j, \omega^j)$: that is, the valuation it announces is $t_j$. Let $(A, P)$ be the outcome of the strategy profile $s = (s_i, s_{-i})$ and $(\hat{A}, \hat{P})$ that of the strategy profile $\hat{s} = (\hat{s}_i, s_{-i})$. There are three possibilities for the winners under $s$ and $\hat{s}$: (1) player $i$ is the winner under both of them; (2) player $i$ is the winner under neither of them; and (3) player $i$ is the winner under $\hat{s}$, but not under $s$. Below, we consider them one by one.

Case 1: $i$ is the winner under both $s$ and $\hat{s}$.

In this case, conditioned on $r = Heads$, $(A_i, P_i) = (\hat{A}_i, \hat{P}_i)$ and $u_i(A, P|r = Heads) = u_i(\hat{A}, \hat{P}|r = Heads)$, since under both $s$ and $\hat{s}$, player $i$ gets his/her favorite subset for free.

Conditioned on $r = Tails$, $A_i = \hat{A}_i = \emptyset$, $\hat{P}_i = \sum_{j:\hat{\alpha}^i_j\neq\emptyset \text{ and } t_j(\hat{\alpha}^i_j)<\hat{\pi}^i_j} \hat{\pi}^i_j - \hat{\delta}_i$, and $P_i = \sum_{j:\alpha^i_j\neq\emptyset \text{ and } t_j(\alpha^i_j)<\pi^i_j} \pi^i_j - \delta_i$, where $\hat{\delta}_i$ is player $i$'s reward under $\hat{s}$ and $\delta_i$ is that under $s$. By the definition of $\hat{\omega}^i$, the set $\{j : \hat{\alpha}^i_j \neq \emptyset \text{ and } t_j(\hat{\alpha}^i_j) < \hat{\pi}^i_j\}$ is empty, so $\hat{P}_i = -\hat{\delta}_i$. As $REV(\hat{\omega}^i) > REV(\omega^i)$, by definition we have $\hat{\delta}_i > \delta_i$, which implies $\hat{P}_i < -\delta_i \leq P_i$. Accordingly, $u_i(\hat{A}, \hat{P}|r = Tails) > u_i(A, P|r = Tails)$.

In sum, we have $u_i(\hat{A}, \hat{P}) > u_i(A, P)$ in Case 1.

Case 2: $i$ is the winner under neither $s$ nor $\hat{s}$.

In this case, the winner $w$ is the same under both strategy profiles. Conditioned on $r = Heads$, $A_i = \hat{A}_i = \emptyset$ and $P_i = \hat{P}_i = 0$, thus $u_i(A, P|r = Heads) = u_i(\hat{A}, \hat{P}|r = Heads)$.

Conditioned on $r = Tails$, $i$ gets the set $\alpha^w_i$ under $s$ if and only if he/she gets it under $\hat{s}$, as he/she announces valuation $\theta_i$ under both strategy profiles. That is, $A_i = \hat{A}_i$. Moreover, the only difference in player $i$'s prices is the rewards he/she gets, and $P_i - \hat{P}_i = -\delta_i + \hat{\delta}_i > 0$. Accordingly, $u_i(\hat{A}, \hat{P}|r = Tails) > u_i(A, P|r = Tails)$.

In sum, we have $u_i(\hat{A}, \hat{P}) > u_i(A, P)$ in Case 2.

Case 3: $i$ is the winner under $\hat{s}$, but not under $s$.

In this case, letting $w$ be the winner under $s$, we have $REV(\omega^i) \leq REV(\omega^w) \leq REV(\hat{\omega}^i)$, and at least one of the inequalities is strict. We compare player $i$'s utilities under $s$ and $\hat{s}$, but conditioned on different outcomes of the random coin. More specifically, we use $r$ to denote the outcome of the coin under $s$ and $\hat{r}$ that under $\hat{s}$.

First, conditioned on $\hat{r} = Heads$, $\hat{A}_i = \operatorname{argmax}_{a\subseteq[m]} \theta_i(a)$ and $\hat{P}_i = 0$; thus:

$$u_i(\hat{A}, \hat{P}|\hat{r} = Heads) = \theta_i(\hat{A}_i).$$

While conditioned on $r = Tails$, we have either $A_i = \emptyset$ and $P_i = -\delta_i$, or $A_i = \alpha_i^w \neq \emptyset$ and $P_i = \pi_i^w - \frac{\epsilon}{n} - \delta_i$; thus:

$$\begin{aligned} u_i(A, P|r = Tails) &\leq \max\{\delta_i, \theta_i(\alpha_i^w) - \pi_i^w + \frac{\epsilon}{n} + \delta_i\} \leq \max\{\delta_i, \theta_i(\hat{A}_i) - 1 + \frac{\epsilon}{n} + \delta_i\} \\ &\leq \theta_i(\hat{A}_i) + \delta_i, \end{aligned}$$

where the second inequality is because $\theta_i(\alpha_i^w) \leq \theta_i(\hat{A}_i)$ and $\pi_i^w \geq 1$, and the third inequality is because both terms in $\max\{\cdot\}$ are less than or equal to $\theta_i(\hat{A}_i) + \delta_i$. Accordingly,

$$u_i(\hat{A}, \hat{P}|\hat{r} = Heads) - u_i(A, P|r = Tails) \geq -\delta_i. \quad (3)$$

Second, conditioned on $\hat{r} = Tails$, $\hat{A}_i = \emptyset$ and $\hat{P}_i = -\hat{\delta}_i$, similar to Case 1 above. Thus:

$$u_i(\hat{A}, \hat{P}|\hat{r} = Tails) = \hat{\delta}_i.$$

While conditioned on $r = Heads$, $A_i = \emptyset$ and $P_i = 0$; thus:

$$u_i(A, P|r = Heads) = 0.$$

Accordingly,

$$u_i(\hat{A}, \hat{P}|\hat{r} = Tails) - u_i(A, P|r = Heads) \geq \hat{\delta}_i. \quad (4)$$

Combining Inequalities 3 and 4 and given that $r$ and $\hat{r}$ are both fair coins, we have:

$$u_i(\hat{A}, \hat{P}) - u_i(A, P) \geq \frac{\hat{\delta}_i - \delta_i}{2} > 0,$$

thus $u_i(\hat{A}, \hat{P}) > u_i(A, P)$ in Case 3, as well.

In sum, $s_i \leq_{U(t)}^{\theta_i} \hat{s}_i$ for all $t \in \mathcal{B}_i$, which implies $s_i \notin \mathcal{C}_i$. Thus, Lemma 2 holds. □

We now analyze the revenue of $M_{Normal}$.

**Proof of Theorem 1.** Arbitrarily fix an auction context $C = (n, m, \theta, \mathcal{B})$ and a strategy profile $s \in \mathcal{C}$. By Lemma 1, we can write $s_i = (\theta_i, \omega^i)$ for each player $i$. Let $(A, P)$ be the outcome of $M_{Normal}$ under $s$. By Lemma 2, $REV(\omega^i) \geq BB_i$ for each $i$, so:

$$R_w = \max_i REV(\omega^i) \geq \max_i BB_i = BB(C).$$

Note that $REV(A, P|r = Heads) = 0$, while:

$$\begin{aligned} &REV(A, P|r = Tails) = \sum_i P_i \\ &= P_w + \sum_{i:\alpha_i^w \neq \emptyset, \theta_i(\alpha_i^w) \geq \pi_i^w} (\pi_i^w - \tfrac{\epsilon}{n} - \delta_i) + \sum_{i:\alpha_i^w \neq \emptyset, \theta_i(\alpha_i^w) < \pi_i^w} (-\delta_i) + \sum_{i:\alpha_i^w = \emptyset, i \neq w} (-\delta_i) \\ &= \left( \sum_{i:\alpha_i^w \neq \emptyset, \theta_i(\alpha_i^w) < \pi_i^w} \pi_i^w \right) - \delta_w + \sum_{i:\alpha_i^w \neq \emptyset, \theta_i(\alpha_i^w) \geq \pi_i^w} (\pi_i^w - \tfrac{\epsilon}{n} - \delta_i) \\ &+ \sum_{i:\alpha_i^w \neq \emptyset, \theta_i(\alpha_i^w) < \pi_i^w} (-\delta_i) + \sum_{i:\alpha_i^w = \emptyset, i \neq w} (-\delta_i) \\ &\geq \sum_{i:\alpha_i^w \neq \emptyset} \pi_i^w - \sum_i \tfrac{\epsilon}{n} - \sum_i \delta_i = R_w - \epsilon - \sum_i \delta_i > R_w - \epsilon - \sum_i \tfrac{\epsilon}{n} = R_w - 2\epsilon \geq BB(C) - 2\epsilon. \end{aligned}$$

Combining the two cases together, we have $REV(A, P) > \frac{BB(C)}{2} - \epsilon$, and Theorem 1 holds. □

## 5. Conservative Distinguishable Implementation

Our main mechanism, together with an auction context, specifies an extensive game with perfect information, chance moves and simultaneous moves [47][6]. For such a mechanism $M$, we denote the set of all pure strategy profiles by $S = S_1 \times \cdots \times S_n$, the history of a strategy profile $s$ by $H(s)$ and, again, the outcome of $s$ by $M(s)$. If $\sigma$ is a mixed strategy profile, then $H(\sigma)$ and $M(\sigma)$ are the corresponding distributions.

Even for extensive games of complete information, the literature has several notions of rationality, with different epistemic foundations and predictions about the players' strategies. Since our setting is of incomplete information without Bayesian beliefs, it is important to define a proper solution concept in order to analyze mechanisms in such settings. Iterated eliminations of dominated strategies and their epistemic characterizations have been the focus of many studies in epistemic game theory. In [30], the authors define distinguishable dominance, prove that it is order independent with respect to surviving histories and characterize it with extensive-form rationalizability [31,32,48]. In some sense, distinguishable dominance is the counterpart of strict dominance in extensive-form games. We incorporate this solution concept with the players' possibilistic beliefs.

**Definition 4.** *Let $C = (n, m, \theta, \mathcal{B})$ be an auction context, $M$ an extensive-form mechanism, $i$ a player, $t_i$ a valuation of $i$ and $T = T_i \times T_{-i}$ a set of pure strategy profiles. A strategy $s_i \in T_i$ is distinguishably-dominated by another strategy $\sigma_i \in \Delta(T_i)$ with respect to $t_i$ and $T$, in symbols $s_i \prec_T^{t_i} \sigma_i$, if:*

1. *$\exists s_{-i} \in T_{-i}$ distinguishing $s_i$ and $\sigma_i$: that is, $H(s_i, s_{-i}) \neq H(\sigma_i, s_{-i})$; and*
2. *$u_i(t_i, M(s_i, s_{-i})) < u_i(t_i, M(\sigma_i, s_{-i}))$ $\forall s_{-i} \in T_{-i}$ distinguishing $s_i$ and $\sigma_i$.*

Intuitively, $s_i$ is distinguishably dominated by $\sigma_i$ if it leads to a smaller utility for $i$ than $\sigma_i$, when played against any $s_{-i}$, except those $s_{-i}$ that produce the same history with $s_i$ and with $\sigma_i$: when such an $s_{-i}$ is used, not only player $i$ has the same utility under $s_i$ and $\sigma_i$, but also nobody can distinguish whether $i$ is using $s_i$ or $\sigma_i$ by observing the history of the game.

For each player $i$, we denote by $DU_i(t_i, T)$ the set of strategies in $T_i$ that are not distinguishably dominated with respect to $t_i$ and $T$ and by $DU_i$ the set $DU_i(\theta_i, S)$. Having seen how to incorporate the iterated elimination of weakly-dominated strategies into our setting, the readers should find the following definition a natural analog.

**Definition 5.** *Let $C = (n, m, \theta, \mathcal{B})$ be an auction context, $M$ a mechanism and $i$ a player. The set of conservatively distinguishably-rational strategies of player $i$ is:*

$$\mathcal{CD}_i \triangleq DU_i \setminus \{s_i : \exists \sigma_i \in \Delta(DU_i) \text{ s.t. } \forall t \in \mathcal{B}_i,\ s_i \prec_{DU(t)}^{\theta_i} \sigma_i\},$$

*where $DU(t) \triangleq \times_j DU_j(t_j, S)$ for any $t \in \Theta$. The set of conservatively distinguishably-rational strategy profiles is $\mathcal{CD} = \mathcal{CD}_1 \times \cdots \times \mathcal{CD}_n$.*

*Mechanism $M$ conservatively distinguishably implements a revenue benchmark $f$ if, for any auction context $C$ and any strategy profile $s \in \mathcal{CD}$, $REV(M(s)) \geq f(C)$.*

A player $i$ may further refine $\mathcal{CD}_i$, but doing so requires more than mutual belief of rationality. We thus do not consider any further refinements.

[6] Section 6.3 of [47] defines extensive games with perfect information and chance moves, as well as extensive games with perfect information and simultaneous moves. It is easy to combine the two to define extensive games with all three characteristics. Such a game can be described by a "game tree". A decision node is an internal node, where the players take actions or chance moves. A terminal node is a leaf, where an outcome is specified. The history of a strategy profile is the probability distribution over paths from the root to the leaves determined by this profile. The outcome of a strategy profile is the probability distribution over outcomes at the leaves determined by this profile.

## 6. The Best-Belief Mechanism

Now, we construct and analyze our best-belief mechanism $M_{BB}$. Similar to the normal-form mechanism, it is parameterized by $n$, $m$ and a constant $\epsilon \in (0,1]$. In the description below, Steps 1–3 correspond to decision nodes, while Steps a–e are again "steps taken by the mechanism".

The best-belief mechanism, $M_{BB}$:

a: Set $A_i := \emptyset$ and $P_i := 0$ for each player $i$.

1: Each player $i$, publicly and simultaneously with the other players, announces:

(1) a subset $\xi_i$ of the goods; and

(2) an outcome $\omega^i = (\alpha^i, \pi^i)$, such that: $\alpha^i_i = \emptyset$, and for each player $j$, $\pi^i_j$ is zero whenever $\alpha^i_j = \emptyset$ and is a positive integer otherwise.

b: Set $R_i := REV(\omega^i)$ for each player $i$ and $w := \text{argmax}_i R_i$ with ties broken lexicographically.

2: Publicly flip a fair coin, and denote the result by $r$.

c: If $r = Heads$, then $A_w := \xi_w$, and halt.

3: (Note that $r = Tails$ when this step is reached.)

Each player $i$, such that $\alpha^w_i \neq \emptyset$ publicly and simultaneously announces YES or NO.

d: For each player $i$ announcing NO, $P_w := P_w + \pi^w_i$.

For each player $i$ announcing YES, $A_i := \alpha^w_i$ and $P_i := \pi^w_i - \frac{\epsilon}{n}$.

For each player $i$, $P_i := P_i - \delta_i$ with $\delta_i = \frac{\epsilon}{n} \cdot \frac{R_i}{1+R_i}$.

e: The final outcome is $(A, P)$.

### 6.1. Analysis of Our Mechanism

As before, given a context $(n, m, \theta, \mathcal{B})$ and an outcome $\omega$, we use $u_i(\omega)$ instead of $u_i(\theta_i, \omega)$ for player $i$'s utility under $\omega$. We have the following.

**Theorem 2.** *For any context $(n, m, \theta, \mathcal{B})$ and constant $\epsilon \in (0,1]$, mechanism $M_{BB}$ conservatively distinguishably implements the revenue benchmark $\frac{BB}{2} - \epsilon$.*

Different from the normal-form mechanism, here, a player does not report his/her true valuation. Instead, the use of his/her valuation is divided into two parts: a subset of the goods, which will be his/her favorite subset as we will see in the proof; and a simple "yes or no" answer to the take-it-or-leave-it offer suggested by the winner. All of the other information about his/her true valuation is redundant and has been removed from the player's report. This can be done because the mechanism is extensive and the players give their answers directly after seeing the offers; thus, the seller does not need to deduce their answers from their reported valuations. We again start by proving the following two lemmas. Some ideas are similar to those for Lemmas 1 and 2; thus, the details have been omitted.

**Lemma 3.** *For any context $(n, m, \theta, \mathcal{B})$, constant $\epsilon$, player $i$ and strategy $s_i$, if $s_i \in DU_i$, then, according to $s_i$, in Step 3 of $\mathcal{M}_{BB}$, $i$ announces YES if and only if $\theta_i(\alpha^w_i) \geq \pi^w_i$*[7]*.*

---

[7] That is, $i$ will announce YES or NO as above at every decision node corresponding to Step 3, which is reachable (with positive probability) by $s_i$ together with some strategy sub-profile $s_{-i}$, where $i$ is an acting player.

**Proof.** We only prove the "if" direction, as the "only if" direction is totally symmetric. Assume that, according to $s_i$, $i$ announces NO at some reachable decision node $d$ of $i$ where $\theta_i(\alpha_i^w) \geq \pi_i^w$. We refer to such a node $d$ as a deviating node. Consider the following strategy $s_i'$:

- $s_i'$ announces the same $\xi_i$ and $\omega^i$ as $s_i$ in Step 1; and
- according to $s_i'$, in Step 3, $i$ announces YES if and only if $\theta_i(\alpha_i^w) \geq \pi_i^w$.

Below, we show that $s_i \prec_S^{\theta_i} s_i'$, where $S$ is the set of all strategy profiles of $M_{BB}$.

For any deviating node $d$, since $d$ is reachable by $s_i$, there exists a strategy sub-profile $s_{-i} \in S_{-i}$, such that the history $H(s_i, s_{-i})$ reaches $d$ with positive probability. In fact, by the construction of the mechanism, the probability is exactly 1/2: when $r = Tails$. For any such $s_{-i}$, by the construction of $s_i'$, the history $H(s_i', s_{-i})$ also reaches $d$ with probability 1/2. By definition, $i$ announces YES at $d$ under $s_i'$ and NO under $s_i$; thus, $H(s_i, s_{-i}|r = Tails) \neq H(s_i', s_{-i}|r = Tails)$ and $s_{-i}$ distinguishes $s_i$ and $s_i'$.

Indeed, for any strategy sub-profile $s_{-i}$, it distinguishes $s_i$ and $s_i'$ if and only if $H(s_i, s_{-i})$ reaches a deviating node $d$ (with probability 1/2). Arbitrarily fixing such an $s_{-i}$ and the corresponding deviating node $d$, it suffices to show:

$$u_i(M_{BB}(s_i, s_{-i})) < u_i(M_{BB}(s_i', s_{-i})). \tag{5}$$

Because $i \neq w$ under $(s_i, s_{-i})$ when $r = Tails$ (that is, when $d$ is reached), $i \neq w$ under $(s_i, s_{-i})$ when $r = Heads$, as well, since $w$ is the same in the two cases. Moreover, because $s_i'$ announces the same $\xi_i$ and $\omega^i$ as $s_i$ in Step 1, we have $H(s_i, s_{-i}|r = Heads) = H(s_i', s_{-i}|r = Heads)$ and $u_i(M_{BB}(s_i, s_{-i})|r = Heads) = u_i(M_{BB}(s_i', s_{-i})|r = Heads) = 0$.

Similar to Lemma 1, $u_i(M_{BB}(s_i, s_{-i})|r = Tails) = \delta_i$, as $i$ announces NO at $d$ under $s_i$. Furthermore, $u_i(M_{BB}(s_i', s_{-i})|r = Tails) = \theta_i(\alpha_i^w) - P_i = \theta_i(\alpha_i^w) - \pi_i^w + \frac{\epsilon}{n} + \delta_i \geq \frac{\epsilon}{n} + \delta_i > \delta_i$, as $\theta_i(\alpha_i^w) \geq \pi_i^w$ at $d$, and $i$ announces YES at $d$ under $s_i'$. Therefore, $u_i(M_{BB}(s_i', s_{-i})|r = Tails) > u_i(M_{BB}(s_i, s_{-i})|r = Tails)$, which implies Equation (5). Accordingly, $s_i \prec_S^{\theta_i} s_i'$, $s_i \notin DU_i$, and Lemma 3 holds. □

**Lemma 4.** *For any context $(n, m, \theta, \mathcal{B})$, constant $\epsilon$, player $i$ and strategy $s_i$, if $s_i \in \mathcal{CD}_i$, then, according to $s_i$, player $i$ announces $\omega^i$ in Step 1 with $REV(\omega^i) \geq BB_i$.*

**Proof.** Arbitrarily fix a strategy $s_i \in DU_i$ according to which, in Step 1, $i$ announces $\xi_i$, and $\omega^i = (\alpha^i, \pi^i)$ with $REV(\omega^i) < BB_i$. By Lemma 3, according to $s_i$, in Step 3, $i$ announces YES if and only if $\theta_i(\alpha_i^w) \geq \pi_i^w$. Consider the following strategy $\hat{s}_i$:

- In Step 1, $i$ announces $\hat{\xi}_i$, and $\hat{\omega}^i = (\hat{\alpha}^i, \hat{\pi}^i)$, such that:
  - $\theta_i(\hat{\xi}_i) = \max_{A \subseteq \{1,\ldots,m\}} \theta_i(A)$;
  - $REV(\hat{\omega}^i) = \max_{(A,P) \in \Omega:\ A_i = \emptyset \text{ and } P_j \leq t_j(A_j)\ \forall j \neq i, \forall t \in \mathcal{B}_i} REV(A, P)$; and
  - $\hat{\pi}_j^i > 0$ whenever $\hat{\alpha}_j^i \neq \emptyset$.
- In Step 3, $i$ announces YES if and only if $\theta_i(\alpha_i^w) \geq \pi_i^w$.

By definition, $REV(\hat{\omega}^i) = BB_i > REV(\omega^i)$, which implies that $\hat{s}_i$ and $s_i$ differ in Step 1: the root of the game tree. Thus, any strategy sub-profile $s_{-i}$ distinguishes them. We show that for all

$t \in \mathcal{B}_i, s_i \prec^{\theta_i}_{DU(t)} \hat{s}_i$[8]. To do so, arbitrarily fixing a valuation profile $t \in \mathcal{B}_i$ and a strategy sub-profile $s_{-i} \in \times_{j \neq i} DU_j(t_j, S)$, it suffices to show:

$$u_i(M_{BB}(s_i, s_{-i})) < u_i(M_{BB}(\hat{s}_i, s_{-i})). \tag{6}$$

Let $\delta_i$ and $\hat{\delta}_i$ be the rewards of player $i$ in Step d, under $(s_i, s_{-i})$ and $(\hat{s}_i, s_{-i})$, respectively. Because $REV(\hat{\omega}^i) > REV(\omega^i)$, we have:

$$\delta_i < \hat{\delta}_i. \tag{7}$$

Similar to Lemma 2, we distinguish three cases.

Case 1. $i$ is the winner under both $(s_i, s_{-i})$ and $(\hat{s}_i, s_{-i})$.

In this case, on the one hand,

$$u_i(M_{BB}(s_i, s_{-i})|r = Heads) = \theta_i(\xi_i) \leq \theta_i(\hat{\xi}_i) = u_i(M_{BB}(\hat{s}_i, s_{-i})|r = Heads),$$

where the inequality is by the definition of $\hat{\xi}_i$.

On the other hand, $u_i(M_{BB}(s_i, s_{-i})|r = Tails) = -(\sum_{j: j \text{ announces NO in } (s_i, s_{-i})} \pi^i_j - \delta_i) \leq \delta_i$ and $u_i(M_{BB}(\hat{s}_i, s_{-i})|r = Tails) = -(\sum_{j: j \text{ announces NO in } (\hat{s}_i, s_{-i})} \hat{\pi}^i_j - \hat{\delta}_i)$. For any player $j$, such that $\hat{\alpha}^i_j \neq \emptyset$, because $t \in \mathcal{B}_i$, by the construction of $\hat{\omega}^i$, we have $\hat{\pi}^i_j \leq t_j(\hat{\alpha}^i_j)$. Because $\hat{s}_j \in DU_j(t_j, S)$, by Lemma 3, $j$ announces YES in Step 3 under $(\hat{s}_i, s_{-i})$. Accordingly, $\sum_{j: j \text{ announces NO in } (\hat{s}_i, s_{-i})} \hat{\pi}^i_j = 0$ and:

$$u_i(M_{BB}(\hat{s}_i, s_{-i})|r = Tails) = \hat{\delta}_i > \delta_i = u_i(M_{BB}(s_i, s_{-i})|r = Tails),$$

where the inequality is by Equation (7). In sum, Equation (6) holds in Case 1.

Case 2. $i$ is the winner under neither $(s_i, s_{-i})$ nor $(\hat{s}_i, s_{-i})$.

Letting $w$ be the winner under both strategy profiles, we have $u_i(M_{BB}(s_i, s_{-i})|r = Heads) = u_i(M_{BB}(\hat{s}_i, s_{-i})|r = Heads) = 0$. Moreover, conditioned on $r = Tails$, by the construction of $\hat{s}_i$, player $i$ announces the same thing under $(s_i, s_{-i})$ and $(\hat{s}_i, s_{-i})$. Thus, the only difference between $i$'s allocation and price under the two strategy profiles is the rewards: one is $\delta_i$, and the other is $\hat{\delta}_i$. Accordingly, $u_i(M_{BB}(s_i, s_{-i})|r = Tails) - u_i(M_{BB}(\hat{s}_i, s_{-i})|r = Tails) = \delta_i - \hat{\delta}_i < 0$, where the inequality is by Equation (7). In sum, Equation (6) holds in Case 2.

Case 3. $i$ is the winner under $(\hat{s}_i, s_{-i})$, but not under $(s_i, s_{-i})$.

In this case, let $w$ be the winner under $(s_i, s_{-i})$ and $r$ and $\hat{r}$ be the outcomes of the coins under $(s_i, s_{-i})$ and $(\hat{s}_i, s_{-i})$, respectively. Similar to Lemma 2, we have:

$$u_i(M_{BB}(s_i, s_{-i})|r = Tails) \leq \max\{\delta_i, \theta_i(\alpha^w_i) - \pi^w_i + \frac{\epsilon}{n} + \delta_i\} \leq \theta_i(\alpha^w_i) + \delta_i,$$

$$u_i(M_{BB}(s_i, s_{-i})|r = Heads) = 0,$$

$$u_i(M_{BB}(\hat{s}_i, s_{-i})|\hat{r} = Heads) = \theta_i(\hat{\xi}_i),$$

---

[8] Without loss of generality, we can assume $\hat{s}_i \in \mathcal{CD}_i$. Otherwise, by the well-studied properties of distinguishable dominance [30], there exists $\sigma_i \in \Delta(\mathcal{CD}_i)$, such that $\hat{s}_i \prec^{\theta_i}_{DU(t)} \sigma_i$ for all $t \in \mathcal{B}_i$, and we can prove $s_i \prec^{\theta_i}_{DU(t)} \sigma_i$.

and:

$$u_i(M_{BB}(\hat{s}_i, s_{-i})|\hat{r} = Tails) = -\left(\sum_{j:j \text{ announces NO in } (\hat{s}_i, s_{-i})} \hat{\pi}_j^i - \hat{\delta}_i\right) = \hat{\delta}_i.$$

Accordingly,

$$u_i(M_{BB}(\hat{s}_i, s_{-i})) = \frac{\theta_i(\hat{\xi}_i) + \hat{\delta}_i}{2} > \frac{\theta_i(\alpha_i^w) + \delta_i}{2} \geq u_i(M_{BB}(s_i, s_{-i})),$$

and Equation (6) holds in Case 3.

Therefore, $s_i \notin \mathcal{CD}_i$, and Lemma 4 holds. □

**Proof of Theorem 2.** Given Lemmas 3 and 4, the proof of Theorem 2 is almost the same as that of Theorem 1, except that, rather than distinguishing players with $\theta_i(\alpha_i^w) \geq \pi_i^w$ or $\theta_i(\alpha_i^w) < \pi_i^w$, here, we distinguish players announcing YES or NO in Step 3. The details have been omitted. □

Note that the revenue guarantee of the mechanism holds no matter whether the players' beliefs about each other are correct or not. If a player $i$ has low values for the goods and believes the others' values to be high and if the others' true values and beliefs are all low, then player $i$ may end up being the winner and getting a negative utility. However, according to player $i$'s beliefs, his/her utility will always be positive, and it is individually rational for him to participate. This is not too dissimilar to the stock market, where not everybody makes money, but everybody believes he/she will make money when entering. Indeed, the final outcome implemented may not be an ex-post Nash equilibrium and instead is supported by the two-step elimination of dominated strategies.

Furthermore, note that the idea of asking players to report their beliefs about each other has been explored in the Nash implementation literature (see, e.g., [49,50]). However, our mechanism does not assume complete information or common beliefs. Moreover, our mechanism does not try to utilize the winner's true valuations for generating revenue: indeed, the focus here is how to generate revenue by leveraging the players' beliefs. Simply choosing at random this mechanism or the VCG mechanism (or any other mechanism for unrestricted combinatorial auctions that may achieve better revenue in some contexts), one can achieve a good approximation to the best of the two.

Finally, it suffices for the players' values to be numbers within certain precisions, say two decimal digits, so that there is a gap between any two different values. If the values are real numbers, then the rewards in our mechanisms are set to zero, and our results hold under a weaker notion of dominance: that is, the desired strategies are still at least as good as any deviation, but may not be strictly better.

### 6.2. Privacy, Complexity and Collusion in Our Mechanism

Finally, we discuss the resiliency of our mechanism with respect to privacy, complexity and collusion concerns.

#### 6.2.1. Privacy

Our main mechanism achieves our revenue benchmark by eliciting from the players much less information than they possess. In Step 1, a player does not reveal anything about his/her own valuation except a subset of goods, which is supposed to be his/her favorite subset. Nor does he/she reveal his/her full beliefs about the valuations of his/her opponents: he/she only reveals a maximum guaranteed-revenue outcome, according to his/her beliefs.

This is all of the information that is revealed if the coin flipped by the mechanism ends up as heads. If it ends up as tails, then a player $i$ reveals at most a modest amount of information about his/her own true valuation in Step 3. Namely, only if he/she is offered a subset $A$ of goods for a price $p$, he/she reveals that his/her true value for that specific subset is $\geq p$ if he/she answers YES and $<p$ otherwise. In particular, therefore, in our mechanism, a player who is not offered any goods does not

reveal any information about his/her own valuation. This is very far from what may happen in many other auction mechanisms: that is, fully revealing your valuation and receiving no goods.

Because privacy is important to many strategic agents, we hope that trying to preserve it will become a standard goal in mechanism design. Achieving this goal will require putting a greater emphasis on extensive mechanisms, where the players and the mechanism may interact over time[9]. The power of "interaction" for privacy preservation is very well documented in cryptography[10]. This power extends to mechanism design, as well: as we have seen in our case, even three sequential moves can save a considerable amount of privacy compared with the previous normal-form mechanism.

#### 6.2.2. Computation and Communication Efficiency

Our mechanism is highly efficient in both computation and communication. Essentially, it only needs to sum up the prices in each reported outcome $\omega^i$ and figure out which reported outcome has the highest revenue. Moreover, each player only reports a subset of goods and an outcome and perhaps announces YES or NO in Step 3. One might object, however, that our mechanism transfers all of the hard computation to the players themselves. This is indeed true, but our mechanism also gives them the incentives to approximate this hard computation.

As we have recalled in our Introduction, approximation (1) may be necessary to compute a reasonable outcome when finding "the best one" is computationally hard, but (2) may also distort incentives. Our mechanism instead ensures that approximation is aligned with incentives. Indeed, our mechanism entrusts the players to propose outcomes, but ensures, as per Lemma 4, that each player wishes to become the winner. Thus, our mechanism makes it in a player's own interest to use the best computationally-efficient approximation algorithm he/she knows, in order to propose a high-revenue outcome. Of course, the best algorithm known by a player may not be the best in the literature, in terms of its approximation ratio to the optimal outcome. In this case, the mechanism's revenue is at least half of the highest revenue the players are capable of computing. To our best knowledge, this is the first mechanism that gives the buyers incentives to perform computationally-efficient approximations. Incentive-compatible and computationally-efficient approximation on the seller's side has also been studied, but again for valuations of restricted forms, such as single-minded players [1], single-value players [2], auctions of multiple copies of the same good [53], etc. By contrast, we do not impose any such restrictions.

#### 6.2.3. Collusion

Collusion is traditionally prohibited (e.g., by using solution concepts that only consider individual deviations in game theory) and punished (e.g., by laws). However, it continues to exist. We thus wish to point out that our mechanism offers a reasonable form of protection against collusion. Namely, when at least some players are independent, denoting by $I$ the set of independent player, it guarantees at least half of the revenue benchmark $BB' \triangleq \max_{i \in I} BB_i$.

Thus, our mechanism is not responsible for generating any revenue if all players are collusive, but must generate revenue at least half of $BB'$ otherwise. This guarantee holds in a strong collusion model: that is, even when collusive players are capable of making side payments and coordinating their actions via secret and enforceable agreements, and the independent players have no idea that collusion is afoot. The only constraint is that every coalition is rational, that is, its members act so

---

[9] It is well known that every extensive mechanism can be transformed to an "equivalent" normal-form game, but this equivalence does not extend to privacy. Indeed, in an extensive mechanism $M$, a player $i$ reveals information only if a decision node of $i$ is reached, and in an execution of $M$, only some of these nodes are reached. Transforming $M$ into the normal form instead asks $i$ to reveal how he/she would like to act at any possible decision node involving him.

[10] Interaction is indeed at the base of zero-knowledge proofs [51,52], where a mistrusted prover can convince a skeptical verifier that a theorem statement is true without revealing any additional information.

to maximize the sum of their individual utilities. In this case, an independent player $i$, reporting in Step 1 an outcome $\omega$ offering a player $j$ a subset of the goods $X$ for a price $p$, need not worry whether $j$ is independent or collusive. If $i$ becomes the winner and the coin toss of the mechanism is tails, then $j$ will answer YES if and only if his/her individual true value for $X$ is greater than or equal to $p$. Accordingly, $i$ will report in Step 1 an outcome whose revenue is at least $BB_i$. If an independent player becomes the winner, then the mechanism will generate at least $BB'/2$ revenue. Else, some collusive player has become the winner; but then, such a player must have reported an outcome with revenue $R \geq BB'$, and the mechanism will generate at least $R/2$ revenue.

Let us point out that the $BB'$ benchmark is actually guaranteed under a weaker requirement of coalition rationality[11].

#### 6.2.4. Social Welfare

Note that each player $i$ has a "truthful" strategy: to report $\theta_i$ and the outcome $\hat{\omega}^i$ as defined in the proof of Lemma 4, whose revenue is exactly $BB_i$. Since the price suggested by $\hat{\omega}^i$ for each player $i' \neq i$ is no more than the true value of $i'$ for the suggested subset of goods for him/her, the players all say YES when $i$ is the winner, and player $i$'s utility is non-negative. Under the truthful strategy profile, the social welfare of the final outcome is at least $\frac{BB}{2}$. When the players overbid and report outcomes whose revenue is higher than the corresponding $BB_i$'s, the social welfare may be smaller than the revenue.

### 6.3. *Variants of Our Mechanism*

Our mechanism sets aside a "budget" of $\epsilon > 0$ for rewarding the players and achieves the benchmark $BB/2 - \epsilon$. We note that our analysis also holds if the mechanism chooses his/her reward budget to be not an absolute value $\epsilon$, but an $\epsilon$ fraction of the revenue it collects. In such a case, however, its guaranteed revenue will be $(1-\epsilon)BB/2$.

Furthermore, for simplicity, we have assumed that the seller/designer knows nothing about the players. However, it is easy to accommodate the case in which he/she too has some beliefs about the players' valuations. For instance, in keeping with our overall approach, let $\omega'$ be the highest revenue outcome among all of the outcomes $(A, P)$ for which he/she is sure that $\theta_i(A_i) \geq P_i$ for all $i$. Then, he/she can use $\omega'$ as a "reserve outcome" as follows. If, in Step 1, the revenue of the outcome reported by the winner is at least that of $\omega'$, then he/she keeps on running our mechanism; otherwise, roughly speaking, he/she makes himself the "winner" and continues with $\omega'$ being the outcome reported by the winner.

**Acknowledgments:** We thank David Easley, Shafi Goldwasser, Avinatan Hassidim, Robert Kleinberg, Eric Maskin, Paul Milgrom, Rafael Pass, Ron Rivest, Paul Valiant and Avi Wigderson for comments and encouragement and several anonymous reviewers for helpful suggestions about the presentation of our results. The authors were partially supported by Office of Naval Research (ONR) Grant No. N00014-09-1-0597. The first author is partially supported by NSF CAREER Award No. 1553385.

**Author Contributions:** Jing Chen and Silvio Micali contributed equally to this work. Jing Chen and Silvio Micali wrote the paper.

**Conflicts of Interest:** The authors declare no conflict of interest.

[11] That is, when the members of a coalition act so as to maximize a different function of their individual utilities. All we need is a mild "monotonicity" condition, informally described as follows. Consider a coalition C and two outcomes $\omega$ and $\omega'$, such that (1) a member $i$ of C is offered a set $A_i$ for a price $P_i$ in outcome $\omega$ and no goods for price $P_i'$ in $\omega'$; and (2) every other member $j$ of C is offered the same set of goods $A_j$ for the same price $P_j$ in both outcomes. Then, the only rationality condition that we require from C is that it prefers $\omega$ to $\omega'$ if and only if $\theta_i(A_i) - P_i \geq -P_i'$. Under this model, in Step 3 of our mechanism, each coalition can delegate the YES or NO decisions to its members as if they were independent. Thus, again, an independent player need not worry whether another player is independent or collusive.

## References


1. Lehmann, D.; O'Callaghan, L.; Shoham, Y. Truth revelation in approximately efficient combinatorial auctions. *J. ACM* **2002**, *49*, 577–602.
2. Babaioff, M.; Lavi, R.; Pavlov, E. Single-Value combinatorial auctions and implementation in undominated strategies. In Proceedings of the 17th Annual ACM-SIAM Symposium on Discrete Algorithms (SODA 2006), Miami, FL, USA, 22–26 January 2006; pp. 1054–1063.
3. Hart, S.; Nisan, N. Approximate revenue maximization with multiple items. In Proceedings of the 13th ACM Conference on Electronic Commerce (EC), Valencia, Spain, 4–8 June 2012; p. 656.
4. Vickrey, W. Counterspeculation, auctions, and competitive sealed tenders. *J. Financ.* **1961**, *16*, 8–37.
5. Clarke, E. Multipart pricing of public goods. *Public Choice* **1971**, *11*, 17–33.
6. Groves, T. Incentives in teams. *Econometrica* **1973**, *41*, 617–631.
7. Abreu, D.; Matsushima, H. Virtual Implementation in iteratively undominated actions: Complete information. *Econometrica* **1992**, *60*, 993–1008.
8. Glazer, J.; Perry, M. Virtual implementation in backwards induction. *Games Econ. Behav.* **1996**, *15*, 27–32.
9. Chen, J.; Hassidim, A.; Micali, S. Robust perfect revenue from perfectly informed players. In Proceedings of the Innovations in Theoretical Computer Science (ITCS), Beijing, China, 5–7 January 2010; pp. 94–105.
10. Chen, J.; Micali, S. Mechanism design with possibilistic beliefs. *J. Econ. Theory* **2015**, *156*, 77–102.
11. Harsanyi, J. Games with incomplete information played by "Bayesian" players, I–III. *Manag. Sci.* **1967–1968**, *14*, 159–182, 320–334, 486–502.
12. Aumann, R. Agreeing to disagree. *Ann. Stat.* **1976**, *4*, 1236–1239.
13. Myerson, R. Optimal auction design. *Math. Oper. Res.* **1981**, *6*, 58–73.
14. Ausubel, L.; Milgrom, P. The lovely but lonely Vickrey auction. In *Combinatorial Auctions*; Cramton, P., Shoham, Y., Steinberg, R., Eds.; MIT Press: Cambridge, MA, USA, 2006; pp. 17–40.
15. Aumann, R. Backward Induction and Common Knowledge of Rationality. *Games Econ. Behav.* **1995**, *8*, 6–19.
16. Milgrom, P. *Putting Auction Theory to Work*; Cambridge University Press: Cambridge, UK, 2004.
17. Cramton, P.; Shoham, Y.; Steinberg, R. (Eds.) *Combinatorial Auctions*; MIT Press: Cambridge, MA, USA, 2006.
18. Klemperer, P. *Auctions: Theory and Practice*; Princeton University Press: Princeton, NJ, USA, 2004.
19. Milgrom, P.; Ausubel, L.; Levin, J.; Segal, I. *Incentive Auction Rules Option and Discussion*; Technical Report, FCC-12-118A2, Federal Communications Commission: Washington, DC, USA, 2012.
20. Milgrom, P.; Segal, I. Deferred-Acceptance Auctions and Radio Spectrum Reallocation. In Proceedings of the 15th ACM Conference on Economics and Computation (EC), Palo Alto, CA, USA, 8–12 June 2014; pp. 185–186.
21. Kash, I.A.; Murty, R.; Parkes, D.C. Enabling Spectrum Sharing in Secondary Market Auctions. *IEEE Trans. Mob. Comput.* **2014**, *13*, 556–568.
22. Cramton, P.; Lopez, H.; Malec, D.; Sujarittanonta, P. *Design of the Reverse Auction in the Broadcast Incentive Auction*; Working Paper; University of Maryland: College Park, MD, USA, 2015.
23. Nguyen, T.D.; Sandholm, T. Multi-Option Descending Clock Auction. In Proceedings of the 15th International Conference on Autonomous Agents and Multiagent Systems (AAMAS), Singapore, 9–13 May 2016; pp. 1461–1462.
24. Milgrom, P.; Weber, R. A Theory of Auctions and Competitive Bidding, II. In *The Economic Theory of Auctions*; Klemperer, P., Ed.; Edward Elgar: Cheltenham, UK, 2000; Volume I, pp. 179–194.
25. Parkes, D. Iterative Combinatorial Auctions. In *Combinatorial Auctions*; Cramton, P., Shoham, Y., Steinberg, R., Eds.; MIT Press: Cambridge, MA, USA, 2006; Chapter 2.
26. Likhodedov, A.; Sandholm, T. Approximating Revenue-Maximizing Combinatorial Auctions. In Proceedings of the 20th National Conference on Artificial Intelligence (AAAI), Pittsburgh, PA, USA, 9–13 July 2005; pp. 267–274.
27. Balcan, M.; Blum, A.; Mansour, Y. *Single Price Mechanisms for Revenue Maximization in Unlimited Supply Combinatorial Auctions*; Technical Report, CMU-CS-07-111; Carnegie Mellon University: Pittsburgh, PA, USA, 2007.
28. Micali, S.; Valiant, P. *Resilient Mechanisms for Unrestricted Combinatorial Auctions*; Technical Report, MIT-CSAIL-TR-2008-067; Massachusetts Institute of Technology: Cambridge, MA, USA, 2008.

29. Jackson, M. Implementation in undominated strategies: A look at bounded mechanisms. *Rev. Econ. Stud.* **1992**, *59*, 757–775.
30. Chen, J.; Micali, S. The order independence of iterated dominance in extensive games. *Theor. Econ.* **2013**, *8*, 125–163.
31. Pearce, D.G. Rationalizable strategic behavior and the problem of perfection. *Econometrica* **1984**, *52*, 1029–1050.
32. Battigalli, P. On rationalizability in extensive games. *J. Econ. Theory* **1997**, *74*, 40–61.
33. Chen, J.; Micali, S.; Pass, R. Tight revenue bounds with possibilistic beliefs and level-$k$ rationality. *Econometrica* **2015**, *83*, 1619–1639.
34. Chen, J.; Micali, S.; Valiant, P. Robustly leveraging collusion in combinatorial auctions. In Proceedings of the Innovations in Theoretical Computer Science (ITCS), Beijing, China, 5–7 January 2010; pp. 81–93.
35. Izmalkov, S.; Lepinski, M.; Micali, S. Perfect Implementation. *Games Econ. Behav.* **2011**, *71*, 121–140.
36. Parkes, D.; Rabin, M.; Shieber, S.; Thorpe, C. Practical secrecy-preserving, verifiably correct and trustworthy auctions. *Electron. Commer. Res. Appl.* **2008**, *7*, 294–312.
37. Dwork, C.; Roth, A. *The Algorithmic Foundations of Differential Privacy*; Foundations and Trends in Theoretical Computer Science; NOW Publishers: Breda, The Netherlands, 2014.
38. Green, J.; Laffont, J. On coalition incentive compatibility. *Rev. Econ. Stud.* **1979**, *46*, 243–254.
39. Schummer, J. Manipulation through bribes. *J. Econ. Theory* **2000**, *91*, 180–198.
40. Chen, J.; Micali, S. Collusive dominant-strategy truthfulness. *J. Econ. Theory* **2012**, *147*, 1300–1312.
41. Moulin, H.; Shenker, S. Strategyproof sharing of submodular costs: budget balance versus efficiency. *Econ. Theory* **2001**, *18*, 511–533.
42. Jain, K.; Vazirani, V. Applications of approximation algorithms to cooperative games. In Proceedings of the 33rd ACM Symposium on Theory of Computing (STOC), Heraklion, Greece, 6–8 July 2001; pp. 364–372.
43. Feigenbaum, J.; Papadimitriou, C.; Shenker, S. Sharing the cost of multicast transmissions. *J. Comput. Syst. Sci.* **2001**, *63*, 21–41.
44. Laffont, J.; Martimort, D. Mechanism design with collusion and correlation. *Econometrica* **2000**, *68*, 309–342.
45. Goldberg, A.; Hartline, J. Collusion-resistant mechanisms for single-parameter agents. In Proceedings of the Symposium on Discrete Algorithms (SODA), Vancouver, BC, Canada, 23–25 January 2005; pp. 620–629.
46. Che, Y.; Kim, J. Robustly collusion-proof implementation. *Econometrica* **2006**, *74*, 1063–1107.
47. Osborne, M.; Rubinstein, A. *A Course in Game Theory*; MIT Press: Cambridge, MA, USA, 1994.
48. Shimoji, M.; Watson, J. Conditional dominance, rationalizability, and game forms. *J. Econ. Theory* **1998**, *83*, 161–195.
49. Maskin, E. Nash equilibrium and welfare optimality. *Rev. Econ. Stud.* **1999**, *66*, 23–38.
50. Jackson, M.; Palfrey, T.; Srivastava, S. Undominated Nash implementation in bounded mechanisms. *Games Econ. Behav.* **1994**, *6*, 474–501.
51. Goldwasser, S.; Micali, S.; Rackoff, C. The knowledge complexity of interactive proof-systems. *SIAM J. Comput.* **1989**, *18*, 186–208.
52. Goldreich, O.; Micali, S.; Wigderson, A. Proofs that yield nothing but their validity or all languages in NP have zero-knowledge proofs. *J. ACM* **1991**, *38*, 691–729.
53. Dobzinski, S.; Dughmi, S. On the power of randomization in algorithmic mechanism design. In Proceedings of the 54th Sympsium on Foundations of Computer Science (FOCS), Atlanta, GA, USA, 25–27 October 2009; pp. 505–514.

# Section 3: Philosophy

 *games* 

*Article*

# Community-Based Reasoning in Games: Salience, Rule-Following, and Counterfactuals

**Cyril Hédoin**

University of Reims Champagne-Ardenne, 51571 Reims, France; cyril.hedoin@univ-reims.fr;
Tel.: +33-326-918-720

Academic Editor: Paul Weirich
Received: 5 August 2016; Accepted: 9 November 2016; Published: 16 November 2016

**Abstract:** This paper develops a game-theoretic and epistemic account of a peculiar mode of practical reasoning that sustains focal points but also more general forms of rule-following behavior which I call community-based reasoning (CBR). It emphasizes the importance of counterfactuals in strategic interactions. In particular, the existence of rules does not reduce to observable behavioral patterns but also encompasses a range of counterfactual beliefs and behaviors. This feature was already at the core of Wittgenstein's philosophical account of rule-following. On this basis, I consider the possibility that CBR may provide a rational basis for cooperation in the prisoner's dilemma.

**Keywords:** community-based reasoning; epistemic logic; game theory; rule-following; counterfactuals

---

## 1. Introduction

The study of strategic interactions is traditionally based on the mathematical specification of a game which includes the set of players, the set of (pure) strategies, the payoff functions and possibly an information structure. The latter specifies what each player knows and believes about the characteristics of the game and also about the other players' rationality. As Thomas Schelling [1] has argued long ago, such a mathematical specification of a game however is most of the time insufficient to predict and to explain the players' actual behavior. Schelling's point is that the mathematical specification fails to acknowledge the fact that the players' practical reasoning is actually building on (aesthetic, cultural, psychological ... ) features that help them to make choices and to coordinate. He particularly emphasizes the role played by salience and focal points in the players' reasoning process.

Following a recent literature reflecting on social conventions and the role of reasoning and common belief (e.g., [2–5]), this paper develops a game-theoretic and epistemic account of a peculiar mode of practical reasoning that sustains focal points but also more general forms of rule-following behavior which I call community-based reasoning (henceforth, CBR). The CBR notion emphasizes the fact that the working of focal points and rules depends on the ability for each player to infer correctly others' behavior on the basis of a mutually believed state of affairs. Since each other player's behavior also depends in principle on his own inference, this leads to an iterative and nested chains of inferences about how each player reasons from a given state of affairs. A player's practical reasoning is community-based when it uses the belief that all the players are members of the same community as a basis for this iterative and nested chains of inferences. Formally, CBR can be modeled through an epistemic game where the players share a theory of the game they are playing. The contribution of this paper is twofold: first, I suggest that CBR provides a useful and insightful perspective to formalize Schelling's account of focal points but also Saul Kripke's [6] "collectivist solution" to Wittgenstein's so-called rule-following paradox. The latter strengthens a claim already made by Giacomo Sillari [7]. Second, I argue that a formalization of CBR brings interesting insights over the issue of epistemic and causal (in)dependence in normal form games. In particular, I consider the possibility that CBR may provide a rational basis for cooperation in the prisoner's dilemma.

The paper is organized as follows. Section 2 characterizes CBR as a mode of practical reasoning that individuals may rationally use in specific circumstances. Section 3 provides a formal account of CBR on the basis of a game-theoretic and epistemic framework. Section 4 argues that CBR offers interesting insights into characterizing the nature of rule-following behavior and more generally the working of institutions. Section 5 claims that CBR is an instance where epistemic dependence holds between the players' beliefs in spite of causal independence between their actions. On this basis, Section 6 contemplates whether CBR provides a rational justification for cooperation in the prisoner's dilemma. Section 7 briefly concludes.

## 2. CBR as a Mode of Practical Reasoning

In this paper, I will be exclusively concerned with normal form games, which correspond to interactions where the players choose simultaneously (or cannot observe each other's choice). Game theory is the tool par excellence used in virtually all the behavioral sciences to study strategic interactions, i.e., decision problems where the consequences of each agent's behavior are a function of the behavior of other agents. This tool can be used with different aims in mind, such as for instance characterizing the necessary or sufficient conditions for a specific solution concept to be implemented. Here, my purpose is different: the goal is to use a game-theoretic framework in order to reflect over the way more or less rational individuals reason in some kind of strategic interactions to achieve coordination or to cooperate. This purpose was already at the core of Thomas Schelling's [1] early use of game theory to study strategic behavior. As it is well-known, Schelling emphasizes the fact that the sole mathematical description of the game (i.e., the matrix with the utility numbers embedded in) is insufficient to explain and to predict the players' behavior. The reason is that obviously, the players' practical reasoning in a strategic interaction does not merely rely on what corresponds to the mathematical features used by the game theorist to define a game. Aesthetic, cultural, psychological and imaginative features also enter as inputs in the players' reasoning process. Moreover, the latter does not necessarily proceed along a logical and deductive path; various forms of inductive reasoning may be used to infer practical conclusions about what one should do from a set of premises corresponding to one's information. The point is thus that a proper explanation of the way people behave in many strategic interactions requires to enriching the game-theoretic models with additional features. The latter must particularly capture the players' beliefs and/or knowledge and account for the reasoning process used to ultimately make a strategic choice.

Consider Figures 1 and 2 which depict two basic games. Figure 1 depicts the familiar coordination game with two Nash equilibria in pure strategies (Right, Right) and (Left, Left), and a third one in mixed strategies where each player plays each strategy with probability one-half. Assuming that the players know each other's preferences, are rational and that these facts are commonly known,[1] it is nevertheless impossible for the game theorist to predict what each player will do as a fundamental indeterminacy remains. Moreover, this indeterminacy extends to the player themselves, at least if we take for granted that their only information corresponds to the matrix and that they use the best-reply reasoning constitutive of the Nash equilibrium solution concept.

| | | Bob | |
|---|---|---|---|
| | | Right | Left |
| Ann | Right | 1 ; 1 | 0 ; 0 |
| | Left | 0 ; 0 | 1 ; 1 |

**Figure 1.** The coordination game.

[1] I leave these notions informally stated here. See the next section for a more formal statement.

| | | Bob | |
|---|---|---|---|
| | | Heads | Tails |
| Ann | Heads | 3 ; 2 | 0 ; 0 |
| | Tails | 0 ; 0 | 2 ; 3 |

**Figure 2.** The heads-tails game.

Schelling's solution to this difficulty is now well-known to most economists and game theorists: in practice, coordination may be achieved thanks to the existence of focal points toward which the players' expectations are converging. The precise nature of focal points is mostly left undefined by Schelling but their function is pretty clear: as everyone recognizes that a particular outcome (i.e., strategy profile) is salient for all, each player expects that everyone expects all the other players to play along this outcome. Since in a coordination game the players' interests are perfectly aligned, this gives one a decisive reason to also play along this outcome. However, focal points may also foster coordination in cases where the players have partially conflicting interests, i.e., in so-called "mixed-motive games." The heads-tails game (Figure 2) is an instance of this kind of games discussed by Schelling. In spite of the symmetry of the game and the fact that the players have conflicting preferences over the Nash equilibrium to be implemented (here Ann prefers (Heads, Heads) while Bob prefers (Tails, Tails)), Schelling's informal experiments indicate that almost three-quarters of the players choose "Heads". Interestingly, the proportion is almost the same among the players in the role of Bob than among the players in the role of Ann. A plausible explanation of this fact is that "Heads" sounds as more salient than "Tails" for almost all players in such a way that everyone expects others to play "Heads." Of course, under such an expectation, it is clearly rational to play "Heads" even for Bob.

Some years later, the philosopher David Lewis [8] used a similar idea to account for the emergence and the nature of conventions. In particular, Lewis suggested that conventions originate through the "force of the precedent", i.e., a particular form of salience of the history of plays in a given strategic interaction. As for Schelling, there is no clear expression of the nature of this salience in Lewis's account. However, the most natural reading of both Schelling's and Lewis's writings is that salience and focal points have mostly psychological groundings. Plainly, salience is essentially natural as far as it derives from cognitive mechanisms which are themselves the ultimate product of our evolutionary history [9]. This reading is not undisputable however and a plausible (and complementary) alternative is that focal points depend on cultural factors.[2] More precisely, the meaning of environmental cues for a given person arguably depends on the context in which she is embedded and, for strategic interactions, on the identity of the other persons with whom she is interacting. In this sense, salience can be said to be community-based [4]: the degree of salience of a strategy, an outcome or more generally of an event[3] is a function of the community in which the interaction is taking place. Consider for instance the market panic game depicted by Figure 3 ([4,10]):

| | | Bob | |
|---|---|---|---|
| | | Sell | Do not sell |
| Ann | Sell | 5 ; 5 | 6 ; 0 |
| | Do not Sell | 0 ; 6 | 10 ; 10 |

**Figure 3.** The market panic game.

[2] Of course, as culture is itself shaped by our evolutionary history, in particular through the genetically and cognitively-based learning mechanisms that have been selected for, salience is surely ultimately natural. This does not undermine the distinction made in the text between natural and community-based forms of salience however. The contrary would imply that speaking of culture is meaningless for anyone endorsing naturalism and materialism, which is surely not the case.

[3] As it will appear in the next section, playing a given strategy or implementing a given outcome can be ultimately formalized as an event in a set-theoretic framework. The looseness of the statement in the text is thus unproblematic.

The matrix corresponds to the classical assurance game which has once again two Nash equilibria in pure strategies, i.e., (Sell, Sell) and (Do not sell, Do not sell). Suppose now that as Ann is watching her TV she notices a speech by the Chairman of the Federal Reserve Board stating that we are close to a financial meltdown. A plausible reasoning for Ann is then the following: "it is highly probable that Bob and other agents on the financial markets have also noticed the Chairman's speech; moreover, it is also highly probable that Bob and others will infer that everyone has noticed the Chairman's speech. Finally, on this basis, I think it is highly probable that Bob and others will believe that everyone will choose 'Sell'. Therefore, given my preferences, I should also choose 'Sell'". Note that there are two steps in Ann's reasoning based on her listening to the Chairman's speech. First, she infers from her listening to the Chairman's speech that others have also probably listened to the speech. Second, she infers from the first two conclusions that others will sell and therefore that she should also sell.[4] Now, acknowledging the fact that Ann may have heard of several other speeches by many different persons claiming that a financial meltdown will occur, what is that makes the Chairman's speech salient in such a way that she noticed it and made the inferences stated above?

An obvious answer to the question above is of course that the salience of the Chairman's speech comes from the privileged institutional status of the Chairman and from the fact that Ann, Bob and others are members of a community that acknowledges this status. It is important then to note that this salience based on community-membership not only explains why the Chairman's speech is noticed in the very first place but also the inferences that lead to Ann's practical conclusion. Actually, I would argue that this is the fact that Ann, Bob and others are members of the same community that allows them to make these inferences and therefore makes the Chairman's speech salient in the very first place. In other words, the Chairman's speech is salient because, contrary to most other potentially observable events, it allows people to reach a firm practical conclusion about what they should do.

Community-based salience is thus grounded on what I will call in the rest of the paper community-based reasoning. CBR is a special kind of practical reasoning which operates on the basis of inferences which are grounded on a belief that I and others are members of the same community. It can be stated in the following generic way:

***Community-Based Reasoning***—Person $P$'s practical reasoning is community-based if, in some strategic interaction $S$, her action $A$ follows from the following reasoning steps:

(a) $P$ believes that her and all other persons $P'$ in $S$ are members of some community $C$.
(b) $P$ believes that some state of affairs $x$ holds.
(c) Given (b), (a) grounds $P$'s belief that all other persons $P'$ in $S$ believe that $x$ holds.
(d) From $x$, $P$ inductively infers that some state of affairs $y$ also holds.
(e) Given (c) and (d), (a) grounds $P$'s belief that all other persons $P'$ in $S$ believe that $y$ also holds.
(f) From (e) and given $P$'s preferences, $P$ concludes that $A$ is best in $S$.

Community-membership sustains two key steps in CBR: in step (c), community-membership serves as a basis for $P$ to infer that the fact that $x$ holds is mutual belief; in step (e), it serves as a basis for $P$ for inferring that everyone infers $y$ from $x$. The term "grounds" that appears in both (c) and (e) singles out that $P$ is making an assumption about the fact that she and other members of $C$ are sharing both an epistemic accessibility to some states of affairs and some form of inductive inference.[5] Any person $P$ that endorses CBR takes this practical reasoning as valid from her point of view, i.e., not from a

[4] See [11] for an early characterization of the notion of common knowledge in these terms. See also [2–4] for conceptual and formal analysis of this kind of reasoning.

[5] This fundamental feature was already emphasized by Lewis [8] in his account of common knowledge. As he pointed out, the very possibility for a state of affairs to become common knowledge (or common belief) depends on "suitable ancillary premises regarding our rationality, inductive standards, and background information" (p. 53). Lewis' account also relies on the key notion of indication (see [2,7]) which broadly corresponds to the various forms of inductive standards that one may use. The inductive inference mentioned in step (d) can be understood in terms of Lewis' indication notion.

logical point of view but in such a way that $P$ considers that she is justified in endorsing it.[6] Clearly, this implies that community-based reasoners may reach wrong conclusions. The point of course is that nothing per se establishes that the members of the same community actually make the same kinds of inductive inferences or that the fact that $x$ holds is mutual belief. Moreover, CBR is partially self-referential in the sense that if $P$ is the only person to use CBR, she will probably reach conclusions that do not match conclusions reached by others using a different kind of practical reasoning.

Finally, if steps (a) to (e) hold for all the members of the community, it can be shown that $y$ is common belief among the members of the community. Therefore, if everyone's preferences and practical rationality are commonly known (or believed), then everyone's action $A$ will also be commonly known (or believed). The next section provides a game-theoretic and epistemic framework that establishes these results and fully characterizes the nature of CBR.

## 3. A Game-Theoretic and Epistemic Framework

This section characterizes CBR in a game-theoretic framework. To this end, I use semantic epistemic models (s.e.m.) of games. The notion of s.e.m. comes from modal and epistemic logic and is a tool used to represent all the relevant facts about how a given game is played. More specifically, an s.e.m. formalizes the players' knowledge, beliefs and how they reason on their basis. As a consequence, it seems that CBR can be captured within an s.e.m., as I show below.

As usual, a game $G$ is defined as a triple $< N, \{S_i, u_i\}_{i \in N} >$ where $N$ is the set of $n \geq 2$ players, $S_i$ is the set of pure strategies for player $i = 1, \ldots, n$ and $u_i$ is $i$'s utility function representing $i$'s preferences over the possible outcomes. The set of possible outcomes simply correspond to the set of strategy profiles $S = \Pi_{i \in N} S_i$; therefore, for all $i$ we have $u_i$: $S \rightarrow \Re$. In the following, we only need to assume that the players' utility functions are ordinal, i.e., they are unique up to any increasing transformation. Cardinal utility functions would be required if we used probabilistic belief operators as it is common in the literature but entering into these complications is not required here. Players are assumed to be rational in the sense that they play their best-response given their belief by choosing the strategy leading to their most preferred outcomes. Defined as such, a game is only a partial formalization of a strategic interaction as it does not state how the game is actually or would be played. This is done by adding to $G$ an s.e.m. $I$: $< W, w^*, \{C_i, B_i\}_{i \in N} >$ where $W$ is the (finite) set of states of the world (or possible worlds) $w$ with $w^*$ the actual state. A state of the world is a complete specification of everything that is relevant from the point of view of the modeler. It can be seen as a list of true propositions about the strategies chosen by the players, what they know and believe, and so on.[7] Therefore, the actual state $w^*$ specifies how the game is actually played while all the other states $w'$ indicate how the game would have been played or could have been played. $C_i$: $W \rightarrow S_i$ is player $i$'s decision function; it indicates what strategy $i$ is playing in each possible world. Finally, $B_i$ is a binary relation called an accessibility relation which specifies for any given state $w$ which are the states $w'$ that player $i$ considers as possible. Therefore, $wB_iw'$ means that at $w$ player $i$ considers $w'$ to be possible. I denote $B_i(w) = \{w' \in W$: $wB_iw'\}$ the set of states $w'$ that are accessible from $w$.[8] The tuple $I$: $<W, w^*, \{C_i, B_i\}_{i \in N} >$ corresponds to what can be called the (semantic) model of game $G$.

The binary relations $B_i$ may satisfy different sets of properties which will determine their substantive meaning. A standard approach in economics consists in using the knowledge-belief semantic structures pioneered by Robert Aumann ([13–15]). This approach relies however on strong

---

6 See Sugden [12] for a similar account of the validity of a scheme of practical reasoning. Validity is not to be understood as an assertion made by the game theorist but rather as the consequence of the fact that a player is actually endorsing it. What I am tacitly assuming is that one cannot endorse community-based reasoning if he regards it as invalid.

7 In logic, it is usual to write explicitly as a counterpart to the semantic models a syntax, i.e., a list of propositions that are derived from a set of axioms on the basis of an alphabet. The semantics is derived from the syntax by building a state space on the basis of a truth value function that indicates for each state whether a given proposition is true or false.

8 Equivalently, one may also specify a possibility operator $B_i$: $W \rightarrow 2^W$ mapping any state $w$ into a set of states $w'$. Then $B_i(w) = B_i(w)$.

assumptions about the players' epistemic abilities as it assumes that the players have introspective access to their knowledge.[9] Moreover, the representation of beliefs requires the definition of a probability measure over the state space that has been regarded by some authors as problematic [19]. A less demanding approach is to define the accessibility relations in purely doxastic terms, i.e., as representing the players' beliefs. This is obtained by assuming that the accessibility relations are serial, transitive, and Euclidean [20]:

*Seriality*: for all $w \in W$, $\exists w'$: $wB_iw'$.
*Transitivity*: for all $w, w', w'' \in W$, if $wB_iw'$ and $w'B_iw''$, then $wB_iw''$.
*Euclideaness*: for all $w, w', w'' \in W$, if $wB_iw'$ and $wB_iw''$, then $w'B_iw''$.

Seriality defines a consistency requirement over the players' beliefs as it guarantees that in each state, the players consider at least one state as possible. Transitivity and Euclideaness guarantee that the players have introspective access to their beliefs, i.e., when they believe something, they believe this and when they do not believe something, they also believe this.[10] On this basis, we can define two important notions of doxastic necessity and doxastic possibility [21]. Consider any proposition $p$: according to player $i$, $p$ is doxastically necessary at state $w$ if and only if $p$ is true in all $w' \in B_i(w)$. Correspondingly, $p$ is doxastically possible at state $w$ if and only if $p$ is true in at least one $w' \in B_i(w)$. I denote $[p]$ the set of states $w$ in which $p$ is true and call $[p]$ an *event*. An event is thus any subset of $W$ and therefore $2^W$ is the set of events. Now, we can define a set of non-probabilistic operators $B_i$ with respect to any event $[p]$ in the following way:

$$B_i[p] = \{w \mid B_i(w) \subseteq [p]\}. \tag{1}$$

Expression (1) indicates that player $i$ believes event $[p]$ at $w$ (i.e., believes that proposition $p$ is true) if and only $p$ is doxastically necessary at $w$. Note that this definition implies

$$B_i[\neg p] = \{w \mid B_i(w) \cap [p] = \emptyset\}. \tag{2}$$

Expression (2) states that $i$ believes the event $[\neg p]$ at $w$ (i.e., believes that proposition $p$ is false) if and only if $p$ is doxastically impossible at $w$. It should be noted that both $B_i[p]$ and $B_i[\neg p]$ are themselves well-defined events as they correspond to set of states where the proposition "$i$ believes that $p$ is true (false)" is true.[11]

Throughout the paper, I make the natural assumption that the players have a doxastic access to what they are doing, i.e., they have the (right) belief that they play any strategy $s_i$ that they are actually playing: i.e., for all $w' \in B_i(w)$, $C_i(w) = C_i(w')$. If we denote $[s_i]$ the event that $i$ plays strategy $s_i$, then this corresponds to the following condition:

$$\text{For all } w \in W \text{ and all players } i,\ B_i[s_i]. \tag{3}$$

A last piece in our basic framework is needed. We can derive a communal accessibility relation $B^*$ defined as the transitive closure of the set of individual accessibility relations $\{B_i\}_{i \in N}$. Therefore, we

[9] Bacharach [16] provides a useful discussion of the so-called Aumann's structures. For a discussion of the problems related to the formalization of knowledge in logic and game theory, see e.g., [17,18].

[10] Accessibility relations with these properties correspond to modal operators satisfying the axioms of the KD45 system of modal logic in the underlying syntax. It is generally regarded as the most relevant way to account for beliefs in the perspective of doxastic logic [18].

[11] The intermediary case is when $B_i(w)$ and $[p]$ intersect: $i$ does not believe $[p]$ is actually the case, but he does not believe that $[p]$ is impossible either. This is clearly unproblematic: I may perfectly not believe that France will necessarily win the Euro championship of football without believing that France cannot win the competition. In other words, this corresponds to cases where while not believing something as necessary, I also believe that I may be wrong and that this something is indeed true.

have $wB^*w'$ if and only if we can design a finite sequence of worlds $w_1, \ldots, w_m$ with $w_1 = w$ and $w_m = w'$ such that for each $k$ from 1 to $m - 1$, we have at least one player $j$ for which $w_k B_j w_{k+1}$. Correspondingly, we have $B^*(w) = \{w': wB^*w'\}$. The common belief operator $B^*$ is then defined in the standard way:

$$B^*[p] = \{w \mid B^*(w) \subseteq [p]\}. \tag{4}$$

Expression (4) means that the event $[p]$ is common belief among the $n$ players: each player believes that each player believes that ... $p$ is true. In the following, I will assume that all the features of any game $G$ are common belief among the players, which implies that the discussion is restricted to games of complete information.

The resulting model $I$: $< W, w^*, \{C_i, B_i\}_{i \in N} >$ provides a complete description of how the game $G$ is played and what the player believes at each possible world. Therefore, it is not only an account of what actually happens but also of what could have happened. As I explain below, this makes s.e.m. a particularly useful tool to discuss the role of counterfactuals in the players' reasoning process. It is worth noting that $I$ is the theorist's model of the game, not necessarily the players' own model which, in this sense, would have to be "commonly known" as pointed out by Aumann and Brandenburger [22] (p. 132). Arguably, this can be regarded as problematic in light of Schelling's claim emphasized above that we should not conflate the mathematical description of the game with the way the players are actually framing the strategic interaction. The point is however that the model of the game is a tool to represent the players' reasoning process through which they reach the practical conclusion regarding their strategy choice. What this representation implies about how individuals are "really" reasoning depends on one's philosophical commitments over the nature of the relationship between a model and the real world but also over the nature of the intentional states (beliefs, preferences) that are accounted for in the game's model. This issue is well beyond the scope of the paper. However, it is important to acknowledge that in the current framework, we will be able to account for the players' reasoning process through several elements. First, the assumption that the players are rational at all states $w$, i.e., they choose their best strategy given their preferences and their beliefs. Second, the characterization of the players' (common) beliefs at each state $w$. Third, the players' actual and counterfactual choices at each state $w$. As I argue in the next section, we can account for CBR in such a framework. I also argue that such an account helps to foster a better understanding of the nature of salience and more generally of the phenomenon of rule-following behavior.

## 4. CBR and Rule-Following Behavior

On the basis of the above framework, I will develop in this section an account of community-based salience and rule-following behavior. In particular, I shall suggest that the salience of an event may arise from the fact that the members of a population are following a rule. In turn, this rule-following behavior is grounded on CBR.

First, I add to the framework of the previous section a rationality assumption according to which a player $i$ is rational at $\omega$ if and only if:

(R) For every $w' \in B_i(w)$, there is no strategy $s_i' \neq C_i(\omega)$ such that$u_i(s_i', C_{-i}(w')) \geq u_i(C_i(w), C_{-i}(w'))$, with $C_{-i}(w) = s_{-i} = (s_1, \ldots, s_{i-1}, s_{i+1}, \ldots, s_n)$.

Expression (R) is a statement about the players' epistemic rationality as it indicates that player $i$ chooses her best strategy given her belief about what others are doing. Another way to state this characterization of rationality is that a rational player never plays a strategy that he believes is strictly dominated by another one. As I will point out below, (R) does not imply that a rational player actually makes her best choice as her belief about others' choices may be mistaken. In particular, no restriction is placed on the fact that a player may believe that others' choices are somehow (causally) dependent on her choice. On this basis, we denote $[r_i]$ the event that $i$ is rational, i.e., $[r_i] = \{w \mid \text{(R) is true}\}$.

Before being able to characterize CBR in model semantic terms, we need to introduce an additional proposition according to which all the players in the game are members of some community. As pointed out in Section 2, this feature is an essential component of CBR as it grounds two key inferences in steps (c) and (e). We simply denote $c$ the proposition that "everyone in population $N$ is a member of the same community $C$" and $[c]$ the corresponding event. Steps (a)–(e) of CBR can now be semantically expressed by the following conditions:

(CBR1) For all $i \in N$, $B_i[e] \cap B_i[c] \subseteq B_i[B_{-i}[e]]$.

(CBR2) For all $i, j \in N$, $B_i[B_{-i}[e]] \cap B_i[c] \subseteq B_i[r_{-i}] \cap B_i[B_j[s_{-j}]] \subseteq B_i[s_{-i}]$ with $[s_{-i}]$ the event that strategy profile $s_{-i} = (s_1, \ldots, s_{i-1}, s_{i+1}, \ldots, s_n)$ is played.

(CBR3) For all $i \in N$, $B_i[s_{-i}] \cap [r_i] \subseteq [s_i]$.

(CBR4) For all $i, j \in N$ and any event $[f]$, $(\{B_i[e] \cap B_i[c]\} \subseteq B_i[f]) \subseteq B_i[B_j[e] \subseteq B_j[f]]$.

Consider each condition in turn. (CBR1) corresponds to steps (a)–(c) of CBR: $i$'s beliefs that event $[e]$ holds and that everyone is a member of the same community (event $[c]$) allow her to infer that everyone else believes that $[e]$ holds. (CBR2) is a complex condition that captures steps (d) and (e). It is easier to understand if we divide it into two parts:

(CBR2.1) $B_i[B_{-i}[e]] \cap B_i[c] \subseteq B_i[B_j[s_{-j}]]$.

(CBR2.2) $B_i[r_{-i}] \cap B_iB_j[s_{-j}] \subseteq B_i[s_{-i}]$.

The first inclusion relation corresponding to (CBR2.1) captures the fact that community-membership grounds $i$'s inference from her belief that others believe that event $[e]$ holds to the event that each $j$ has specific belief about what everyone else will do. The second inclusion relation corresponding to (CBR2.2) indicates that, in the context of strategic interactions, player $i$ needs to believe that the other players are rational to infer what they will be doing given that they believe that $[e]$ holds. In other words, this is the combination of the beliefs in community-membership and rationality that ground the inference from event $B_{-i}[e]$ to event $[s_{-i}]$ for any player $i$. It is important because it makes explicit the fact that CBR must rely on a belief that others are rational in some well-defined sense. Indeed, without rationality, CBR is only an instance of theoretical (or doxastic) reasoning, i.e., a reasoning scheme that operates only at the level of beliefs. A rationality principle is needed to make this reasoning scheme a practical one, i.e., one also concerned with action. Moreover, we need to add the fourth condition (CBR4) that was implicit in the scheme of CBR stated in Section 2: each player must be implicitly assuming that others are also community-based reasoners. Condition (CBR4) corresponds to the assumption of symmetric reasoning that underlies Lewis' account of conventions ([2,23,24]). It expresses the idea that each player believes that others are community-based reasoners and that on this basis they make the same (inductive) inferences from event $[e]$.[12] The combination (CBR1)–(CBR3) clearly entails $B_i[e] \cap B_i[c] \subseteq [s_i]$ and the addition of (CBR4) implies that the event $[s]$ that the strategy profile $s = (s_1, \ldots, s_n)$ is played is common belief:

(CB) $B_N[e] \cap B_N[c] \subseteq B^*[s]$ with $[s]$ the event that strategy profile $s = (s_1, \ldots, s_n)$ is played and $B_N = \cap_i B_i$ the mutual belief operator.[13]

---

[12] As a referee as pointed out, the very use of *s.e.m.* to account for salience and rule-following behavior does not allow to formalize inductive inferences as the relationships between all propositions (events) are logical ones. As I briefly discuss below, this depends on how the inclusion relation is interpreted. Semantically, the latter defines a relation between the truth-value of two propositions that has the status of a logical implication *from the modeler's point of view*. However, I contend that the semantics does not set constraints regarding the kind of inferences that the *players* are making to derive the truth-value of a proposition from the truth-value of another. That is, the inclusion relation may perfectly reflects the fact that the players are making logical/deductive inferences or inductive inferences. A way to avoid any ambiguity would be to make the underlying syntax explicit and to distinguish between the standard material implication and a Lewis-like indication relation. See for instance [25,26].

[13] The proof is relatively straightforward. Here is a sketch: consider any player $i \neq j$. According to (CBR1), we have $B_i[e] \cap B_i[c] \subseteq B_i[B_j[e]]$ and therefore, combined with (CBR2), we have $B_i[B_j[e]] \cap B_i[c] \subseteq B_i[r_j] \cap B_i[B_j[s_{-j}]] \subseteq B_i[s_j]$ for any $j$. Applying this to each $j$ and using (CBR3), we obtain $B_i[B_{-i}[e]] \cap B_i[c] \subseteq B_i[s]$. As this is true for all $i$ and slightly abusing the notation, we can write $B_N[B_N[e]] \cap B_N[c] \subseteq B_N[s]$ with $B_N$ the mutual belief operator. Combining this latter result with (CBR4) gives $B_N[B_N[B_N[e]] \subseteq B_N[B_N[s] = [f]]$. Use this result in (CBR4) to obtain $(B_N[B_N[B_N[e]] \cap B_N[c] \subseteq B_N[f]) \subseteq B_N[B_N[B_N[B_N[e]]] \subseteq B_N[B_N[f]]$ and replicate the process for any number $k$ of steps to obtain $B^*[s]$.

Expression (CB) states that under the required conditions, the event that the strategy profile $s$ played is common belief in the population. CBR, as formalized through conditions (CBR1)–(CBR4), thus indicates a specific way through which a given proposition (event) can become commonly believed. It is also worth to emphasize two additional points. First, as CBR includes condition (R), it may seem that it leads to (Nash) equilibrium play. This is not true in general as a further condition regarding the causal independence of strategy choices (and the players' beliefs in this independence) is required. Without such requirement, CBR may justify cooperation in the prisoner's dilemma, as I discuss in Section 6. Second, given the definition of the common belief operator $B^*$ according to Expression (4), common belief is here characterized in fixed point terms rather than in iterative terms. This does not make any difference at the substantive level though.[14]

Consider again the market panic game depicted by Figure 3 above played by $n$ players on the basis of the framework of the preceding section, and take a first-person point of view (Ann's point of view). To replicate Ann's reasoning process stated in Section 2, we need to characterize a small set of events. Denote $a$ the proposition that "the Chairman's speech predicts a financial meltdown" and $[a]$ the corresponding event. Denote $(s_{-Ann} = \text{Sell})$ and $(s_{Ann} = \text{Sell})$ the events that everyone except Ann plays "Sell" and that Ann plays "Sell" respectively. Then, what happens in the market panic game following the Chairman's speech from Ann's point of view can be represented through an s.e.m. with the following relationships between these different events.

$$B_{Ann}[a] \subseteq B_{Ann}[B_{-Ann}[a]]. \tag{5a}$$

$$B_{Ann}[B_{-Ann}/[a]] \subseteq B_{Ann}[s_{-Ann} = \text{Sell}]. \tag{5b}$$

$$B_{Ann}[s_{-Ann} = \text{Sell}] \subseteq [s_{Ann} = \text{Sell}]. \tag{5c}$$

Expression (5a) indicates that the fact that Ann believes that the Chairman has delivered a speech implies that she believes that everyone else also listened to the Chairman's speech. Expression (5b) states a second implication: Ann's belief that everyone else also listened to the Chairman's speech entails that she believes that everyone else will decide to sell. Finally, Expression (5c) indicates that this latter belief implies that Ann will also sell. However, stated in this way, Expressions (5a)–(5c) are misguided. Indeed, they seem to imply that the relationships between these different events are purely logical and therefore that Ann's practical reasoning is a deductive one. Of course this is wrong: there is nothing logical in the fact that the Chairman's speech leads everyone except Ann to sell their assets and therefore the relationship between Ann's beliefs indicated by (5b) cannot be grounded on a logical reasoning.[15] The same is obviously true for (5a) and (5c): this is not because of logic that Ann can infer from her listening to the Chairman's speech that others have also listened nor that she should sell from the fact that others are selling. The point is that Ann's practical reasoning relies on other premises that do not make the relationships between the different events logical ones. In Section 2, I suggested that Ann's practical reasoning is community-based. Taking this claim for granted, we can expand the s.e.m. to account for this fact:

$$B_{Ann}[a] \cap B_{Ann}[c] \subseteq B_{Ann}[B_{-Ann}[a]]. \tag{6a}$$

$$B_{Ann}[B_{-Ann}[a]] \cap B_{Ann}[c] \subseteq B_{Ann}[r_{-Ann}] \cap B_{Ann}[B_{-Ann}[s_{-i} = \text{Sell}]] \subseteq B_{Ann}\ [s_{-Ann} = \text{Sell}] \text{ for all } i \neq Ann. \tag{6b}$$

$$B_{Ann}\ [s_{-Ann} = \text{Sell}] \cap [r_{Ann}] \subseteq [s_{Ann} = \text{Sell}]. \tag{6c}$$

Combined with (CBR4), (6a)–(6c) entail the event $[s = \text{Sell}]$ is common belief, i.e., $B_{Ann}[B^*[s]]$. Now, we can see that what makes the event $[a]$ salient for Ann in the very first place in such a

---

[14] See [3] for a comparison of the fixed-point and iterative definitions of the common belief notion.

[15] See [2,4,8] for similar claims that the relationship between the events $[a]$ and $[s_{Ann} = \text{Sell}]$ cannot be reduced to a mere logical implication.

framework is the fact that it is the starting point of the whole reasoning scheme. This is captured by the following principle:

***Community-Based Salience***—The set $\phi_i$ of subjectively community-based salient events in the model $I$ of a game $G$ for a player $i$ corresponds to all events $[e]$ that implies an event $B_i[B^*[s]]$ through conditions (CBR1)–(CBR4) . The set $\phi$ of objectively community-based salient events in the model $I$ of a game $G$ is the set of events that are community-based salient for all $i$. Formally:

- Subjectively salient events: $\phi_i = \{[e]\colon B_i[e] \cap B_i[c] \subseteq B_i[B^*[s]]$ through (CBR1)–(CBR4)$\}$.
- Objectively salient events: $\phi = \cap_i \phi_i$.

Consider now any event $[e] \subseteq \phi$. By assumption, (CB) obtains and therefore $[e] \subseteq B^*[s]$ with $[s]$ the event that strategy profile $s = (s_1, \ldots, s_n)$ is played. This is a pretty interesting result: assuming that the players are indeed all community-based reasoners and that as a consequence they share some inductive inferences mode, an objectively salient event will generate a commonly believed strategy profile in the population. Note that this result does not depend on an assumption of common belief in rationality (only mutual belief in rationality is needed) but that community-based reasoning is common knowledge in an "informal" sense.[16]

To close this section, I now want to suggest that the scope of CBR actually goes beyond the phenomenon of salience but also captures an aspect of rule-following behavior that has been emphasized by some readers of the late writings of Ludwig Wittgenstein, especially his masterpiece Philosophical Investigations ([26,27]). As Wittgenstein's writings make it clear, there is no easy way to account for the nature of rule-following behavior: there is simply no answer to questions like 'what is it to follow a rule?' or 'how can I be sure that I follow this rule and not another one?'. The main difficulty lies in the kind of indeterminacy that no inductive reasoning can resolve: whatever the number of times I or others follow some specific pattern of behavior, there is no way to decide which the underlying rule that sustains the behavior is. Saul Kripke's [6] famous "quaddition" example provides a great illustration: the rule of quaddition functions like the rule of addition for any two numbers that do not sum up to more than 57, and otherwise gives the result of 5. Now, no matter how many times I have added two numbers, as long as they do not have summed up to more than 57 I can never be sure that the rule I was following was the rule of addition rather than the rule of quaddition. Wittgenstein's more general point is that (behavioral) patterns do not have any intrinsic meaning. In terms of practical reasoning, whatever others' past behavior or the various other features that may indicate to me what I should do, there is no way to overcome this kind of indeterminacy regarding the underlying rule that should be followed.

There are of course many numbers of conflicting interpretations of Wittgenstein's account. I believe however that there are at least two key ideas that are related to the notion of CBR in this game-theoretic framework. The first idea is the importance of counterfactual reasoning and makes a direct link with the next section. The second idea is the importance of the community in fostering the normative force of rules in a given population.[17] Consider the latter first. Acknowledging that rule-following behavior always takes place inside some specific community whose very identity corresponds to the rules that its members are following, it is not farfetched to claim that rule-following is essentially community-based in the sense developed above. In some way, this is nothing but a generalization of Lewis's account of the role of salience in the working of conventions to all kinds of institutional objects: social and moral norms, legal rules, customs, and so on. I would therefore argue for the following characterization of rule-following in an s.e.m.:

---

[16] This is due to the fact that condition (CBR4) is part of the s.e.m. used. See Section 3 above for a discussion of this point.

[17] The first idea is essential in David Bloor's [28] dispositional reading of Wittgenstein's account, though it is also essential in Kripke's skeptical paradox. The second idea is clearly suggested by Wittgenstein but has been essentially popularized by Kripke. It is discussed by Sillari [7] who relates it to Lewis's account of conventions and common belief.

***Rule-following behavior***—The players in a game $G$ are following a rule according to some s.e.m. $I$ if and only if, for the behavioral pattern defined by the strategy profile $s = (s_1, \ldots, s_n)$ corresponding to the event $[s]$, there is at least one objectively salient event $[e]$ such that $w^* \in [e] \cap B_N[c] \subseteq B^*[s]$.

Three points are worth emphasizing regarding this characterization. First, the very salience of the event $[e]$ is constitutively (rather than causally) explained by the fact that the member of the community are actually following a rule. On this account, the fact that an event is salient is an indication that some rule holds in the population. This is interesting because most discussions of salience in philosophy and in economics have tended to make salience either the cause of the emergence of some rules (e.g., Schelling and Lewis, at least under some interpretation) or simply the product of some relatively "blind" evolutionary process (e.g., [29]). The latter approach indeed provides an interesting account of the origins of salience in a context of cultural evolution. Game-theoretic models of cultural evolution have been argued however to rely on excessively simplistic and abstract assumptions regarding the underlying learning mechanisms [30–32]. Moreover, even if we grant their relevance, such evolutionary explanations of salience do not account for the reasoning process that individuals may use in cases that require an explicit deliberation. Second, the nature of the inclusion relation in the statement $B_N[e] \cap B_N[c] \subseteq B^*[s]$ remains largely undefined. In some way, this reflects the difficulty to give a full characterization of the nature of rule-following behavior. My account points out however that this inclusion relation depends on CBR and on the fact that this practical reasoning is shared in the population, i.e., the assumption of symmetric reasoning expressed by (CBR4). Much work remains to be done to understand the latent cognitive functions and processes that underlie CBR and the more general abilities of humans to account for others' behavior. On this issue, I conjecture that interesting contributions can be made by combining the resources of game theory with the recent advancements in the so-called "Theory of Mind" literature [33]. In particular, some scholars have argued that the reasoning process of rational individuals in a strategic interaction relies on a simulation mechanism where one assumes that the others are reasoning like her [34,35]. This could indeed provide a justification for assuming that the players are symmetric reasoners as stated by (CBR4).[18] The third and last point concerns the importance of counterfactual reasoning: to follow a rule does not reduce to the actual and observable pattern of behavior. A rule also specifies what would have happened in other circumstances than those that have actually happened. A major interest of s.e.m. is that they are perfectly fitted to deal with this important feature: indeed, in the model of a game, with the exception of the actual state $w^*$, all states describe what would have happened in other circumstances. Of course, to meaningfully state that a rule is followed, we have to be in an actual world which indeed belongs to the salient event leading to condition (CB). Hence, we should impose $w^* \in [e] \subseteq \phi$ as a condition for an s.e.m. to formalize rule-following behavior. However, for all other worlds $w \in [e]$, the model indeed expresses aspects of practical reasoning that do not materialize at the behavioral level. This distinguishes my approach with respect to some recent game-theoretic accounts of institutions in social ontology and economics [36,37]. This also naturally leads to consider a related issue that arises from my account of CBR: the kind of epistemic and causal dependence that is implied.

## 5. CBR and Counterfactuals in Games

The last section has pointed out that a key feature of CBR that sustains salience and rule-following behavior is the condition that players are symmetric reasoners with respect to some salient event $[e]$ (condition (CBR4)). This can be seen as a requirement that the players have a common understanding of the situation they are embedded in and which shares some similarities with Wittgenstein's notion of lebensform (forms of life) [7,26]. The precise nature of this condition has been left undefined as we do not have a full understanding of the cognitive mechanisms through which individuals are able to

[18] As noted by Guala [34], such kind of simulation mechanism was already suggested by Lewis [8] (p. 27.).

replicate others' reasoning. As I suggested above, studies belonging to the Theory of Mind literature may bring insights on this issue in a near future. However, from a game-theoretic and philosophical perspective, the symmetric reasoning condition has interesting connections with two deeply related issues: the role of counterfactuals in strategic interactions and the distinction between causal and epistemic dependence in games.[19] Indeed, I shall argue in this section that the symmetric reasoning condition may be interpreted in different ways. Either it refers to a mere epistemic dependence between the players' beliefs and practical reasoning but with a causal independence, or it underlies some kind of causal dependence, at least from the point of view of the players themselves. Which of these two interpretations is the most convincing matters in prisoner's dilemma type of interactions as the latter may make cooperation rational.

Consider again the first person view of any player, say Ann, in any game $G$ like the market panic game of Figure 3. The reasoning scheme through which Ann reaches the conclusion that it is common belief that everyone will sell depends on (i) Ann's rationality; (ii) Ann's belief in others' rationality; (iii) Ann's belief that everyone is a member of some community and (iv) Ann's belief in the symmetric reasoning that the latter entails. It is not contentious that the latter two beliefs entail, from Ann's point of view, a form of epistemic dependence: Ann's believes that her and others' beliefs are somehow correlated. This is the essence of condition (CBR4). If we generalize to all players, then there is a mutual belief in the population that beliefs are correlated. For each player $i$, believing something implies that others have the same belief; moreover, for any two other players $j$ and $k$, correlation also holds as $j$'s and $k$'s beliefs are both correlated to $i$'s. Ann's rationality is expressed by (R) and the fact that $[r_{Ann}] = W$ in the corresponding s.e.m. and Ann's belief in other rationality is the event $B_{Ann}[r_{-Ann}] = W$. As this is also the case for all other players, there is indeed mutual belief in rationality in the population.[20] This assumption sets however very few constraints on the players' practical reasoning. What it says is that each player maximizes her utility (i.e., chooses the strategy that leads to her most preferred outcome) given her beliefs. Nothing is said however regarding the content of these beliefs and the way they are derived. The latter indeed depends on the properties of CBR. At this point, it might be argued that condition (R) should be strengthened in such a way that the formation of beliefs satisfies a condition of causal independency. The intuition is the following: in normal form games, players choose without knowing what others are doing. More specifically, a foundational assumption is that the players' choices are causally independent: Ann's choice cannot have any causal influence on Bob's choice as well as the converse, as both are choosing independently. While we cannot exclude the possibility that Ann's belief about what Bob is doing can be correlated to Bob's corresponding belief (or someone else's belief for that matter), this cannot be due to a causal relationship. This leads to the notion of causal rationality developed by causal decision theorists [40].

Dealing with this issue necessitates to consider the role played by counterfactuals in CBR. Counterfactuals are generally defined as false subjunctive conditionals. Subjunctive conditionals are conditionals of the form "If I were to do $x$, then $y$ would result". Counterfactuals are then of the form "If I had done $x$, then $y$ would have resulted". There are several ways to capture counterfactuals in an s.e.m.[21] The simplest is to add to the structure $< W, w^*, \{C_i, B_i\}_{i \in N} >$ a selection function $f$: $W \times 2^W \rightarrow 2^W$ [43]. A selection function maps each pair of state $w$ and event $[e]$ into a subset $f(w, [e]) \in [e]$. The latter corresponds to the state $w'$ that is the closest to $w$, where closeness is understood in the standard sense of Stalnaker-Lewis theory of counterfactuals [38]: we assume that all states in $W$ can be ordered with respect to any state $w$ by a binary relation $\preccurlyeq_w$ which is complete, transitive, asymmetric and, centered.[22] Then, $x \preccurlyeq_w y$ reads as the statement that $x$ is closer to $w$ than $y$. This

[19] On these issues, see for instance [17,38,39].

[20] Actually, in the s.e.m. that is sketched here, there is even common belief in rationality as all players are rational at all states $w$. Note however that this assumption is not needed. See footnote 10 above where the mutual belief in rationality is only needed once to derive the result.

[21] Counterfactuals have been dealt with in various ways in the game-theoretic literature. See [41] for a formalization of "hypothetical knowledge" in partition structures. See [42] for an analysis of "conditional beliefs" in doxastic models.

[22] $\preccurlyeq_w$ is asymmetric if $x \preccurlyeq_w y$ and $y \preccurlyeq_w x$ imply $x = y$. It is centered if, for all $x \in W$, $w \preccurlyeq_w x$.

provides a straightforward way to state the truth value of counterfactuals in an s.e.m. Take any two propositions $p$ and $q$ and denote the counterfactual "had $p$ been the case, then $q$ would have been the case" by $p \rightrightarrows q$. The latter is true at $w$ if and only if $q$ is true at the closest world with respect to $w$ where $p$ is true. Denote $min_w[p] = \{w' \mid w' \preccurlyeq_w w''$ for all $w', w'' \in W\}$ the set of the $w$-closest worlds where $p$ is true, i.e., the subset of $[p]$ that is the closest to $w$.[23] The selection function $f(w, [e])$ is thus simply defined in terms of the closeness relation:

$$f(w, [e]) = min_w[e]. \tag{7}$$

Now, the event $[p \rightrightarrows q]$ obviously holds at $w$ if and only if $min_w[p] \in [q]$. Correspondingly, the counterfactual event $[e \rightrightarrows f]$ can be characterized through the selection function: $[e \rightrightarrows f] = \{w \mid f(w, [e]) \in [f]\}$. In this perspective, two further restrictions can be naturally imposed on the selection function $f$ [44]:

$$\text{If } w \in [e], \text{ then } f(w, [e]) = \{w\}. \tag{8a}$$

$$\text{If } f(w, [e]) \in [f] \text{ and } f(w, [f]) \in [e], \text{ then } f(w, [e]) = f(w, [f]). \tag{8b}$$

Condition (8a) simply states that the closest (and indeed identical) world to $w$ is $w$ itself. This is a direct implication of the assumption that $\preccurlyeq_w$ is centred. (8b) says that if the $w$-closest state in $[e]$ is in $[f]$ and the $w$-closest state in $[f]$ is in $[e]$, then the two states must coincide. In the context of causal decision theory, $f(w, [e])$ is thus the state that would be causally true (in an objective sense) at state $w$ if $[e]$ were the case. The combination of the selection function $f$ with the accessibility relations $B_i$ allows to define the set of states that player $i$ believes *could be* causally true at $w$. It corresponds to the union $\cup_{w' \in Bi(w)} f(w', [e])$. Therefore, the belief that the counterfactual $e \rightrightarrows f$ is true corresponds to the event

$$B_i\,[e \rightrightarrows f] = \left\{w \middle| \cup_{w' \in Bi(w)} f\,(w', \,[e]) \subseteq [f]\right\} \tag{9}$$

In a game-theoretic context, the set of states that $i$ believes could be causally true at $w$ is determined by the following counterfactual reasoning: "If I were to play $s_i'$ instead of $s_i$, then I believe that others would play some strategy profile $s_{-i}''$". Then, we have $[e] = [s_i']$ and therefore $\cup_{w' \in Bi(w)} f(w, [s_i'])$, i.e., the set of states that could causally happen according to $i$ if he were to play $s_{-i}$. It is now possible to provide a formal condition of causal independence imposed to the decision functions $\{C_i\}_{i \in n}$:

(CI) For every strategy $s_i$ and for all players $i$, if $w' \in f(w, [s_i])$, then $C_{-i}(w) = C_{-i}(w')$.

Condition (CI) states that the players' decision functions are such that each player's strategy choice is causally independent from other players' choices. On this basis, we can strengthen the rationality condition (R) by requiring that the players maximize given condition (CI):

(CR) Player $i$ is causally rational at $w$ if, for every $w' \in B_i(w)$, there is no strategy $s_i' \neq C_i(\omega)$ such that $u_i(s_i', C_{-i}(f(w', [s_i']))) \geq u_i(C_i(w), C_{-i}(w'))$, and $C_{-i}(f(w', [s_i'])) = C_{-i}(w')$ for all $s_i'$.

Condition (CR) states that a player $i$ is causally rational at $w$ if and only if she plays her best response given the fact that others' choices are causally independent to her choice, i.e., if $i$ were to play any other strategy $s_i'$, others would still play $s_{-i}$. This definition emphasizes an important point: the fact that the model satisfies (CI) is not sufficient to guarantee that the players are causally rational at $w$, even if they play their best response given their beliefs. Indeed, the players themselves must believe that (CI) holds at $w$. If we denote $[ci]$ the event that (CI) holds, then this condition is straightforward to define [43]:

[23] Depending on the specific variant of the Stalnaker-Lewis theory of counterfactual, $min_w[p]$ may be assumed to be a singleton, i.e., there is always one and only one world $w'$ which is the closest to $w$. Without loss of generality, I will assume that this is the case here.

(BCI) Player $i$ believes that players' choices are causally independent at $w$ if and only if:
$B_i(w) \subseteq [ci]$.
This implies that for every strategy $s_i$ and every $w' \in B_i(w)$, if
$w'' \in f(w', [s_i])$, then $C_{-i}(w'') = C_{-i}(w')$.

I have assumed so far that the rationality condition (R) is constitutive of CBR. Moreover, in the context of normal form games, the causal independence condition (CI) should probably be regarded as an undisputable structural constraint as there is absolutely no reason to assume that it could not hold. The central issue concerns the status of condition (BCI) with respect to CBR and thus whether or not CBR implies the stronger condition of causal rationality (CR). The next section will discuss this issue in the context of the famous prisoner's dilemma game.

## 6. CBR in the Prisoner's Dilemma

The prisoner's dilemma (see Figure 4a below) is by far the most studied game by game theorists in all disciplinary contexts: economics, biology, philosophy ... The reason is that it points out the contradiction between what can be called "collective rationality" and "individual rationality". On the former, the players should cooperate as mutual cooperation leads to the Pareto-optimal outcome that (by assumption) maximizes social welfare. On the latter however, mutual defection is unavoidable as defection is the players' dominant strategy. Crucially, in the prisoner's dilemma, it is not even required that the players have a mutual belief in other's rationality because whatever the other is doing or is expecting one to do, playing the dominant strategy appears to be the sole rational choice.

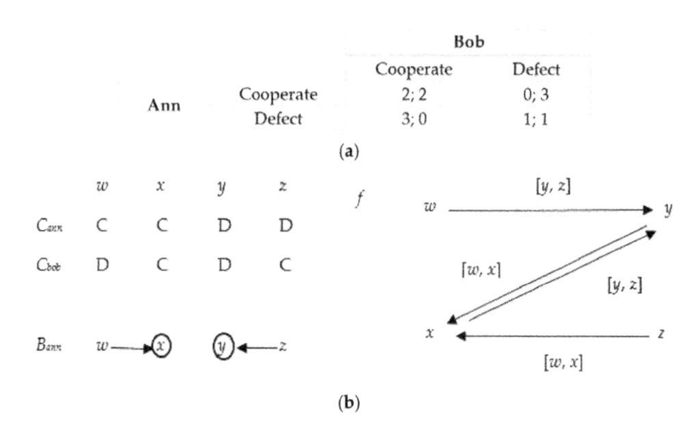


**Figure 4.** (**a**) The prisoner's dilemma; (**b**) A partial epistemic model.

There have been many attempts (mostly by philosophers) to show that cooperation in the prisoner's dilemma may be rational.[24] Economists and game theorists have generally dismissed them as being based on a misunderstanding of both of the rationality principle and of the general purpose of game theory (e.g., [45]). However, once we take counterfactual reasoning into account, things are not so straightforward. Consider for instance the case where condition (R) holds but not (BCI). As a consequence, the players are not causally rational: they maximize their utility but we do not put any constraint on the way they form their beliefs. Suppose for instance that Ann holds the beliefs formalized by the following partial model of the prisoner's dilemma [43] (see Figure 4b for a partial graphical representation): $W = \{w, x, y, z\}$, $C_{Ann}(w) = C_{Ann}(x) = C$, $C_{Ann}(y) = C_{Ann}(z) = D$; $C_{Bob}(x)$

[24] It may be worth insisting that we are only concerned here with the one-shot prisoner's dilemma. Of course, (conditional) cooperation is perfectly rational in the infinitely repeated prisoner's dilemma.

$= C_{Bob}(z) = C$, $C_{Bob}(w) = C_{Bob}(y) = D$; $B_{Ann}(w) = B_{Ann}(x) = \{x\}$, $B_{Ann}(y) = B_{Ann}(z) = \{y\}$; $f(w, [w]) = f(w, [W]) = \{w\}$, $f(x, [x]) = f(x, [W]) = \{x\}$, $f(y, [y]) = f(y, [W]) = \{y\}$, $f(z, [z]) = f(z, [W]) = \{z\}$, $f(w, [y, z]) = f(x, [y, z]) = \{y\}$, $f(y, [w, x]) = f(z, [w, x]) = \{x\}$. Suppose that the "true" state is $w$. In such a model, Ann believes that her and Bob's choices are correlated since she believes that Bob is playing C ($B_{Ann}(w) = \{x\}$) but that if she were to play D (event $\{y, z\}$), then Bob would play D ($f(x, [y, z]) = \{y\}$). As $u_{Ann}(C, C) > u_{Ann}(D, D)$, Ann is indeed rational at $w$ by choosing to cooperate. Clearly however, Ann's beliefs do not satisfy (BCI) since at $w$, we have $B_{Ann}(w) = \{x\}$ and $f(x, [y, z]) = \{y\}$, but $C_{Bob}(x) \neq C_{Bob}(y)$. In other words, at $w$ Ann believes that the counterfactual event $[s_{Ann} = D \rightrightarrows s_{Bob} = D]$ holds while at the same time she believes that Bob will actually cooperate. The problem is thus whether such beliefs can be justified under some kind of practical reasoning. More specifically, does CBR license the conjunction of two beliefs of the kind $B_i[s_{-i}]$ but $B_i[s_i' \rightrightarrows s_{-i}']$ for any $s_i' \neq C_i(w)$?

At first sight, the assumption of symmetric reasoning seems to make beliefs in the correlation of strategy choices plausible: if my (community-based) reasoning leads me to the conclusion that I should cooperate because I believe that if I cooperate you will also cooperate but if I defect you will defect, then the fact that you and I are symmetric reasoners leads me to believe that you have the same belief. However, this is superficial because symmetric reasoning only entails epistemic dependence of the players' beliefs: if I believe that event $[e]$ entails event $[f]$ and that $[e]$ is mutual belief, then I believe that $[f]$ is also mutual belief. Now, suppose that in the prisoner's dilemma the players mutually believe some event $[e]$ (for instance the instructions of the experimenter in an experimentally designed game) indicating that all should cooperate. On this basis, I may perfectly believe that, on the basis of some kind of reasoning that may be community-based, others will reach the conclusion that we all choose to cooperate. Actually, they are wrong because as far as I am concerned and because I am rational, I will choose to defect! If I believe that others are rational, then it will also occur to me that they cannot cooperate and therefore the only commonly believed outcome compatible with the mutual belief in rationality is mutual defection. The point is that community-based reasoning (and thus the underlying symmetric reasoning assumption) does not work in isolation from the rationality principle, as conditions (CBR2.2) and (CBR3) make it clear. Indeed, I have no reason to cooperate unless I have an *independent* belief that causal independence does not hold.

This is not to say that CBR entails condition (BCI) and thus causal rationality. CBR is simply silent regarding the combination of the selection function $f$ and the accessibility relations $B_i$. The point is rather that CBR provides no support against (BCI) and that, as far as condition (CI) seems unobjectionable, there is no reason to reject the use of causal rationality per se. In this case, cooperation in the prisoner's dilemma is not supported by CBR alone. A possibility however is discussed by Bicchieri and Green [21] and is based on the distinction between causal independence and causal necessity. While the players may be causally rational in the sense of (CR), they may hold the belief that it is causally necessary that they play the same strategy because of some "identicality" assumption. Bicchieri and Green add to the s.e.m. a "nomic accessibility" binary relation $C$ where $wCw'$ reads as "$w'$ is causally possible relative to $w$". Then, an event $[e]$ is causally necessary at state $w$ if and only if $\{w': wCw'\} \subseteq [e]$. Denote [C] and [D] the event that everyone cooperates and everyone defects respectively and suppose that all the players in the prisoner's dilemma take the following nomic relation as true: for all $w, w' \in W$, $\{w': wCw'\} \subseteq [C] \cup [D]$. In words, the players consider as a causal necessity that they play identically. Then obviously, the sole fact of causal rationality entails that the players will cooperate. Clearly, it is controversial that the belief in this kind of causal necessity can be defended on the basis of any scheme of practical reasoning. I do not think that community-membership, however we expand and refine CBR, can foster and justify such a belief. I must thus conclude that CBR is unhelpful to commend cooperation as the rational choice in the prisoner's dilemma.[25]

[25] Another possibility, formally isomorphic to Bicchieri and Green's approach is worth mentioning: we may substitute a "deontic accessibility relation" $D$ for Bicchieri and Green's nomic relation by interpreting $wDw'$ as "$w'$ is a deontic possibility relatively to $w$". Now suppose that for any world $w \in [C]$ it is assumed that $\{w': wDw'\} \subseteq [C]$, i.e., if everyone cooperates,

## 7. Conclusions

This paper has presented CBR as a specific scheme of practical reasoning in strategic interactions. Community-based reasoners use the fact that they are the members of the same community as an epistemic resource to generate common belief about everyone's behavior. My main claim in this paper has been twofold: first, CBR plausibly grounds focal points and salience phenomena and more generally may underlie most rule-following behaviors. This first point may be worth testing experimentally. Second, I have argued that CBR emphasizes the importance of counterfactuals in strategic interactions. In particular, the existence of rules does not reduce to observable behavioral patterns but also encompasses a range of counterfactual beliefs and behaviors. On this basis, I have explored the possibility that CBR might rationalize cooperation in the prisoner's dilemma but in the end I remain highly skeptical.

**Conflicts of Interest:** The author declare no conflict of interest.

## References


1. Schelling, T.C. *The Strategy of Conflict*; Harvard University Press: Cambridge, MA, USA, 1981.
2. Cubitt, R.P.; Sugden, R. Common Knowledge, Salience and Convention: A Reconstruction of David Lewis' Game Theory. *Econ. Philos.* **2003**, *19*, 175–210. [CrossRef]
3. Paternotte, C. Being realistic about common knowledge: A Lewisian approach. *Synthese* **2011**, *183*, 249–276. [CrossRef]
4. Sillari, G. A Logical Framework for Convention. *Synthese* **2005**, *147*, 379–400. [CrossRef]
5. Sillari, G. Common Knowledge and Convention. *Topoi* **2008**, *27*, 29–39. [CrossRef]
6. Kripke, S.A. *Wittgenstein on Rules and Private Language: An Elementary Exposition*; Harvard University Press: Cambridge, MA, USA, 1982.
7. Sillari, G. Rule-following as coordination: A game-theoretic approach. *Synthese* **2013**, *190*, 871–890. [CrossRef]
8. Lewis, D.K. *Convention: A Philosophical Study*; John Wiley and Sons: Hoboken, NJ, USA, 2002.
9. Hédoin, C. A Framework for Community-Based Salience: Common Knowledge, Common Understanding and Community Membership. *Econ. Philo.* **2014**, *30*, 365–395. [CrossRef]
10. Levine, D. Neuroeconomics? *Int. Rev. Econ.* **2011**, *58*, 287–305. [CrossRef]
11. Milgrom, P. An Axiomatic Characterization of Common Knowledge. *Econometrica* **1981**, *49*, 219–222. [CrossRef]
12. Sugden, R. The Logic of Team Reasoning. *Philos. Explor.* **2003**, *6*, 165–181. [CrossRef]
13. Aumann, R.J. Agreeing to Disagree. *Ann. Stat.* **1976**, *4*, 1236–1239. [CrossRef]
14. Aumann, R.J. Correlated Equilibrium as an Expression of Bayesian Rationality. *Econometrica* **1987**, *55*, 1–18. [CrossRef]
15. Aumann, R.J. Interactive epistemology I: Knowledge. *Int. J. Game Theory* **1999**, *28*, 263–300. [CrossRef]
16. Bacharach, M. When do we have information partition? In *Mathematical Models in Economics*; University of Oxford: Oxford, UK, 1993; pp. 1–23.
17. Stalnaker, R. Knowledge, Belief and Counterfactual Reasoning in Games. *Econ. Philos.* **1996**, *12*, 133–163. [CrossRef]
18. Stalnaker, R. On Logics of Knowledge and Belief. *Philos. Stud.* **2006**, *128*, 169–199. [CrossRef]
19. Gul, F. A Comment on Aumann's Bayesian View. *Econometrica* **1998**, *66*, 923–927. [CrossRef]
20. Bonanno, G.; Nehring, K. Assessing the truth axiom under incomplete information. *Math. Soc. Sci.* **1998**, *36*, 3–29. [CrossRef]
21. Bicchieri, C.; Green, M.S. Symmetry arguments for cooperation in the prisoner's dilemma. In *The Logic of Strategy*; Kluwer Academic Publishe/Oxford University Press: New York, NY, USA, 1997; pp. 229–249.


---

it is a deontic necessity to cooperate. Such a deontic constraint would entail mutual cooperation in the case each player believes that everyone else is cooperating. A deontic extension of CBR could account for this possibility.

22. Aumann, R.; Brandenburger, A. Epistemic Conditions for Nash Equilibrium. *Econometrica* **1995**, *63*, 1161–1180. [CrossRef]
23. Gintis, H. *The Bounds of Reason: Game Theory and the Unification of the Behavioral Sciences*; Princeton University Press: Princeton, NJ, USA, 2009.
24. Vanderschraaf, P.; Sillari, G. Common Knowledge. Available online: http://plato.stanford.edu/entries/common-knowledge/ (accessed on 11 November 2016).
25. Cubitt, R.P.; Sugden, R. Common reasoning in games: A Lewisian Analysis of common knowledge of rationality. *Econ. Philos.* **2014**, *30*, 285–329. [CrossRef]
26. Hédoin, C. Institutions, Rule-Following and Game Theory. *Econ. Philos.* **2016**. [CrossRef]
27. Wittgenstein, L. *Philosophical Investigations*; John Wiley & Sons: Hoboken, NJ, USA, 2010.
28. Bloor, D. *Wittgenstein, Rules and Institutions*; Routledge: Oxford, UK, 1997.
29. Skyrms, B. *Evolution of the Social Contract*; Cambridge University Press: Cambridge, UK, 1996.
30. Sugden, R. The evolutionary turn in game theory. *J. Econ. Methodol.* **2002**, *8*, 113–130. [CrossRef]
31. Grüne-Yanoff, T. Evolutionary game theory, interpersonal comparisons and natural selection: A dilemma. *Biol. Philos.* **2011**, *26*, 637–654. [CrossRef]
32. Guala, F. The Philosophy of Social Science: Metaphysical and Empirical. *Philos. Compass* **2007**, *2*, 954–980. [CrossRef]
33. Hédoin, C.; Larrouy, L. Game Theory, Institutions and the Schelling-Bacharach Principle: Toward an Empirical Social Ontology. Available online: http://www.gredeg.cnrs.fr/working-papers/GREDEG-WP-2016-21.pdf (accessed on 11 November 2016).
34. Guala, F. *Understanding Institutions: The Science and Philosophy of Living Together*; Princeton University Press: Princeton, NJ, USA, 2016.
35. Morton, A. Game Theory and Knowledge by Simulation. *Ratio* **1994**, *7*, 14–25. [CrossRef]
36. Hindriks, F.; Guala, F. Institutions, rules, and equilibria: A unified theory. *J. Inst. Econ.* **2015**, *11*, 459–480. [CrossRef]
37. Smit, J.P.; Buekens, F.; du Plessis, S. What Is Money? An Alternative to Searle's Institutional Facts. *Econ. Philos.* **2011**, *27*, 1–22. [CrossRef]
38. Board, O. The Equivalence of Bayes and Causal Rationality in Games. *Theory Decis.* **2006**, *61*, 1–19. [CrossRef]
39. Stalnaker, R. Belief revision in games: Forward and backward induction1. *Math. Soc. Sci.* **1998**, *36*, 31–56. [CrossRef]
40. Weirich, P. Causal Decision Theory. Available online: http://plato.stanford.edu/entries/decision-causal/ (accessed on 11 November 2016).
41. Samet, D. Hypothetical Knowledge and Games with Perfect Information. *Games Econ. Behav.* **1996**, *17*, 230–251. [CrossRef]
42. Battigalli, P.; Siniscalchi, M. Hierarchies of Conditional Beliefs and Interactive Epistemology in Dynamic Games. *J. Econ. Theory* **1999**, *88*, 188–230. [CrossRef]
43. Bonanno, G. Counterfactuals and the Prisoner's Dilemma. In *The Prisoner's Dilemma*; Cambridge University Press: Cambridge, UK, 2015; pp. 133–155.
44. Bonanno, G. Reasoning About Strategies and Rational Play in Dynamic Games. In *Models of Strategic Reasoning*; van Benthem, J., Ghosh, S., Verbrugge, R., Eds.; Springer: Berlin/Heidelberg, Germany, 2015; pp. 34–62.
45. Binmore, K.G. *Game Theory and the Social Contract: Playing Fair*; MIT Press: Cambridge, MA, USA, 1994.

MDPI

*Article*

# Probabilistic Unawareness

**Mikaël Cozic**

Université Paris-Est Créteil (EA LIS), Institut Universitaire de France & IHPST (UMR 8590); 61, avenue du général de Gaulle, 94000 Créteil, France; mikael.cozic@u-pec.fr; Tel.: +33-145-171-141

Academic Editor: Paul Weirich
Received: 9 September 2016; Accepted: 10 November 2016; Published: 30 November 2016

**Abstract:** The modeling of awareness and unawareness is a significant topic in the doxastic logic literature, where it is usually tackled in terms of full belief operators. The present paper aims at a treatment in terms of partial belief operators. It draws upon the modal probabilistic logic that was introduced by Aumann (1999) at the semantic level, and then axiomatized by Heifetz and Mongin (2001). The paper embodies in this framework those properties of unawareness that have been highlighted in the seminal paper by Modica and Rustichini (1999). Their paper deals with full belief, but we argue that the properties in question also apply to partial belief. Our main result is a (soundness and) completeness theorem that reunites the two strands—modal and probabilistic—of doxastic logic.

**Keywords:** unawareness; epistemic logic; probabilistic logic

**JEL Classification:** C70; C72; D80

---

## 1. Introduction

Full (or categorical) beliefs are doxastic attitudes, like those ascribed when one says:

Pierre believes that $\phi$.

Modal logic provides a way of modeling full beliefs. It is well known that it suffers from two main cognitive idealizations. The first one is logical omniscience: a family of properties such as the closure of beliefs under logical consequence (from the premise that $\phi$ implies $\psi$, infer that $B\phi$ implies $B\psi$, also known as the rule of monotonicity) or substitutability of logically equivalent formulas (from the premise that $\phi$ is equivalent to $\psi$, infer that $B\phi$ is equivalent to $B\psi$, also known as the rule of equivalence). The second cognitive idealization is full awareness, which is more difficult to characterize precisely. As a first approximation, let us say that, according to this assumption, the agent is supposed to have a full understanding of the underlying space of possibilities and of the propositions that can be built upon them.

Logicians and computer scientists have devoted much attention to the weakening of logical omniscience. This literature is surveyed in [1], and in particular the two main extant solutions: structures with subjective, logically-impossible states[1] and awareness© structures introduced by R. Fagin and J. Halpern [4][2]. The very same awareness© structures are used to weaken the full awareness assumption. More recently, game theorists have become interested in weakening awareness in epistemic and doxastic logic and in related formalisms ([5–10])[3]. For a recent and detailed survey of models of unawareness, see [14].

---

[1] See [2,3].
[2] In the whole paper we use "awareness©" to denote the model of [4], to be distinguished from the attitude of awareness.
[3] There is also a literature that studies games with unaware players. See, for instance, [11–13].

Doxastic logic is a rather coarse-grained model of doxastic attitudes, because it excludes partial beliefs, i.e., the fact that an agent believes that it is unlikely that $\phi$ or very likely that $\psi$. The main formalism for partial beliefs makes use of probabilities in their subjective or epistemic interpretation, where probability values stand for degrees of belief. There is a noteworthy contrast between modal doxastic logic and those probabilistic models: whereas the former make beliefs explicit (part of the formal language), they are left implicit in the latter. However, one may enrich the syntax with explicit partial belief operators. For instance, R. Aumann has introduced in [15] an operator $L_\alpha\phi$ interpretable as

the agent believes at least to degree $\alpha$ that $\phi$

A possible-world semantics is given for these operators (which is inspired by [16])[4]. This semantics has been axiomatized by [19] under the form of a weak (soundness and) completeness theorem. This probabilistic logic is the true counterpart of Kripkean epistemic logic for degrees of beliefs, and it is the framework of this paper.

This probabilistic logic suffers from the same cognitive idealizations as doxastic logic: logical omniscience and full awareness. In a preceding paper [20], we dealt with the problem of logical omniscience in probabilistic logic. Our proposal was mainly based on the use of so-called impossible states, i.e., subjective states where the logical connectives can have a non-classical behavior. The aim of the present paper is to enrich probabilistic logic with a modal logic of unawareness. Our main proposal is a generalization of Aumman's semantics that uses impossible states like those of [5] and provably satisfies a list of intuitive requirements. Our main result is a weak completeness theorem like the one demonstrated by [19], but adapted to the richer framework that includes awareness. To our knowledge, [21] is the closest work to ours: in this paper, the state-space model with "interactive unawareness" introduced in [8] is extended to probabilistic beliefs in order to deal with the issue of speculative trade. One of the differences with our paper is that their framework is purely set-theoretical, whereas we rely on a formal language.

The remainder of the paper proceeds as follows. In Section 2, we try to provide some intuitions about the target attitudes, awareness and unawareness. Section 3 presents briefly probabilistic logic, and notably the axiom system of [19] (that will be called 'system $HM$'). In Section 4, we vindicate a slightly modified version of the Generalized Standard Structures of [5]. Section 5 contains the main contribution of the paper: a logic for dealing with unawareness in probabilistic logic[5]. Our axiom system (named 'system $HM_U$') enriches the probabilistic logic with an awareness operator and accompanying axioms. Section 6 concludes.

## 2. Awareness and Unawareness

### 2.1. Some Intuitions

Unawareness is a more elusive concept than logical omniscience. This section gives insights on the target phenomena and puts forward properties that a satisfactory logic of (un)awareness should embody. Following the lead of [5], we may say that there is unawareness when

- there is "ignorance about the state space"
- "some of the facts that determine which state of nature occurs are not present in the subject's mind"
- "the agent does not know, does not know that she does not know, does not know that she does not know that she does not know, and so on..."

---

[4] This is not the only way to proceed. Fagin, Halpern and Moses introduced in [17] an operator $w(\phi)$ and formulas $a_1 w(\phi_1) + ... + a_n w(\phi_n) \geq c$ interpretable as "the sum of $a_1$ times the degree of belief in $\phi_1$ and...and $a_n$ times the degree of belief in $\phi_n$ is at least as great as $c$". For a recent survey of probabilistic logic, see [18].

[5] B. Schipper in [22] pointed out that a similar result has been stated in an unpublished paper by T. Sadzik; see [23]. The framework is slightly different from ours.

Here is an illustrative example. Pierre plans to rent a house for the next holiday, and from the observer's point of view, there are three main factors relevant to his choice:

- $p$: the house is no more than 1 km far from the sea
- $q$: the house is no more than 1 km far from a bar
- $r$: the house is no more than 1 km far from an airport

There is an intuitive distinction between the two following doxastic states:

State (i): Pierre is undecided about $r$'s truth: he neither believes that $r$, nor believes that $\neg r$; there are both $r$-states and $\neg r$-states that are epistemically accessible to him.

State (ii): the possibility that $r$ does not come up to Pierre's mind. Pierre does not ask himself: 'is there an airport no more than 1 km far from the house?".

The contrast between the two epistemic states can be rendered in terms of a state space with either a fine or coarse grain[6]. The observer's set of possible states is:
where each state is labeled by the sequence of literals that are true in it. This state is also Pierre's in doxastic State (i). The doxastic State (ii), on the other hand, is:
Some states in the initial state space have been fused with each other; those that differ only in the truth value they assign to the formula the agent is unaware of, namely $r$.[7]

### 2.2. Some Principles in Epistemic Logic

More theoretically, what properties should one expect awareness to satisfy? In what follows:

- $B\phi$ means "the agent believes that $\phi$,
- $A\phi$ means "the agent is aware that $\phi$".

Here is a list of plausible properties for the operators $B$ and $A$:

| | |
|---|---|
| $A\phi \leftrightarrow A\neg\phi$ | (symmetry) |
| $A(\phi \wedge \psi) \leftrightarrow A\phi \wedge A\psi$ | (distributivity over $\wedge$) |
| $A\phi \leftrightarrow AA\phi$ | (self-reflection) |
| $\neg A\phi \rightarrow \neg A \neg A\phi$ | (U-introspection) |
| $\neg A\phi \rightarrow \neg B\phi \wedge \neg B \neg B\phi$ | (plausibility) |
| $\neg A\phi \rightarrow (\neg B)^n \phi \; \forall n \in \mathbb{N}$ | (strong plausibility) |
| $\neg B \neg A\phi$ | (BU-introspection) |

Natural as they are, these properties cannot be jointly satisfied in Kripkean doxastic logic. This has been recognized by [6], who show that it is impossible to have both:

(i) a non-trivial awareness operator that satisfies plausibility, U-introspection and BU-introspection and
(ii) a belief operator that satisfies either necessitation or the rule of monotonicity[8].

---

6 This is close to the "small world" concept of [24]. In Savage's language, "world" means the state space or set of possible worlds, itself.

7 Once again, the idea is already present in [24]: "...a smaller world is derived from a larger by neglecting some distinctions between states". The idea of capturing unawareness with the help of coarse-grained or subjective state spaces is widely shared in the literature; see, for instance, [8] or [9]. By contrast, in the framework of first-order epistemic logic, unawareness is construed as unawareness of some objects in the domain of interpretation by [25]. This approach is compared with those based on subjective state spaces in [26].

8 For a recent elaboration on the results of [6], see [27].

Of course, the standard belief operator of epistemic logic does satisfy both necessitation and the rule of monotonicity. The main challenge is therefore to build a logic of belief and awareness that supports the above intuitive principles. Since necessitation and the rule of monotonicity are nothing but forms of logical omniscience, it becomes a major prerequisite to weaken the latter. Indeed, both the generalized standard structures of [5] and the awareness© structures of [4] do weaken logical omniscience.

### 2.3. Some Principles in Probabilistic Logic

Probabilistic logics are both lesser known than and not so well unified as modal doxastic logics. The syntactic framework, in particular, varies from one to another. The logic on which this paper is based relies on a language that is quite similar to that of doxastic logic, and it therefore can be seen as constituting a probabilistic modal logic: its primary doxastic operators are $L_a$, where $a$ is a rational number between zero and one ("the agent believes at least to degree $a$ that...")[9] We can express the relevant intuitive principles for $L_a$ as:

| | |
|---|---|
| $A\phi \leftrightarrow A\neg\phi$ | (symmetry) |
| $A(\phi \wedge \psi) \leftrightarrow A\phi \wedge A\psi$ | (distributivity over $\wedge$) |
| $A\phi \leftrightarrow AA\phi$ | (self-reflection) |
| $\neg A\phi \rightarrow \neg A\neg A\phi$ | (U-introspection) |
| $\neg A\phi \rightarrow \neg L_a\phi \wedge \neg L_a\neg L_a\phi$ | (plausibility) |
| $\neg A\phi \rightarrow (\neg L_a)^n\phi \ \forall n \in \mathbb{N}$ | (strong plausibility) |
| $\neg L_a\neg A\phi$ | ($L_a$U-introspection) |
| $L_0\phi \leftrightarrow A\phi$ | (minimality) |

Seven of these eight principles are direct counterparts of those put forward for modal doxastic logic, minimality being the exception. On the one hand, if an agent believes to some degree (however small) that $\phi$, then he or she is aware of $\phi$. This conditional is intuitive for a judgmental rather than for a purely behavioral conception of partial beliefs, according to which degrees of beliefs are causal determinants of behavior, which may or may not be consciously grasped by the agent[10]. The reverse conditional roughly means that an agent aware of $\phi$ has some degree of belief toward $\phi$. This directly echoes Bayesian epistemology. These eight principles may be seen as a set of requirements for a satisfactory probabilistic logic[11].

## 3. Probabilistic (Modal) Logic

This section briefly reviews the main concepts of probabilistic logic following [15,19][12]. Probabilistic logic is of course related to the familiar models of beliefs where the doxastic states are represented by a probability distribution on a state space (or on the formulas of a propositional language), but the doxastic operator is made explicit here. Syntactically, as we have already said, this means that the language is endowed with a family of operators $L_a$. Semantically, there are sets of possible states (or "events") corresponding to the fact that an agent believes (or does not believe) such and such formula at least to such and such degree.

### 3.1. Language

**Definition 1** (probabilistic language). *The set of formulas of a probabilistic language $\mathcal{L}^L(At)$ based on a set $At$ of propositional variables is defined by:*

$$\phi ::= p|\bot|\top|\neg\phi|\phi \wedge \phi|L_a\phi$$

[9] By contrast, in [28,29], one considers formulas like $a_1w(\phi_1) + ... + a_nw(\phi_n) \geq b$, where $a_1, ..., a_n, b$ are integers, $\phi_1, ...., \phi_n$ are propositional formulas and $w(\phi)$ is to be interpreted as the probability of $\phi$.

[10] For a philosophical elaboration on this distinction, see [30].

[11] A similar list of properties is proven in [21].

[12] Economists are leading contributors to the study of explicit probabilistic structures because they correspond to the so-called type spaces, which are basic to games of incomplete information since [16]. See [31].

*where $p \in At$ and $a \in [0,1] \cap \mathbb{Q}$.*

From this, one may define two derived belief operators:

- $M_a\phi = L_{1-a}\neg\phi$ (the agent believes at most to degree $a$ that $\phi$)
- $E_a\phi = M_a\phi \wedge L_a\phi$ (the agent believes exactly to degree $a$ that $\phi$)[13]

### 3.2. Semantics

Probabilistic structures (PS), as introduced by [15], aim at interpreting the formal language we just defined. They are the true probabilistic counterpart of Kripke structures for epistemic logic. In particular, iterated beliefs are allowed because a probability distribution is attributed to each possible state, very much like a Kripkean accessibility relation. We follow the definition of [19]:

**Definition 2** (probabilistic structures). *A probabilistic structure for $\mathcal{L}^L(At)$ is a four-tuple $\mathcal{M} = (S, \Sigma, \pi, P)$ where:*

*(i)* *$S$ is a state space*
*(ii)* *$\Sigma$ is a $\sigma$-field of subsets of $S$*
*(iii)* *$\pi : S \times At \rightarrow \{0,1\}$ is a valuation for $S$ s.t. $\pi(.,p)$ is measurable for every $p \in At$*
*(iv)* *$P : S \rightarrow \Delta(S,\Sigma)$ is a measurable mapping from $S$ to the set of probability measures on $\Sigma$ endowed with the $\sigma$-field generated by the sets*

$$\{\mu \in \Delta(S,\Sigma) : \mu(E) \geq a\} \; \forall E \in \Sigma, a \in [0,1].$$

**Definition 3.** *The satisfaction relation, labeled $\models$, extends $\pi$ to every formula of the language according to the following conditions:*

*(i)* $\mathcal{M},s \models p$ *iff* $\pi(p,s) = 1$
*(ii)* $\mathcal{M},s \models \phi \wedge \psi$ *iff* $\mathcal{M},s \models \phi$ *and* $\mathcal{M},s \models \psi$
*(iii)* $\mathcal{M},s \models \neg\phi$ *iff* $\mathcal{M},s \not\models \phi$
*(iv)* $\mathcal{M},s \models L_a\phi$ *iff* $P(s)([[\phi]]) \geq a$

As usual, $[[\phi]]$ denotes the set of states where $\phi$ is true, or the proposition expressed by $\phi$. From a logical point of view, one of the most striking features of probabilistic structures is that compactness does not hold. Let $\Gamma = \{L_{1/2-1/n}\phi : n \geq 2, n \in \mathbb{N}\}$ and $\psi = \neg L_{1/2}\phi$. For each finite $\Gamma' \subset \Gamma$, $\Gamma' \cup \{\psi\}$ is satisfiable, but $\Gamma \cup \{\psi\}$ is not. As a consequence, an axiomatization of probabilistic structures will provide at best a weak completeness theorem.

### 3.3. Axiomatization

Explicit probabilistic structures were not axiomatized in [15]. To deal with this issue, an axiom system was proposed in [19] that is (weakly) complete for these structures. We coin it the system $HM$[14].

[13] Since $a$ is a rational number and the structures will typically include real-valued probability distributions, it may happen in some state that for no $a$, it is true that $E_a\phi$. It happens when the probability assigned to $\phi$ is a real, but non-rational number.
[14] The work in [19] calls this system $\Sigma+$.

| System $HM$ |
|---|
| (PROP) Instances of propositional tautologies<br>(MP) From $\phi$ and $\phi \rightarrow \psi$, infer $\psi$ |
| (L1) $L_0\phi$<br>(L2) $L_a\top$<br>(L3) $L_a\phi \rightarrow \neg L_b\neg\phi$ $(a+b>1)$<br>(L4) $\neg L_a\phi \rightarrow M_a\phi$<br>(DefM) $M_a\phi \leftrightarrow L_{1-a}\neg\phi$<br>(RE) From $\phi \leftrightarrow \psi$ infer $L_a\phi \leftrightarrow L_a\psi$ |
| (B) From $((\phi_1, ..., \phi_m) \leftrightarrow (\psi_1, ..., \psi_n))$ infer<br>$((\bigwedge_{i=1}^{m} L_{ai}\phi_i) \wedge (\bigwedge_{j=2}^{n} M_{bj}\psi_j) \rightarrow L_{(a1+...+am)-(b1+...+bn)}\psi_1)$ |

The inference rule (B) deserves attention. The content and origin of (B) is explained in [19], so we can be brief. The pseudo-formula $((\phi_1, ..., \phi_m) \leftrightarrow (\psi_1, ..., \psi_n))$ is an abbreviation for:

$$\bigwedge_{k=1}^{max(m,n)} \phi^{(k)} \leftrightarrow \psi^{(k)}$$

where:

$$\phi^{(k)} = \bigvee_{1 \le < l_1 ... < l_k \le m} (\phi_{l_1} \wedge ... \wedge \phi_{l_k})$$

(if $k > m$, by convention $\phi^{(k)} = \perp$). Intuitively, $\phi^{(k)}$ "says" that at least $k$ of the formulas $\phi_i$ is true. The meaning of (B) is simpler to grasp when it is interpreted set-theoretically. Associate $(E_1, ..., E_m)$ and $(F_1, ..., F_n)$, two sequences of events, with the sequences of formulas $(\phi_1, ..., \phi_m)$ and $(\psi_1, ..., \psi_n)$. Then, the premise of (B) is a syntactical rendering of the idea that the sum of the characteristic functions is equal, i.e., $\sum_{i=1}^{m} \mathbf{I}_{E_i} = \sum_{j=1}^{n} \mathbf{I}_{F_j}$. If $P(E_i) \ge \alpha_i$ for $i = 1, ..., n$ and $P(F_j) \le \beta_j$ for $j = 2, ..., m$, then $P(F_1)$ has to "compensate", i.e.,

$$P(F_1) \ge (\alpha_1 + ... + \alpha_n) - (\beta_2 + ... + \beta_m)$$

The conclusion of (B) is a translation of this "compensation". It is very powerful from the probabilistic point of view and plays a crucial role in the (sophisticated) completeness proof. In comparison with the modal doxastic logic, one of the issues is that it is not easy to adapt the usual proof method, i.e., that of canonical models. More precisely, with Kripke logics, there is a natural accessibility relation on the canonical state space. Here, we need to prove the existence of a canonical probability distribution from relevant mathematical principles. This step is linked to a difficulty that is well known in the axiomatic study of quantitative and qualitative probability: how to ensure a (numerical) probabilistic representation for a finite structure of qualitative probability. A principle similar to (B) has been introduced in an set-theoretical axiomatic framework by [32] and imported into a probabilistic modal logic (with a qualitative binary operator) by [33].

## 4. A Detour by Doxastic Logic without Full Awareness

Our probabilistic logic without full awareness is largely an adaptation of the generalized standard structures (GSS) introduced by [5] to deal with unawareness in doxastic logic. Actually, we will slightly modify the semantics of [5] and obtain a partial semantics for unawareness. Before giving our own logical system, we remind the reader of GSSs of the case of doxastic logic.

### *4.1. Basic Heuristics*

Going back to the motivating example, suppose that the "objective" state space is based on the set of atomic formulas $At = \{p, q, r\}$ as in Figure 1. Suppose furthermore that the actual state is $s = pqr$ and that in this state, Pierre believes that $p$, is undecided about $q$ and is unaware of $r$. In the language used in [5] , one would say that the actual state is "projected" to a subjective state $\rho(s) = pq$ of a "subjective" state space based on the set of atomic formulas that the agent is aware of, i.e., $p$ and $q$ (Figure 2). In Kripkean doxastic logic, the agent's accessibility relation selects, for each possible state $s$, the set of states $R(s)$ that are epistemically possible for him or her. GSSs define an accessibility relation on this subjective state space. In Figure 3, projection is represented by a dotted arrow and accessibility by solid arrows.

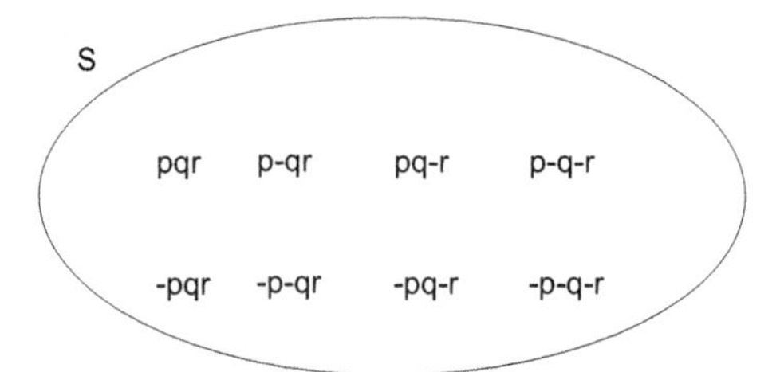


**Figure 1.** An objective state space.

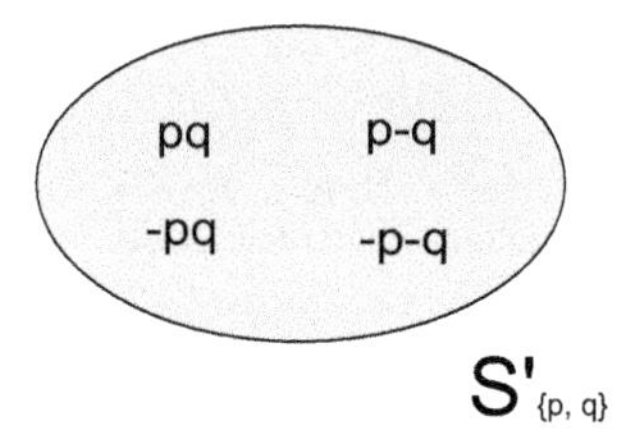


**Figure 2.** A subjective state space.

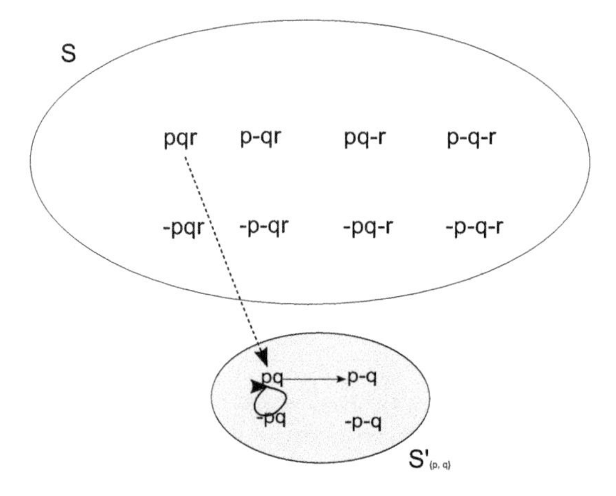


**Figure 3.** Projection of an objective state in a subjective state space.

This picture does not represent all projections between objective states and subjective states, and it corresponds to only one subjective state space. Generally, there are as many subjective state spaces $S'_X$ as there are subsets $X$ of the set $At$ on which the objective state space is based[15]. It is crucial to specify in the right way the conditions on the projection $\rho(.)$ from objective to subjective states. Suppose that another objective state $s'$ is projected to $pq$, as well; then, two conditions should be obtained. First, $s = pqr$ and $s'$ should agree on the atomic formulas the agent is aware of; so, for instance, $s'$ could not be $\neg pqr$, since the agent is aware of $p$. Second, the states accessible from $s$ and $s'$ should be the same. Another natural assumption is that all of the states accessible from a given state are located in the same "subjective" state space (see (v)(2) in Definition 4 below). Figure 4 pictures a GSS more faithfully than the preceding one.

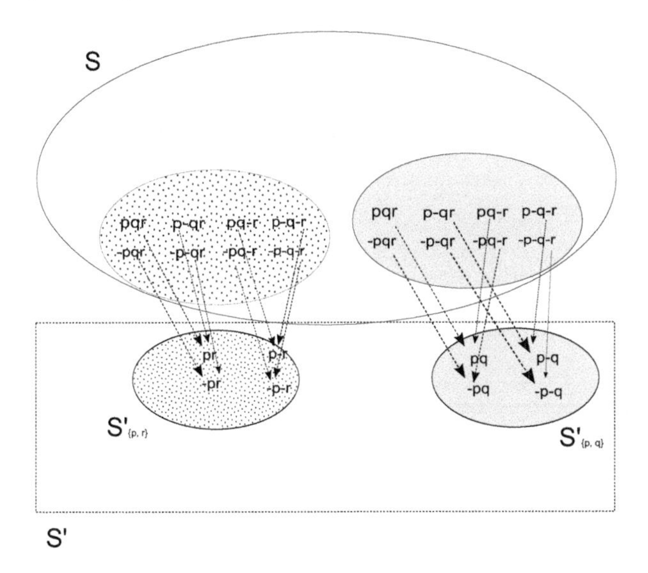


**Figure 4.** Partial picture of a generalized standard structure (GSS).

## 4.2. Generalized Standard Structures

The next definition is a precise rendering of the intuitive ideas. We have followed [7] rather than [5], notably because he makes clear that GSSs can be seen as structures with impossible states.

**Definition 4.** *A GSS is a t-tuple* $\mathcal{M} = (S, S', \pi, R, \rho)$*:*

*(i)* $S$ *is a state space*
*(ii)* $S' = \bigcup_{X \subseteq At} S'_X$ *(where* $S'_X$ *are disjoint) is a (non-standard) state space*
*(iii)* $\pi : S \times At \to \{0, 1\}$ *is a valuation for* $S$
*(iv)* $R : S \to \wp(S')$ *is an accessibility relation for* $S$
*(v)* $\rho : S \to S'$ *is an onto map s.t.*

*(1) if* $\rho(s) = \rho(t) \in S'_X$*, then (a) for each atomic formula* $p \in X$*,* $\pi(s, p) = \pi(t, p)$ *and (b)* $R(s) = R(t)$ *and*

*(2) if* $\rho(s) \in S'_X$*, then* $R(s) \subseteq S'_X$

*One can extend R and π to the whole state space:*

[15] In an objective state space, the set of states need not reflect the set of possible truth-value assignments to propositional variables (as is the case in our example).

(vi) $\pi' : S' \times At \rightarrow \{0,1\}$ *is a valuation for* $S'$ *s.t. for all* $X \subseteq At$ *for all* $s' \in S'_X$, $\pi'(s',p) = 1$ *iff (a)* $p \in X$ *and (b) for all* $s \in \rho^{-1}(s'), \pi(s,p) = 1$. *We note* $\pi^* = \pi \cup \pi'$.
(vii) $R' : S' \rightarrow \wp(S')$ *is an accessibility relation for* $S$ *s.t. for all* $X \subseteq At$ *for all* $s' \in S'_X$, $R'(s') = R(s)$ *for some* $s \in \rho^{-1}(s')$. *We note* $R^* = R \cup R'$.

In comparison with Kripke structures, a modification is introduced as regards negation. In a subjective state $s' \in S'_X$, for a negated formula $\neg\phi$ to be true, it has to be not only that $\phi$ is not true, but also that $\phi$ belongs to the sub-language induced by $X$. Semantic partiality follows: it may be the case that in some $s'$, neither $\phi$, nor $\neg\phi$ is true (this is why subjective states are impossible states). The main reason why, following [7], we introduce this semantics is that it is a very simple way of inducing the "right" kind of partiality. This will be shown in the next subsection. In the sequel, $\mathcal{L}^{BA}(X)$ denotes the language containing the operators $B$ (full beliefs) and $A$ (awareness) and based on the set $X$ of propositional variables.

**Definition 5.** *The satisfaction relation for GSS is defined for each* $s^* \in S^* = S \cup S'$:

(i) $\mathcal{M}, s^* \vDash p$ *iff* $\pi^*(s^*,p) = 1$
(ii) $\mathcal{M}, s^* \vDash \phi \wedge \psi$ *iff* $\mathcal{M}, s^* \vDash \phi$ *and* $\mathcal{M}, s^* \vDash \psi$
(iii) $\mathcal{M}, s^* \vDash \neg\phi$ *iff* $\mathcal{M}, s^* \nvDash \phi$ *and either* $s^* \in S$, *or* $s^* \in S'_X$ *and* $\phi \in \mathcal{L}^{BA}(X)$
(iv) $\mathcal{M}, s^* \vDash B\phi$ *iff for each* $t^* \in R^*(s^*)$, $\mathcal{M}, t^* \vDash \phi$
(v) $\mathcal{M}, s^* \vDash A\phi \Leftrightarrow \mathcal{M}, s^* \vDash B\phi \vee B\neg B\phi$

**Example 1.** *In Figure 3, let us consider* $s = pqr$. $\mathcal{M}, \rho(s) \nvDash r$ *given Clause (vi) of the definition of GSS. However,* $\mathcal{M}, \rho(s) \nvDash \neg r$ *given Clause (iii) of the definition of the satisfaction relation. Clause (iv) of the satisfaction relation and Clause (v)(2) of the GSS imply that* $\mathcal{M}, \rho(s) \nvDash Br$ *and that* $\mathcal{M}, s \nvDash Br$. *However, Clause (iii) of the satisfaction relation again implies that* $\mathcal{M}, \rho(s) \nvDash \neg Br$. *The same holds in the state accessible from* $\rho(s)$. *Therefore,* $\mathcal{M}, s \nvDash B\neg Br$. *By Clause (v) of the satisfaction relation, this implies that* $\mathcal{M}, s \vDash \neg Ar$.

### *4.3. Partial Generalized Standard Structures*

The preceding definition characterizes awareness in terms of beliefs: Pierre is unaware of $\phi$ if, and only if, he does not believe that $\phi$ and does not believe that he does not believe that $\phi$. This is unproblematic when one studies, as [5] do, partitional structures, i.e., structures where the accessibility relation is an equivalence relation and, thus, induces a partition of the state space. Game theorists rely extensively on this special case, which has convenient properties. In this particular case, the fact that an agent does not believe that $\phi$ and does not believe that she/he does not believe that $\phi$ implies that she/he does not believe that she/he does not believe...that she/he does not believe that $\phi$-, at all levels of iteration[16]. However, without such an implication, the equivalence between the fact that an agent is unaware of $\phi$ and the fact that she/he does not believe that $\phi$ and does not believe that she/he does not believe that $\phi$ is dubious, at least in one of the two directions.

We therefore need a more general characterization of awareness and unawareness. Our proposal proceeds from the following observation: in the definition of satisfaction for GSS, the truth-conditions for negated formulas introduce (semantic) partiality. If $p \notin X$ and $s^* \in S'_X$, then neither $\mathcal{M}, s^* \vDash p$ nor $\mathcal{M}, s^* \vDash \neg p$ obtain. Let us indicate by $\mathcal{M}, s^* \Uparrow \phi$ that the formula $\phi$ is undefined at $s^*$ and by $\mathcal{M}, s^* \Downarrow \phi$ that it is defined. The following is true:

**Fact 1.** *Let* $\mathcal{M}$ *be a GSS and* $s^* \in S'_X$ *for some* $X \subseteq At$. *Then:*

[16] Intuitively: if the agent does not believe that $\phi$, this means that not every state of the relevant partition's cell makes $\phi$ true. If $\phi$ belonged to the sub-language associated with the relevant subjective state space, since the accessibility relation is partitional, this would imply that $\neg B\phi$ would be true in every state of the cell. However, by hypothesis, the agent does not believe that she/he does not believe that $\phi$. We have therefore to conclude that $\phi$ does not belong to the sub-language of the subjective state space (see Fact 1 below). Hence, no formula $B\neg B\neg B...\neg B\phi$ can be true.

$$\mathcal{M}, s^* \Downarrow \phi \textit{ iff } \phi \in \mathcal{L}^{BA}(X)$$

**Proof.** see Appendix A.2. □

We suggest to keep the underlying GSS, but change the definition of (un)awareness. Semantical partiality, stressed above, is an attractive guide. In our introductory example, one would like to say that the possible states that Pierre conceives of do not "answer" the question "Is is true that $r$?", whereas they do answer the questions "Is it true that $p$?" and "Is it true that $q$?". In other words, the possible states that Pierre conceives of make neither $r$, nor $\neg r$ true. Awareness can be defined semantically in terms of partiality:

$$\mathcal{M}, s \vDash A\phi \text{ iff } \mathcal{M}, \rho(s) \Downarrow \phi$$

Of course, the appeal of this characterization depends on the already given condition: if $\rho(s) \in S'_X$, then $R(s) \subseteq S'_X$. Let us call a partial GSS a GSS where the truth conditions of the (un)awareness operator are in terms of partiality:

**Fact 2.** *Symmetry, distributivity over $\wedge$, self-reflection, U-introspection, plausibility and strong plausibility are valid under partial GSSs. Furthermore, BU-introspection is valid under serial partial GSS.*

**Proof.** This is left to the reader. □

## 5. Probabilistic Logic without Full Awareness

### 5.1. Language

**Definition 6** (Probabilistic language with awareness). *The set of formulas of a probabilistic language with awareness $\mathcal{L}^{LA}(At)$ based on a set $At$ of propositional variables is defined by:*

$$\phi ::= p|\bot|\top|\neg\phi|\phi \wedge \phi|L_a\phi|A\phi$$

*where $p \in At$ and $a \in [0,1] \cap \mathbb{Q}$.*

### 5.2. Generalized Standard Probabilistic Structures

Probabilistic structures make a full awareness assumption, exactly in the same way that Kripke structures do. An obvious way to weaken this assumption is to introduce in the probabilistic setting the same kind of modification as the one investigated in the previous section. The probabilistic counterpart of generalized standard Structures are the following generalized standard probabilistic structures (GSPS):

**Definition 7** (Generalized standard probabilistic structure). *A generalized standard probabilistic structure for $\mathcal{L}^{LA}(At)$ is a t-tuple*
$\mathcal{M} = (S, (S'_X)_{X \subseteq At}, (\Sigma'_X)_{X \subseteq At}, \pi, (P'_X)_{X \subseteq At})$ *where:*

*(i)* $S$ *is a state space.*
*(ii)* $S'_X$ *where* $X \subseteq At$ *are disjoint "subjective" state spaces. Let* $S' = \bigcup_{X \subseteq At} S'_X$.
*(ii)* *For each* $X \subseteq At$, $\Sigma'_X$ *is a σ-field of subsets of* $S'_X$.
*(iii)* $\pi : S \times At \to \{0,1\}$ *is a valuation.*
*(iv)* $P'_X : S'_X \to \Delta(S'_X, \Sigma'_X)$ *is a measurable mapping from* $S'_X$ *to the set of probability measures on* $\Sigma'_X$ *endowed with the σ-field generated by the sets* $\{\mu \in \Delta(S'_X, \Sigma'_X) : \mu(E) \geq a\}$ *for all* $E \in \Sigma'_X$, $a \in [0,1]$.
*(v)* $\rho : S \to S'$ *is an onto map s.t. if* $\rho(s) = \rho(t) \in S'_X$, *then for each atomic formula* $p \in X$, $\pi(s,p) = \pi(t,p)$. *By definition,* $P^*(s^*) = P^*(\rho(s^*))$ *if* $s \in S$ *and* $P^*(s^*) = P'_X(s^*)$ *if* $s^* \in S'_X$.

(vi) $\pi' : S' \times At \rightarrow \{0,1\}$ *extends* $\pi$ *to* $S'$ *as follows: for all* $s' \in S'_X$, $\pi'(s',p) = 1$ *iff* $p \in X$ *and for all* $s \in \rho^{-1}(s')$, $\pi(s,p) = 1$[17]. *For every* $p \in At$, $s' \in S'_X$, $\pi'(.,p)$ *is measurable w.r.t.* $(S'_X, \Sigma'_X)$.

Two comments are in order. First, Clause (iv) does not introduce any special measurability condition on the newly-introduced awareness operator (by contrast, there is still a condition for the doxastic operators). The reason is that in a given subjective state space $S'_X$, for any formula $\phi$, either there is awareness of $\phi$ everywhere, and in this case $[[A\phi]] \cap S'_X = S'_X$, or there is never awareness of $\phi$, and in this case $[[A\phi]] \cap S'_X = \emptyset$. These two events are of course already in any $\Sigma'_X$. Second, Clause (v) imposes conditions on the projection $\rho$. With respect to GSSs, the only change is that we do not require something like: if $\rho(s) \in S'_X$, then $R(s) \subseteq S'_X$. The counterpart would be that if $\rho(s) \in S'_X$, then $Supp(P(s)) \subseteq S'_X$. However, this is automatically satisfied by the definition.

**Example 2.** *In Figure 5, the support of the probability distribution associated with* $s = pqr$ *is* $\{\rho(s) = pq, p\neg q\}$, $P'_{\{p,q\}}(\rho(s))(pq) = a$ *and* $P'_{\{p,q\}}(\rho(s))(p\neg q) = 1 - a$.

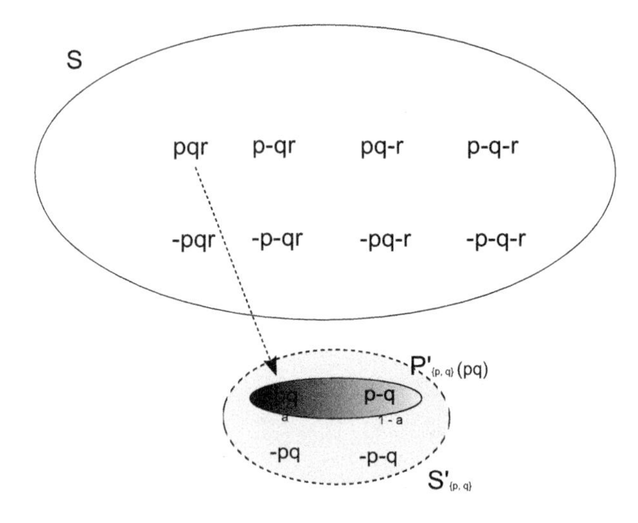


**Figure 5.** Projection of an objective state to a subjective state space in a probabilistic setting.

**Definition 8** (Satisfaction relation for GSPS). *The satisfaction relation for GSPS is defined for each* $s^* \in S^* = S \cup S'$:

(i) $\mathcal{M}, s^* \vDash p$ *iff* $\pi(s^*, p) = 1$
(ii) $\mathcal{M}, s^* \vDash \phi \wedge \psi$ *iff* $\mathcal{M}, s^* \vDash \phi$ *and* $\mathcal{M}, s^* \vDash \psi$
(iii) $\mathcal{M}, s^* \vDash \neg\phi$ *iff* $\mathcal{M}, s^* \nvDash \phi$ *and either* $s^* \in S$, *or* $s^* \in S'_X$ *and* $\phi \in \mathcal{L}^{LA}(X)$
(iv) $\mathcal{M}, s^* \vDash L_a\phi$ *iff* $P^*(s^*)([[\phi]]) \geq a$ *and* $\mathcal{M}, \rho(s) \Downarrow \phi$
(v) $\mathcal{M}, s^* \vDash A\phi$ *iff* $\mathcal{M}, \rho(s^*) \Downarrow \phi$

The following fact is the counterpart for GSPS of what was proven above for GSS.

**Fact 3.** *Let* $\mathcal{M}$ *be a GSPS and* $s' = \rho(s) \in S'_X$ *for some* $s \in S$. *Then:*

$$\mathcal{M}, s' \Downarrow \phi \text{ iff } \phi \in \mathcal{L}^{LA}(X)$$

[17] It follows from Clause (v) above that this extension is well defined.

**Proof.** The proof is analogous to the one provided for Fact 1 above. □

One can show that all of the properties mentioned in Section 2.3 are valid under GSPSs.

**Proposition 1.** *For all GSPS $\mathcal{M}$ and all standard states $s \in S$, the following formulas are satisfied:*

| | |
|---|---|
| $A\phi \leftrightarrow A\neg\phi$ | *(symmetry)* |
| $A(\phi \wedge \psi) \leftrightarrow A\phi \wedge A\psi$ | *(distributivity over $\wedge$)* |
| $A\phi \leftrightarrow AA\phi$ | *(self-reflection)* |
| $\neg A\phi \rightarrow \neg A\neg A\phi$ | *(U-introspection)* |
| $\neg A\phi \rightarrow \neg L_a\phi \wedge \neg L_a\neg L_a\phi$ | *(plausibility)* |
| $\neg A\phi \rightarrow (\neg L_a)^n\phi \; \forall n \in \mathbb{N}$ | *(strong plausibility)* |
| $\neg L_a \neg A\phi$ | *($L_a U$-introspection)* |
| $L_0\phi \leftrightarrow A\phi$ | *(minimality)* |

**Proof.** This is left to the reader. □

### 5.3. Axiomatization

Proposition 1 suggests that GSPSs provide a plausible analysis of awareness and unawareness in a probabilistic setting. To have a more comprehensive understanding of this model, we need to investigate its logical properties. It turns out that an axiom system can be given that is weakly complete with respect to GSPS. We call it system $HM_U$[18].

System $HM_U$

(PROP) Instances of propositional tautologies
(MP) From $\phi$ and $\phi \rightarrow \psi$, infer $\psi$

(A1) $A\phi \leftrightarrow A\neg\phi$
(A2) $A(\phi \wedge \psi) \leftrightarrow A\phi \wedge A\psi$
(A3) $A\phi \leftrightarrow AA\phi$
(A4$_L$) $A\phi \leftrightarrow AL_a\phi$
(A5$_L$) $A\phi \rightarrow L_1 A\phi$

(L1$_U$) $A\phi \leftrightarrow L_0\phi$
(L2$_U$) $A\phi \rightarrow L_a(\phi \vee \neg\phi)$
(L3) $L_a\phi \rightarrow \neg L_b \neg\phi \; (a + b > 1)$
(L4$_U$) $(\neg L_a\phi \wedge A\phi) \rightarrow M_a\phi$

(RE$_U$) From $\phi \leftrightarrow \psi$ and $Var(\phi) = Var(\psi)$, infer $(L_a\phi \leftrightarrow L_a\psi)$

(B$_U$) From $((\phi_1, ..., \phi_m) \leftrightarrow (\psi_1, ..., \psi_n))$, infer:
$((\bigwedge_{i=1}^{m} L_{ai}\phi_i) \wedge (\bigwedge_{j=2}^{n} M_{bj}\psi_j)) \rightarrow (A\psi_1 \rightarrow L_{(a1+...+am)-(b1+...+bn)}\psi_1))$

Some comments are in order. (a) Axioms (A1)–(A5$_L$) concern the awareness operator and its relationship with the doxastic operators. The subscript "$_L$" indicates an axiom that involves a probabilistic doxastic operator, to be distinguished from its epistemic counterpart indicated by "$_B$" in the axiom system for doxastic logic reproduced in Appendix A.3. The other axioms and inference rules were roughly part of probabilistic logic without the awareness operator appearing in them. Subscript "$_U$" indicates a modification with respect to the System $HM$ due to the presence of the

[18] In what follows, $Var(\phi)$ denotes the set of propositional variables occurring in $\phi$.

awareness operator. (b) Axiom ($L1_U$) substitutes $A\phi \leftrightarrow L_0\phi$ for $L_0\phi$. This means that in our system, the awareness operator can be defined in terms of the probabilistic operator. However, this does not imply that a non-standard semantics is not needed: if ones defines $A\phi$ as $L_0\phi$, $A\phi$ is valid for any $\phi$ in standard semantics[19]. (c) The relationship between logical omniscience and full awareness comes out clearer in the restriction we have to impose on the rule of equivalence (RE)[20]. It is easy to see why in semantical terms why the rule of equivalence no longer holds universally. Suppose that $L_{1/2}p$ holds at some state of some model. From propositional logic, we know that $p \equiv (p \wedge q) \vee (p \wedge \neg q)$. However, if the agent is not aware of $q$, it is not true that $L_{1/2}((p \wedge q) \vee (p \wedge \neg q))$. (d) For the same kind of reason, the inference rule (B) no longer holds universally. Consider for instance $\phi_1 = (p \vee \neg p)$, $\phi_2 = (r \vee \neg r)$ $\psi_1 = (q \vee \neg q)$ and $\psi_2 = (p \vee \neg p)$. Additionally, suppose that that the agent is aware of $p$ and $r$, but not of $q$. Clearly, the premise of (B), i.e., $((\phi_1, \phi_2) \leftrightarrow (\psi_1, \psi_2))$ is satisfied. Furthermore, the antecedent of (B)'s conclusion, i.e., $L_1\phi_1$ and $M_1\psi_2$ is satisfied, as well. However, since the agent is unaware of $q$, we cannot conclude what we should conclude were (B) valid, i.e., that $L_1\psi_1$.

We are now ready to formulate our main result.

**Theorem 1** (Soundness and completeness of $HM_U$). *Let $\phi \in \mathcal{L}^{LA}(At)$. Then:*

$$\models_{GSPS} \phi \text{ iff } \vdash_{HM_U} \phi$$

**Proof.** See the Appendix B. □

## 6. Conclusions

This study of unawareness in probabilistic logic could be unfolded later in several directions. First, we did not deal with the extension to a multi-agent framework, an issue tackled recently by [8]; second, we did not investigate applications to decision theory or game theory. However, we would like to end by stressing another issue that is less often evoked and nonetheless conceptually very challenging: the dynamics of awareness. The current framework, as much of the existing work, is static, i.e., it captures the awareness and doxastic states at a given time. It does not tell anything of the fact that, during an inquiry, an agent may become aware of some new possibilities[21].

Let us consider our initial example where Pierre is aware of $p$ and $q$, but not of $r$, and let us suppose that Pierre's partial beliefs are represented by some probability distribution on a subjective state space $S_{\{p,q\}}$. Assume that at some time, Pierre becomes aware of $r$; for instance, someone has asked him whether he thinks that $r$ is likely or not. It seems that our framework can be extended to accommodate the situation: Pierre's new doxastic state will be represented on a state space $S_{\{p,q,r\}}$ appropriately connected to the initial one $S_{\{p,q\}}$ (see Figure 6). Typically, a state $s' = pq$ will be split into two fine-grained states $s_1 = pqr$ and $s_2 = pq\neg r$. However, how should Pierre's partial beliefs evolve? Obviously, a naive Laplacian rule according to which the probability assigned to $s'$ is equally allocated to $s_1$, and $s_2$ will not be satisfactory. Are there rationality constraints capable of determining a new probability distribution on $S_{\{p,q,r\}}$? Or should we represent the new doxastic state of the agent by a set of probability distributions?[22] We leave the answers to these questions for future investigation.

[19] I thank an anonymous referee for suggesting to me to clarify this point.
[20] This restricted rule is reminiscent of the rule $RE_{sa}$ in [5].
[21] Note that he/she could become unaware of some possibilities as well, but we will not say anything about that.
[22] The dynamics of awareness has been studied by [34] in doxastic logic and by [35] in doxastic logic. See also [36].

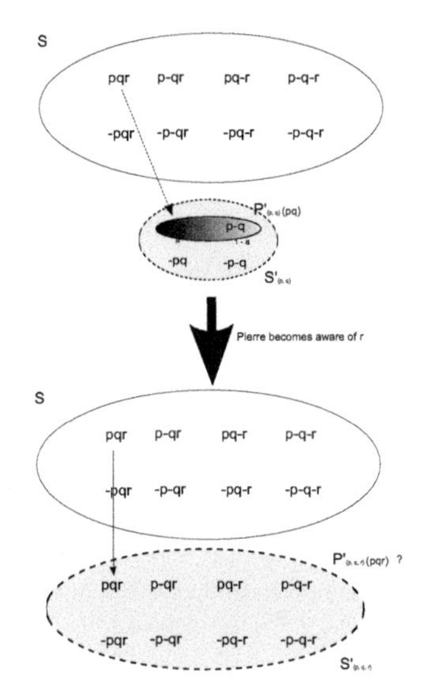


**Figure 6.** The issue of becoming aware.

**Acknowledgments:** I would like to thank for their comments or advices two anonymous referees, C. Dégremont, A. Heifetz, B. Hill, M. Meier, T. Sadzik, B. Schipper and P. Weirich and audiences at the Institut d'Histoire et de Philosophie des Sciences et des Techniques (Paris), the London School of Economics (workshop "Decision, Games and Logic", London), the Institute for Logic, Language and Computation (seminar "Logic and Games", Amsterdam), the Laboratoire d'Algorithmique, Complexité et Logique (Université Paris-Est Créteil). Special thanks are due to Philippe Mongin, who helped me a lot to improve the penultimate version. This work was supported by the Institut Universitaire de France, the ANR-10-LABX-0087 IEC and ANR-10-IDEX-001-02PSL* grants.

**Conflicts of Interest:** The author declares no conflict of interest.

## Appendix A. Doxastic Logic: Proofs and Illustrations

### *Appendix A.1. Illustration*

Let us consider a GSS $\mathcal{M}$ where:

- the actual state is $s \in S$
- $s$ is projected in $s_1 \in S'_X$ for some $X \subseteq At$
- $R(s_1) = \{s_2, s_3\}$, $R(s_2) = \{s_2\}$ and $R(s_3) = \{s_3\}$
- $\mathcal{M}, s_2 \vDash p$ and $\mathcal{M}, s_3 \vDash \neg p$

The relevant part of the model is represented in Figure A1.

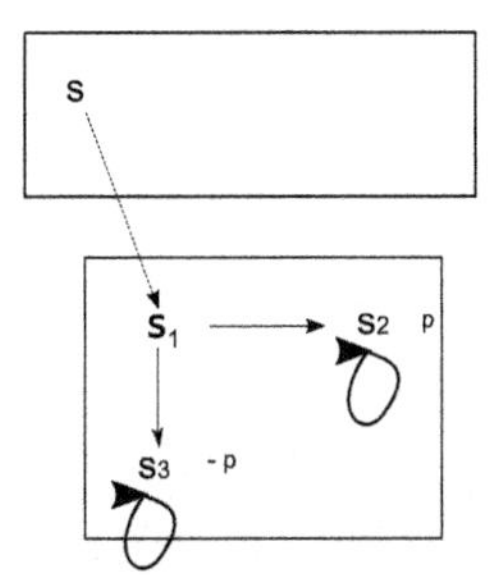


**Figure A1.** Unawareness without partitions.

It is easy to check that $\mathcal{M}, s \vDash Up$ and $\mathcal{M}, s \vDash B(B\neg p \vee Bp)$, since $\mathcal{M}, s_2 \vDash Bp$ and $\mathcal{M}, s_3 \vDash B\neg p$.

*Appendix A.2. Proof of Fact 1*

For greater convenience, we give the proof for partial GSSs, but Fact 1 holds for original GSSs, as well. We have to show that if $\mathcal{M}$ is a partial GSS and $s^* \in S'_X$ for some $X \subseteq At$, then $\mathcal{M}, s^* \Downarrow \phi$ iff $\phi \in \mathcal{L}^{BA}(X)$. The proof is by induction on the complexity of formulas:

- if $\phi := p$, then $\mathcal{M}, s^* \vDash p$ or $\mathcal{M}, s^* \vDash \neg p$ iff ($p \in \mathcal{L}^{BA}(X)$ and $\pi(\rho^{-1}(s^*), p) = 1$) or ($p \in \mathcal{L}^{BA}(X)$ and not $\pi(\rho^{-1}(s^*), p) = 1$) iff $p \in \mathcal{L}^{BA}(X)$.
- if $\phi := \neg\psi$, then $\mathcal{M}, s^* \vDash \phi$ or $\mathcal{M}, s^* \vDash \neg\phi$ iff $\mathcal{M}, s^* \vDash \neg\psi$ or $\mathcal{M}, s^* \vDash \neg\neg\psi$ iff ($\psi \in \mathcal{L}^{BA}(X)$ and $\mathcal{M}, s^* \nvDash \psi$) or ($\neg\psi \in \mathcal{L}^{BA}(X)$ and $\mathcal{M}, s^* \nvDash \neg\psi$) iff ($\psi \in \mathcal{L}^{BA}(X)$ and $\mathcal{M}, s^* \nvDash \psi$) or ($\psi \in \mathcal{L}^{BA}(X)$ and $\mathcal{M}, s^* \vDash \psi$) iff $\psi \in \mathcal{L}^{BA}(X)$ iff $\neg\psi \in \mathcal{L}^{BA}(X)$
- if $\phi := \psi_1 \wedge \psi_2$, then $\mathcal{M}, s^* \vDash \phi$ or $\mathcal{M}, s^* \vDash \neg\phi$ iff ($\mathcal{M}, s^* \vDash \psi_1$ and $\mathcal{M}, s^* \vDash \psi_2$) or ($\psi_1 \wedge \psi_2 \in \mathcal{L}^{BA}(X)$ and ($\mathcal{M}, s^* \nvDash \psi_1$ or $\mathcal{M}, s^* \nvDash \psi_2$)) iff by IH ($\psi_1 \wedge \psi_2 \in \mathcal{L}^{BA}(X)$ and $\mathcal{M}, s^* \vDash \psi_1$ and $\mathcal{M}, s^* \vDash \psi_2$) or ($\psi_1 \wedge \psi_2 \in \mathcal{L}^{BA}(X)$ and not ($\mathcal{M}, s^* \vDash \psi_1$ and $\mathcal{M}, s^* \vDash \psi_2$)) iff $\psi_1 \wedge \psi_2 \in \mathcal{L}^{BA}(X)$
- if $\phi := B\psi$, then $\mathcal{M}, s^* \vDash \phi$ or $\mathcal{M}, s^* \vDash \neg\phi$ iff (for each $t^* \in R^*(s^*)$, $\mathcal{M}, t^* \vDash \phi$) or ($B\psi \in \mathcal{L}^{BA}(X)$ and $\mathcal{M}, s^* \nvDash B\psi$) iff, by the induction hypothesis and since each $t^* \in R^*(s^*)$ belongs to $S_X$ - ($B\psi \in \mathcal{L}^{BA}(X)$ $\mathcal{M}, s^* \vDash B\psi$) or ($B\psi \in \mathcal{L}^{BA}(X)$ and $\mathcal{M}, s^* \nvDash B\psi$) iff $B\psi \in \mathcal{L}^{BA}(X)$
- if $\phi := A\psi$, then $\mathcal{M}, s^* \vDash \phi$ or $\mathcal{M}, s^* \vDash \neg\phi$ iff $\mathcal{M}, s^* \Downarrow \psi$ or ($A\psi \in \mathcal{L}^{BA}(X)$ and $\mathcal{M}, s^* \nvDash A\psi$) iff (by Induction Hypothesis) $\psi \in \mathcal{L}^{BA}(X)$ or ($A\psi \in \mathcal{L}^{BA}(X)$ and $\mathcal{M}, s^* \Uparrow \psi$) iff (by the induction hypothesis) $\psi \in \mathcal{L}^{BA}(X)$ or ($A\psi \in \mathcal{L}^{BA}(X)$ and $\psi \notin \mathcal{L}^{BA}(X)$) iff $\psi \in \mathcal{L}^{BA}(X)$ iff $A\psi \in \mathcal{L}^{BA}(X)$.

*Appendix A.3. An Axiom System for Partial GSSs*

We may obtain a complete axiom system for serial partial GSS thanks to [37] who relates GSS and awareness© structures. Actually, one obtains a still closer connection with serial partial GSS. Let us first restate the definition of awareness© structures.

**Definition A1.** *An awareness© structure is a t-tuple $(S, \pi, R, A)$ where*

*(i) S is a state space,*
*(ii) $\pi : At \times S \to \{0, 1\}$ is a valuation,*
*(iii) $R \subseteq S \times S$ is an accessibility relation,*
*(iv) $A : S \to Form(\mathcal{L}^{BA}(At))$ is a function that maps every state in a set of formulas ("awareness© set").*

The new condition on the satisfaction relation is the following:
$\mathcal{M}, s \vDash B\phi$ iff $\forall t \in R(t)$ $\mathcal{M}, t \vDash \phi$ and $\phi \in A(s)$
Let us say that an awareness© structure $\mathcal{M} = (S, R, \mathcal{A}, \pi)$ is propositionally determined (p.d.) if (1) for each state $s$, $\mathcal{A}(s)$ is generated by some atomic formulas $X \subseteq At$, i.e., $\mathcal{A}(s) = \mathcal{L}^{BA}(X)$, and (2) if $t \in R(s)$, then $\mathcal{A}(s) = \mathcal{A}(t)$.

**Proposition A1** (Adapted from Halpern 2001 Theorem 4.1).

1. *For every serial p.d. awareness© structure $\mathcal{M}$, there exists a serial partial GSS $\mathcal{M}'$ based on the same state space S and the same valuation $\pi$ s.t. for all formulas $\phi \in \mathcal{L}^{BA}(At)$ and each possible state s*
   $\mathcal{M}, s \vDash_{a©} \phi$ *iff* $\mathcal{M}', s \vDash_{pGSS} \phi$
2. *For every serial partial GSS $\mathcal{M}$, there exists a serial p.d. awareness© structure $\mathcal{M}'$ based on the same state space S and the same valuation $\pi$ s.t. for all formulas $\phi \in \mathcal{L}^{BA}(At)$ and each possible state s*
   $\mathcal{M}, s \vDash_{pGSS} \phi$ *iff* $\mathcal{M}', s \vDash_{a©} \phi$

An axiom system has been devised in [37] that is (sound and) complete with respect to p.d. awareness© structures. An axiom system for serial p.d. awareness© structures can be devised by enriching this axiom system with:

$$(\text{D}_U)\ B\phi \rightarrow (\neg B\neg\phi \wedge A\phi)$$

The resulting axiom system, coined $KD_U$, is this one:

| System $KD_U$ |
|---|
| (PROP) Instances of propositional tautologies |
| (MP) From $\phi$ and $\phi \rightarrow \psi$, infer $\psi$ |
| (K)$B\phi \wedge B(\phi \rightarrow \psi) \rightarrow B\psi$ |
| (Gen)From $\phi$, infer $A\phi \rightarrow B\phi$ |
| ($\text{D}_U$) $B\phi \rightarrow (\neg B\neg\phi \wedge A\phi)$ |
| (A1) $A\phi \leftrightarrow A\neg\phi$ |
| (A2) $A(\phi \wedge \psi) \leftrightarrow (A\phi \wedge A\psi)$ |
| (A3) $A\phi \leftrightarrow AA\phi$ |
| ($\text{A4}_B$) $AB\phi \leftrightarrow A\phi$ |
| ($\text{A5}_B$) $A\phi \rightarrow BA\phi$ |
| (Irr)If no atomic formulas in $\phi$ appear in $\psi$, from $U\phi \rightarrow \psi$, infer $\psi$ |

The following derives straightforwardly from Proposition 2.

**Proposition A2** (Soundness and completeness theorem). *Let $\phi \in \mathcal{L}^{BA}(At)$. Then:*

$$\models_{spGSS} \phi \textit{ iff } \vdash_{KD_U} \phi$$

## Appendix B. Probabilistic Logic: Proof of the Completeness Theorem for $HM_U$

**Proof.** ($\Leftarrow$). Soundness is easily checked and is left to the reader. ($\Rightarrow$). We have to show that if $\models_{GSPS} \phi$, then $\vdash_{HM_U} \phi$. The proof relies on the well-known method of filtration. First, we define a restricted language $\mathcal{L}^{LA}_{[\phi]}$ as in [19]: $\mathcal{L}^{LA}_{[\phi]}$ contains:

- as atomic formulas, only $Var(\phi)$, i.e., the atomic formulas occurring in $\phi$,
- only probabilistic operators $L_a$ belonging to the finite set $Q(\phi)$ of rational numbers of the form $p/q$, where $q$ is the smallest common denominator of indexes occurring in $\phi$ and
- only formulas of epistemic depth smaller than or equal to that of $\phi$ (an important point is that we stipulate that the awareness operator $A$ does not add any epistemic depth to a formula: $dp(A\psi) = dp(\psi)$).

As we will show, the resulting language $\mathcal{L}^{LA}_{[\phi]}$ is finitely generated: there is a finite subset $\mathfrak{B}$ of $\mathcal{L}^{LA}_{[\phi]}$ called a base, such that $\forall\psi \in \mathcal{L}^{LA}_{[\phi]}$, there is a formula $\psi'$ in the base, such that $\vdash_{HM_U} \psi \leftrightarrow \psi'$. In probabilistic structures, it is easy to construct such a base[23]. The basic idea is this:

(1) consider $D_0$, the set of all the disjunctive normal forms built from $B_0 = Var(\phi)$, the set of propositional variables occurring in $\phi$.
(2) $B_k$ is the set of formulas $L_a\psi$ for all $a \in Q(\phi)$ where $\psi$ is a disjunctive normal form built with "atoms" coming from $B_0$ to $B_{k-1}$.
(3) the construction has to be iterated up to the epistemic depth $n$ of $\phi$, hence to $B_n$. The base $\mathfrak{B}$ is $D_n$, i.e., the set of disjunctive normal forms built with "atoms" from $B_0$ to $B_n$.

---

23 The work in [19] leaves the construction implicit.

Obviously, $\mathfrak{B}$ is finite. It can be shown by induction that $\mathcal{L}^{LA}_{[\phi]}$ is finitely generated by $\mathfrak{B}$. For formulas with a Boolean connective as top connective, this is obvious. For formulas of the form $L_a\psi$, it comes from substitutability under logically-equivalent formulas: by the induction hypothesis, there is a formula $\psi'$ equivalent to $\psi$ in $\mathfrak{B}$. Therefore, there is in $\mathfrak{B}$ a formula equivalent to $L_a\psi'$. However, since $\vdash_{HM} \psi \leftrightarrow \psi'$, it follows that $\vdash_{HM} L_a\psi \leftrightarrow L_a\psi'$. We will now show how to unfold these ideas formally.

**Definition B1.** *Let $X = \{\psi_1, ...\psi_m\}$ be a finite set of formulas. $DNF(X)$ is the set of disjunctive normal forms that can be built from X, i.e., the set of all possible disjunctions of conjunctions of the form $e_1\psi_1 \wedge ... \wedge e_m\psi_m$ where $e_i$ is a blank or $\neg$. The members of X are called the atomic components of $DNF(X)$.*

**Definition B2.** *The base $\mathfrak{B}$ for a language $\mathcal{L}^{L}_{[\phi]}$ where $dp(\phi) = n$ is defined as $D_n$ in the following doubly-inductive construction:*

*(i) $B_0 = Var(\phi)$ ($B_0$ is the set of atomic components of epistemic depth 0)*
*(i') $D_0 = DNF(B_0)$ ($D_0$ is the set of disjunctive normal forms based on $B_0$)*

*(ii) $B_k = \{L_a\psi : \psi \in D_{k-1}\}$*
*(ii') $D_k = DNF(\bigcup_{l=0}^{k} B_l)$.*

**Notation B1.** *Let $\psi \in DNF(X)$ and $X \subseteq Y$. The expansion of $\psi$ in $DNF(Y)$ is the formula obtained by the replacement of each conjunction $e_1\psi_1 \wedge ... \wedge e_m\psi_m$ occurring in $\psi$ by a disjunction of all possible conjunctions built from $e_1\psi_1 \wedge ... \wedge e_m\psi_m$ by adding literals of atomic components in $Y - X$.*

For instance, consider $X = \{p\}$ and $Y = \{p, q\}$: the DNF $p$ is expanded in $(p \wedge q) \vee (p \wedge \neg q)$.

**Fact B1.**

*(i) $\forall k, \forall \psi \in D_k \cup B_k, dp(\psi) = k$.*
*(ii) For each $\psi \in D_k$, in each $D_l$, $l > k$, there is a formula $\psi'$, which is equivalent to $\psi$*

**Proof.** (i) is obvious; (ii) follows from the fact that any formula $\psi \in D_k$ can be expanded in $\psi'$ from $D_l$, $l > k$ and that (by propositional reasoning) $\psi$ and $\psi'$ are equivalent. □

It can be proven that $\mathfrak{B} = D_n$ is a finite base for $\mathcal{L}^{L}_{[\phi]}$. First, for each $\psi \in \mathcal{L}^{L}_{[\phi]}$, there is a formula $\psi'$ in $D_l$ s.t. $\vdash_{HM} \psi \leftrightarrow \psi'$ where $dp(\psi) = l$. Since $\psi'$ can be expanded in a logically-equivalent formula $\psi'' \in D_n$, it is sufficient to conclude that for each $\psi \in \mathcal{L}^{L}_{[\phi]}$, there is an equivalent formula in the base.

(i) $\psi := p$: $\psi$ is obviously equivalent to some DNF in $D_0$ and $dp(p) = 0$.
(ii) $\psi := (\chi_1 \wedge \chi_2)$: by the induction hypothesis, there is $\chi_1'$ equivalent to $\chi_1$ in $D_{dp(\chi_1)}$ and $\chi_2'$ equivalent to $\chi_2$ in $D_{dp(\chi_2)}$. Suppose w.l.o.g.that $dp(\chi_2) > dp(\chi_1)$ and, therefore, that $dp(\psi) = dp(\chi_2)$. Then, $\chi_1'$ can be expanded in $\chi_1'' \in D_{dp(\chi_2)}$. Obviously, the disjunction of the conjunctions occurring both in $\chi_1''$ and $\chi_2'$ is in $D_{dp(\chi_2)}$ and equivalent to $\psi$.
(iii) $\psi := L_a\chi$: by IH, there is $\chi'$ equivalent to $\chi$ in $D_{dp(\chi)}$. Note that $dp(\chi) < n = dp(\phi)$. By construction, $L_a(\chi') \in B_{dp(\chi)+1}$. Consequently, there will be in $D_{dp(\chi)+1}$ a DNF $\psi'$ equivalent to $L_a(\chi')$. Since $dp(\chi) + 1 \leq n$, this DNF can be associated by expansion to a DNF in the base $D_n$. Furthermore, since $\vdash_{HM} \chi \leftrightarrow \chi'$ and $\vdash_{HM} L_a\chi' \leftrightarrow \psi'$, it follows by the rule of equivalence that $\vdash_{HM} L_a\chi \leftrightarrow \psi'$.

□

There are changes needed to deal with unawareness. (1) First, the awareness operator $A$ has to be included. This is not problematic given that for any formula $\psi$, $\vdash_{HM_U} A\psi \leftrightarrow \bigwedge_m Ap_m$ where $Var(\psi) = \{p_1, ..., p_m, ...p_M\}$. Consequently, the only modification is to include any formula $Ap$ with $p \in Var(\phi)$

in $B_0$. (2) With unawareness, it is no longer true that if $\vdash_{HM_U} \psi \leftrightarrow \chi$, then $\vdash_{HM_U} L_a\psi \leftrightarrow L_a\chi$. For instance, it is not true that $\vdash_{HM_U} L_a p \leftrightarrow L_a((p \wedge q) \vee (p \wedge \neg q))$: the agent may be unaware of $q$. Nevertheless, the rule holds restrictedly: under the assumption that $Var(\psi) = Var(\chi)$, then if $\vdash_{HM_U} \psi \leftrightarrow \chi$, then $\vdash_{HM_U} L_a\psi \leftrightarrow L_a\chi$ (RE$_U$). We can use this fact to make another change to the basis: instead of considering only the disjunctive normal forms built from the whole set $Var(\phi)$, we consider the disjunctive normal forms built from any non-empty subset $X \subseteq Var(\phi)$.

**Definition B3.** *Let* $X \subseteq Var(\phi)$;

*(i)* $B_0^X = X \cup \{Ap : p \in X\}$
*(i')* $D_0^X = DNF(B_0)$

*(ii)* $B_k^X = \{L_a\psi : \psi \in D_{k-1}\}$ *and*
*(ii')* $D_k^X = DNF(\bigcup_{l=0}^{k} B_l^X)$.

**Fact B2.**

*(i)* $\forall k \leq n$, $\psi \in D_k^X$ *where* $X \subseteq Var(\phi)$, $dp(\psi) = k$.
*(ii)* $\forall X \subseteq Var(\phi)$, $\forall \psi \in D_k^X$, $\forall D_l^X, l > k$, *there is a formula* $\psi' \in D_l^X$, *which is equivalent to* $\psi$.
*(iii)* $\forall X \subseteq Y \subseteq Var(\phi)$, *if* $\psi \in D_k^X$, *then there is a formula* $\psi'$, *which is equivalent to* $\psi$ *in* $D_k^Y$.
*(iv)* $\forall X \subseteq Var(\phi)$, $\forall \psi \in D_k^X$, $Var(\psi) = X$.

**Proof.** (i)–(ii) are similar to the classical case; (iii) is straightforwardly implied by Clause (ii') of Definition 12; (iv) is obvious. □

We are now ready to prove that $\mathfrak{B} = \bigcup_{X \subseteq Var(\phi)} D_n^X$ is a basis for $\mathcal{L}_{[\phi]}^{LA}$. We will actually show that for any $\psi \in L_{[\phi]}^{LA}$ with $dp(\psi) = k$, there are $X \subseteq Var(\phi)$ and $\psi' \in D_k^X$ s.t. $\vdash_{HM_U} \psi \leftrightarrow \psi'$ and $dp(\psi) = dp(\psi') = k$ and $Var(\psi) = Var(\psi')$ (Induction Hypothesis, IH).

(i) $\psi := p$: $\psi$ is obviously equivalent to some DNF $\psi'$ in $D_0^{\{p\}}$. Clearly, $dp(\psi) = dp(\psi')$ and $Var(\psi) = Var(\psi')$.
(ii) $\psi := (\chi_1 \wedge \chi_2)$: by IH,
- there is $\chi_1'$ s.t. $\vdash_{HM_U} \chi_1 \leftrightarrow \chi_1'$ and $Var(\chi_1) = Var(\chi_1') = X_1$ and $\chi_1' \in D_{dp(\chi_1)}^{X_1}$
- there is $\chi_2'$ s.t. $\vdash_{HM_U} \chi_2 \leftrightarrow \chi_2'$ and $Var(\chi_2) = Var(\chi_2') = X_2$ and $\chi_2' \in D_{dp(\chi_2)}^{X_2}$

Let us consider $X' = X_1 \cup X_2$ and suppose without loss of generality that $dp(\chi_2) > dp(\chi_1)$. One may expand $\chi_1'$ from $D_{dp(\chi_1)}^{X_1}$ to $D_{dp(\chi_1)}^{X'}$ and expand the resulting DNF to $\chi_1'' \in D_{dp(\chi_2)}^{X'}$. On the other hand, $\chi_2'$ may be expanded to $\chi_2'' \in D_{dp(\chi_2)}^{X'}$. $\psi'$ is the disjunction of the conjunctions common to $\chi_1''$ and $\chi_2''$. Obviously, $dp(\psi) = dp(\psi')$ and $Var(\psi) = Var(\psi')$.
(iii) $\psi := A\chi$: by IH, there is $\chi'$ equivalent to $\chi$ in $D_{dp(\chi)}^X$ with $Var(\chi) = Var(\chi')$. $A\chi'$ is equivalent to $\bigwedge_m Ap_m$ where $Var(\chi') = \{p_1, ..., p_m, ...p_M\}$. Each $Ap_m$ is in $B_0^X$, so by expansion in $D_{dp(\chi)}^X$, there is a DNF equivalent to it and, therefore, a DNF equivalent to $\bigwedge_m Ap_m$.
(iv) $\psi := L_a\chi$: by IH, there is $\chi'$ equivalent to $\chi$ in $D_{dp(\chi)}^X$ with $dp(\chi) = dp(\chi')$ and $Var(\chi) = Var(\chi')$. Note that $dp(\chi) < n = dp(\phi)$. By construction, $L_a(\chi') \in B_{dp(\chi)+1}^X$. Consequently, there will be in $D_{dp(\chi)+1}^X$ a DNF $\psi'$ logically equivalent to $L_a(\chi')$. Since $dp(\chi) + 1 \leq n$, there will be in the base a formula $\psi''$ logically equivalent to $\psi'$. Furthermore, since $\vdash_{HM_U} \chi \leftrightarrow \chi'$ and $Var(\chi) = Var(\chi')$ and $\vdash_{HM_U} L_a\chi' \leftrightarrow \psi''$, it follows that $\vdash_{HM_U} L_a\chi \leftrightarrow \psi''$.

We will now build: (1) the objective state space; (2) the subjective states spaces and the projection $\rho$; and (3) the probability distributions.

(1) The objective state space:

The objective states of the $\phi$-canonical structure are the intersections of the maximally-consistent sets of formulas of the language $\mathcal{L}^{LA}(At)$ and the restricted language $\mathcal{L}^{LA}_{[\phi]}$:

$$S^{\phi} = \{\Gamma \cap \mathcal{L}^{LA}_{[\phi]} : \Gamma \text{ is a maximal } HM_U\text{-consistent set}\}$$

First, let us notice that the system $HM_U$ is a "modal logic" ([38], p. 191): a set of formulas (1) that contains every propositional tautologies; (2) such that the Lindenbaum lemma holds.

**Definition B4.**

(i) *A formula $\phi$ is deducible from a set of formulas $\Gamma$, symbolized $\Gamma \vdash_{HM_U} \phi$, if there exists some formulas $\psi_1, ..., \psi_n$ in $\Gamma$ s.t. $\vdash_{HM_U} (\psi_1 \wedge ... \wedge \psi_n) \rightarrow \phi$.*
(ii) *A set of formulas $\Gamma$ is $HM_U$-consistent if it is false that $\Gamma \vdash_{HM_U} \bot$*
(iii) *A set of formulas $\Gamma$ is maximally $HM_U$-consistent if (1) it is $HM_U$-consistent and (2) if it is not included in a $HM_U$-consistent set of formulas.*

**Lemma B1** (Lindenbaum Lemma). *If $\Gamma$ is a set of $HM_U$-consistent formulas, then there exists an extension $\Gamma^+$ of $\Gamma$ that is maximally $HM_U$-consistent.*

**Proof.** See, for instance, [38] (p.199). □

**Notation B2.** *For each formula $\psi \in \mathcal{L}^{LA}_{[\phi]}$, let us note $[\psi] = \{s \in S^{\phi} : \psi \in s\}$*

**Lemma B2.** *The set $S^{\phi}$ is finite.*

**Proof.** This Lemma is a consequence of the fact that $\mathcal{L}^{LA}_{[\phi]}$ is finitely generated.

(a) Let us say that two sets of formulas are $\Delta$-equivalent if they agree on each formula that belongs to $\Delta$. $S^{\phi}$ identifies the maximal $HM_U$-consistent sets of formulas that are $\mathcal{L}^{LA}_{[\phi]}$-equivalent. $S^{\phi}$ is infinite iff there are infinitely many maximal $HM_U$-consistent sets of formulas that are not pairwise $\mathcal{L}^{LA}_{[\phi]}$-equivalent.
(b) If $\mathfrak{B}$ is a base for $\mathcal{L}^{LA}_{[\phi]}$, then two sets of formulas are $\mathcal{L}^{LA}_{[\phi]}$-equivalent iff they are $\mathfrak{B}$-equivalent. Suppose that $\Delta_1$ and $\Delta_2$ are not $\mathcal{L}^{LA}_{[\phi]}$-equivalent. This means w.l.o.g. that there is a formula $\psi$ s.t. (i) $\psi \in \Delta_1$, (ii) $\psi \notin \Delta_2$ and (iii) $\psi \in \mathcal{L}^{LA}_{[\phi]}$. Let $\psi' \in \mathfrak{B}$ be a formula s.t. $\vdash_{HM_U} \psi \leftrightarrow \psi'$. Clearly, $\psi' \in \Delta_1$ and $\psi' \in \mathcal{L}^{LA}_{[\phi]}$ and $\neg\psi' \in \Delta_2$. Therefore, $\Delta_1$ and $\Delta_2$ are not $\mathfrak{B}$-equivalent. The other direction is obvious.
(c) Since $\mathfrak{B}$ is finite, there are only finitely many maximal $HM_U$-consistent sets of formulas that are not pairwise $\mathfrak{B}$-equivalent. Therefore, $S^{\phi}$ is finite.

□

(2) The subjective state spaces and the projection $\rho(.)$:

As it might be expected, the subjective state associated with an objective state $\Gamma \in S^{\phi}$ will be determined by the formulas that the agent is aware of in $\Gamma$.

**Definition B5.** *For any set of formulas $\Gamma$, let $Var(\Gamma)$ be the set of atomic formulas that occur in the formulas that belong to $\Gamma$. For any $\Gamma \in S^{\phi}$, let:*

(i) $A^+(\Gamma) = \{\psi : A\psi \in \Gamma\}$ *and* $A^-(\Gamma) = \{\psi : \neg A\psi \in \Gamma\}$
(ii) $a^+(\Gamma) = \{p \in Var(\mathcal{L}^{LA}_{[\phi]}) : Ap \in \Gamma\}$ *and* $a^-(\Gamma) = \{p \in Var(\mathcal{L}^{LA}_{[\phi]}) : \neg Ap \in \Gamma\}$.

**Lemma B3.** *Let $\Gamma \in S^{\phi}$.*

(i) $A^+(\Gamma) = \mathcal{L}^{LA}_{[\phi]}(a^+(\Gamma))$

*(ii)* $A^+(\Gamma) \cup A^-(\Gamma) = \mathcal{L}^{LA}_{[\phi]}$

**Proof.** (i) Follows from (A1)-(A4$_L$); (ii) follows from (i) and the fact that since $\Gamma$ comes from a maximal consistent set, $\neg\psi \in \Gamma$ iff $\psi \notin \Gamma$. □

One may group the sets that have the same awareness profile into equivalence classes: $|\Gamma|_a = \{\Delta \in S^\phi\colon a^+(\Delta) = a^+(\Gamma)\}$. The sets that belong to the same equivalence class $|\Gamma|_a$ will be mapped in the same subjective state space $S'_{|\Gamma|_a}$. We are now ready to define the projection $\rho$ and these subjective states.

**Definition B6.** *The projection* $\rho : S^\phi \to \bigcup_{\Gamma \in S^\phi} S'_{|\Gamma|_a}$ *is defined by:*

$$\rho(\Gamma) = \Gamma \cap A^+(\Gamma)$$

*where* $S'_{|\Gamma|_a} = \{\Delta \cap A^+(\Gamma) \cap \mathcal{L}^{LA}_{[\phi]} : \Delta$ *is a maximal* $HM_U$*-consistent set and* $a^+(\Delta) = a^+(\Gamma)\}$.

Note that in the particular case where the agent is unaware of every formula, $A^+(\Gamma) = \emptyset$. Therefore, each objective state where the agent is unaware of every formula will be projected in the same subjective state $\emptyset \in S'_\emptyset = \{\emptyset\}$. More importantly, one can check that $\rho$ is an onto map: suppose that $\Lambda \in S'_{|\Gamma|_a}$ where $\Gamma \in S^\phi$. By definition, for some $\Delta$ (a maximal $HM_U$-consistent set), $\Lambda = \Delta \cap A^+(\Gamma) \cap \mathcal{L}^{LA}_{[\phi]}$ and $a^+(\Delta) = a^+(\Gamma)$. As a consequence, $A^+(\Delta) = A^+(\Gamma)$, and therefore, $\Lambda = \Delta \cap A^+(\Delta) \cap \mathcal{L}^{LA}_{[\phi]}$. Hence, $\Lambda = \rho(\Delta \cap \mathcal{L}^{LA}_{[\phi]})$. One can show also the following lemma.

**Lemma B4.**

*(i)* *For each* $\Gamma \in S^\phi$, $S'_{|\Gamma|_a}$ *is finite.*
*(ii)* *For each subset* $E \subseteq S'_{|\Gamma|_a}$, *there is* $\psi \in A^+(\Gamma) \cap \mathcal{L}^{LA}_{[\phi]}$ s.t. $E = [\psi]_{S'_{|\Gamma|_a}}$ *where* $[\psi]_{S'_{|\Gamma|_a}}$ *denotes the set of states of* $S'_{|\Gamma|_a}$ *to which* $\psi$ *belongs.*
*(iii)* *For all* $\psi_1, \psi_2 \in A^+(\Gamma) \cap \mathcal{L}^{LA}_{[\phi]}$, $[\psi_1]_{S'_{|\Gamma|_a}} \subseteq [\psi_2]_{S'_{|\Gamma|_a}}$ *iff* $\vdash_{HM_U} \psi_1 \to \psi_2$

**Proof.** (i) Follows trivially since the objective state space is already finite; (ii) let us pick a finite base $\mathfrak{B}_\Gamma$ for $A^+(\Gamma) \cap \mathcal{L}^{LA}_{[\phi]}$. For each element $\beta$ of this base and each $\Delta \in S'_{|\Gamma|_a}$, either $\beta \in \Delta$ or $\neg\beta \in \Delta$. Two sets $\Delta$ and $\Delta' \in S'_{|\Gamma|_a}$ differ at least by one such formula of $\mathfrak{B}_\Gamma$. Let $C(\Delta) = \bigwedge_m e_m \beta_m$ where $\beta_m \in \mathfrak{B}_\Gamma$ and $e_m$ is a blank if $\beta_m \in \Delta$ and $\neg$ if $\beta_m \notin \Delta$. For two distinct sets $\Delta$ and $\Delta'$, $C(\Delta) \neq C(\Delta')$. For each event $E \subseteq S'_{|\Gamma|_a}$, one can therefore consider the disjunction $\bigvee_k C(\Delta_k)$ for each $\Delta_k \in E$. Such a formula belongs to each $\Delta_k$ and only to these $\Delta_k$. (iii) ($\Rightarrow$). For each formula $\psi \in A^+(\Gamma) \cap \mathcal{L}^{LA}_{[\phi]}$ and each $\Delta \in S'_{|\Gamma|_a}$, $\neg\psi \in \Delta$ iff $\psi \notin \Delta$. Therefore, there are two possibilities for any $\Delta$: either $\psi \in \Delta$ or $\neg\psi \in \Delta$. (a) If $\psi_1 \in \Delta$, then by hypothesis $\psi_2 \in \Delta$ and given the construction of the language, $\neg\psi_1 \vee \psi_2 \in \Delta$, hence $\psi_1 \to \psi_2 \in \Delta$. (b) If $\psi_1 \notin \Delta$, then $\neg\psi_1 \in \Delta$, hence $\psi_1 \to \psi_2 \in \Delta$. This implies that for any $\Delta$, $\psi_1 \to \psi_2 \in \Delta$. Given the definition of $S'_{|\Gamma|_a}$ and the properties of maximal consistent sets, this implies that $\vdash_{HM_U} \psi_1 \to \psi_2$. ($\Leftarrow$). Given the construction of the language, if $\psi_1, \psi_2 \in A^+(\Gamma) \cap \mathcal{L}^{LA}_{[\phi]}$, then $\psi_1 \to \psi_2 \in A^+(\Gamma) \cap \mathcal{L}^{LA}_{[\phi]}$. Since $\vdash_{HM_U} \psi_1 \to \psi_2$, for each $\Delta$, $\psi_1 \to \psi_2 \in \Delta$. If $\psi_1 \in \Delta$, clearly $\psi_2 \in \Delta$, as well. Therefore, $[\psi_1]_{S'_{|\Gamma|_a}} \subseteq [\psi_2]_{S'_{|\Gamma|_a}}$. □

(3) The probability distributions:

**Definition B7.** *For* $\Gamma \in S^\phi$ *and* $\psi \in \mathcal{L}^{LA}_{[\phi]}$, *let:*

- $\tilde{a} = \max\{a : L_a\psi \in \Gamma\}$
- $\tilde{b} = \min\{b : M_b\psi \in \Gamma\}$

In the classical case [19], $\tilde{a}$ and $\tilde{b}$ are always defined. This is not so in our structure with unawareness: if the agent is not aware of $\psi$, no formula $L_a\psi$ will be true because of $(A0_U)$ $A\psi \leftrightarrow L_0\psi$. Given (A1) and (DefM), one can derive:

$$\vdash_{HM_U} A\psi \leftrightarrow M_1\psi$$

The construction of the language implies that for any $\Gamma$, $A\psi \in \Gamma$ iff $L_0\psi \in \Gamma$ iff $M_1\psi \in \Gamma$. Therefore, $\tilde{a}$ and $\tilde{b}$ are defined iff $A\psi \in \Gamma$.

**Lemma B5.** *Let us suppose that* $A\psi \in \Gamma$.

*(i)* $\forall c \in Q(\phi)$, $c \leq \tilde{a}$ *implies* $L_c\psi \in \Gamma$, *and* $c \geq \tilde{b}$ *implies* $M_c\psi \in \Gamma$
*(ii) There are only two cases: (i) either* $\tilde{a} = \tilde{b}$ *and* $E_{\tilde{a}}\psi \in \Gamma$ *while* $E_c\psi \notin \Gamma$ *for* $c \neq \tilde{a}$, *(ii) or* $\tilde{a} < \tilde{b}$ *and* $E_c\psi \notin \Gamma$ *for any* $c \in Q(\phi)$.
*(iii)* $\tilde{b} - \tilde{a} \leq \frac{1}{q}$ *(where q is the common denominator to the indexes)*

**Proof.** See [19]; the modifications are obvious. □

**Definition B8.** *Given* $\Gamma \in S^{\Phi}$ *and* $\psi \in \mathcal{L}^{LA}_{[\phi]}$, *if* $A\psi \in \Gamma$, *let* $I^{\Gamma}_{\psi}$ *be either* $\{\tilde{a}\}$ *if* $\tilde{a} = \tilde{b}$ *or* $(\tilde{a}, \tilde{b})$ *if* $\tilde{a} < \tilde{b}$.

Lemma A.5 in [19] can be adapted to show that for each $S'_{|\Gamma|_a}$ and $\Gamma \in S'_{|\Gamma|_a}$, there is a probability distribution $P'_{|\Gamma|_a}(\Gamma)$ on $2^{S'_{|\Gamma|_a}}$, such that

(C) for all $\psi \in \mathcal{L}^{LA}_{[\phi]}$ if $A\psi \in \Gamma$, $P'_{|\Gamma|_a}(\Gamma)([\psi]_{S'_{|\Gamma|_a}}) \in I^{\Gamma}_{\psi}$.

The proof in [19] relies on a theorem by Rockafellar that can be used because of the inference rule (B). It would be tedious to adapt the proof here. One comment is nonetheless important. In our axiom system $HM_U$, the inference rule holds under a restricted form $(B_U)$. Therefore, one could wonder whether this will not preclude adapting the original proof, which relies on the unrestricted version (B). It turns out that the answer is negative. The reason is that the formulas involved in the application of (B) are only representatives for each subset of the state space. We have previously shown how to build these formulas in our case, and they are such that the agent is necessarily aware of them. Therefore, the restriction present in $(B_U)$ does not play any role, and we may define the $\phi$-canonical structure as follows.

**Definition B9.** *The* $\phi$*-canonical structure is the GSPS* $\mathcal{M}_{\phi} = (S^{\Phi}, S^{\Phi'}, (2^{S'_{|\Gamma|_a}})_{\Gamma \in S^{\Phi}}, \pi_{\phi}, (P^{\phi}_{|\Gamma|_a})_{\Gamma \in S^{\Phi}})$ *where:*

*(i)* $S^{\Phi} = \{\Gamma \cap \mathcal{L}^{LA}_{[\phi]} : \Gamma$ *is a maximal* $HM_U$*-consistent set*$\}$
*(ii)* $S^{\Phi'} = \bigcup_{\Gamma \in S^{\Phi}} S'_{|\Gamma|_a}$ *where* $S'_{|\Gamma|_a} = \{\Delta \cap A^{+}(\Gamma) \cap \mathcal{L}^{LA}_{[\phi]} : \Delta$ *is a maximal* $HM_U$*-consistent set, and* $a(\Delta) = a(\Gamma)\}$
*(iii) for each* $\Gamma \in S^{\Phi}$, $\rho(\Gamma) = \Gamma \cap A^{+}(\Gamma)$
*(iv) for all state* $\Gamma \in S^{\Phi} \cup S^{\Phi'}$ *and atomic formula* $p \in At$, $\pi_{\phi}(p, \Gamma) = 1$ *iff* $p \in \Gamma$
*(v) for* $\Gamma \in S^{\Phi}$, $P^{\phi}_{|\Gamma|_a}$ *is a probability distribution on* $2^{S'_{|\Gamma|_a}}$ *satisfying Condition (C)*[24].

We are now ready to state the crucial truth lemma.

**Lemma B6** (Truth lemma). *For every* $\Gamma \in S^{\Phi}$ *and every* $\psi \in, \mathcal{L}^{LA}_{[\phi]}$,

[24] In the particular case where $A^{+}(\Gamma) = \emptyset$, the probability assigns maximal weight to the only state of $S'_{\emptyset}$.

$$\mathcal{M}_{\phi}, \Gamma \vDash \psi \text{ iff } \psi \in \Gamma$$

**Proof.** The proof proceeds by induction on the complexity of the formula.

- $\psi := p$; following directly from the definition of $\pi_{\phi}$.
- $\psi := \neg\chi$. Since $\Gamma$ is a standard state $\mathcal{M}_{\phi}, \Gamma \vDash_{\phi} \neg\chi$ iff $\mathcal{M}_{\phi}, \Gamma \nvDash \chi$ iff (by IH) $\chi \notin \Gamma$. We shall show that $\chi \notin \Gamma$ iff $\neg\chi \in \Gamma$. ($\Rightarrow$) Let us suppose that $\chi \notin \Gamma$; $\chi$ is in $\mathcal{L}^{LA}_{[\phi]}$; hence, given the properties of maximally-consistent sets, $\neg\chi \in \Gamma^{+}$ where $\Gamma^{+}$ is the extension of $\Gamma$ to $\mathcal{L}^{LA}(At)$ (the whole language). Additionally, since $\Gamma = \Gamma^{+} \cap \mathcal{L}^{LA}_{[\phi]}$, $\neg\chi \in \Gamma$. ($\Leftarrow$) Let us suppose that $\neg\chi \in \Gamma$. $\Gamma$ is coherent, therefore $\chi \notin \Gamma$.
- $\psi := \psi_1 \wedge \psi_2$. ($\Rightarrow$). Let us assume that $\mathcal{M}_{\phi}, \Gamma \vDash \psi_1 \wedge \psi_2$. Then, $\mathcal{M}_{\phi}, \Gamma \vDash \psi_1$ and $\mathcal{M}_{\phi}, \Gamma \vDash \psi_2$. By IH, this implies that $\psi_1 \in \Gamma$ and $\psi_2 \in \Gamma$. Given the properties of maximally-consistent sets, this implies in turn that $\psi_1 \wedge \psi_2 \in \Gamma$. ($\Leftarrow$). Let us assume that $\psi_1 \wedge \psi_2 \in \Gamma$. Given the properties of maximally-consistent sets, this implies that $\psi_1 \in \Gamma$ and $\psi_2 \in \Gamma$ and, therefore, by IH, that $\mathcal{M}_{\phi}, \Gamma \vDash \psi_1$ and $\mathcal{M}_{\phi}, \Gamma \vDash \psi_2$.
- $\psi := A\chi$. We know that in any GSPS $\mathcal{M}$, if $s' = \rho(s) \in S'_X$ for some $s \in S$, then $\mathcal{M}, s' \Downarrow \chi$ iff $\chi \in \mathcal{L}^{LA}(X)$. In our case, $s = \Gamma$, $s' = \rho(\Gamma)$ and $X = a^{+}(\Gamma)$. Therefore, $\mathcal{M}_{\phi}, \Gamma \vDash A\chi$ iff $\chi \in \mathcal{L}^{LA}(a^{+}(\Gamma))$. However, given that $A\chi \in \mathcal{L}^{LA}_{[\phi]}$, $\chi \in \mathcal{L}^{LA}(a^{+}(\Gamma))$ iff $A\chi \in \Gamma$.
- $\psi := L_a\chi$. By definition $\mathcal{M}_{\phi}, \Gamma \models L_a\chi$ iff $P_{|\rho(\Gamma)|_a}(\Gamma)([[\chi]]) \geq a$ and $\mathcal{M}_{\phi}, \rho(\Gamma) \Downarrow \chi$.($\Leftarrow$) Let us suppose that $P_{|\rho(\Gamma)|_a}(\Gamma)([[\chi]]) \geq a$ and $\mathcal{M}_{\phi}, \rho(\Gamma) \Downarrow \chi$. Hence, $\tilde{a}$ is well defined. It is clear that $\tilde{a} \geq a$ given our definition of $P_{|\rho(\Gamma)|_a}(\Gamma)$. It is easy to see that $\vdash_{HM_U} L_a\psi \rightarrow L_b\psi$ for $b \leq a$. As a consequence, $L_a\psi \in \Gamma$. ($\Rightarrow$) Let us suppose that $L_a\psi \in \Gamma$. This implies that $A\psi \in \Gamma$ and, therefore, that $\mathcal{M}_{\phi}, \rho(\Gamma) \Downarrow \chi$. By construction, $a \leq \tilde{a}$, and therefore, $P_{|\rho(\Gamma)|_a}(\Gamma)([[\chi]]) \geq a$. Hence, $\mathcal{M}_{\phi}, \Gamma \models L_a\chi$.

□

**Proof.**

- If $\phi := p$, then $\mathcal{M}, s' \vDash p$ or $\mathcal{M}, s' \vDash \neg p$

  iff ($p \in \mathcal{L}^{LA}(X)$ and $\pi^*(s', p) = 1$) or ($p \in \mathcal{L}^{LA}(X)$ and $\pi^*(s', p) = 0$)

  iff $p \in \mathcal{L}^{BA}(X)$

- If $\phi := \neg\psi$, then $\mathcal{M}, s' \vDash \phi$ or $\mathcal{M}, s' \vDash \neg\phi$

  iff $\mathcal{M}, s' \vDash \neg\psi$ or $\mathcal{M}, s' \vDash \neg\neg\psi$

  iff ($\psi \in \mathcal{L}^{LA}(X)$ and $\mathcal{M}, s' \nvDash \psi$) or ($\neg\psi \in \mathcal{L}^{LA}(X)$ and $\mathcal{M}, s' \nvDash \neg\psi$)

  iff ($\psi \in \mathcal{L}^{LA}(X)$ and $\mathcal{M}, s' \nvDash \psi$) or ($\psi \in \mathcal{L}^{LA}(X)$ and $\mathcal{M}, s' \vDash \psi$)

  iff $\psi \in \mathcal{L}^{LA}(X)$

  iff $\neg\psi \in \mathcal{L}^{LA}(X)$

- If $\phi := \psi_1 \wedge \psi_2$, then $\mathcal{M}, s' \vDash \phi$ or $\mathcal{M}, s' \vDash \neg\phi$

  iff ($\mathcal{M}, s' \vDash \psi_1$ and $\mathcal{M}, s' \vDash \psi_2$) or ($\psi_1 \wedge \psi_2 \in \mathcal{L}^{LA}(X)$ and ($\mathcal{M}, s' \nvDash \psi_1$ or $\mathcal{M}, s' \nvDash \psi_2$))

  iff by the induction hypothesis ($\psi_1 \wedge \psi_2 \in \mathcal{L}^{LA}(X)$ and $\mathcal{M}, s' \vDash \psi_1$ and $\mathcal{M}, s' \vDash \psi_2$) or ($\psi_1 \wedge \psi_2 \in \mathcal{L}^{LA}(X)$ and not ($\mathcal{M}, s' \vDash \psi_1$ and $\mathcal{M}, s' \vDash \psi_2$))

  iff $\psi_1 \wedge \psi_2 \in \mathcal{L}^{LA}(X)$

- if $\phi := B\psi$, then $\mathcal{M}, s' \models \phi$ or $\mathcal{M}, s' \models \neg\phi$
  iff (for each $t^* \in R^*(s')$, $\mathcal{M}, t^* \models \phi$) or ($B\psi \in \mathcal{L}^{LA}(X)$) and $\mathcal{M}, s' \not\models B\psi$)
  iff, by the induction hypothesis and since each $t^* \in R^*(s')$ belongs to $S_X$ - ($B\psi \in \mathcal{L}^{LA}(X)$ $\mathcal{M}, s' \models B\phi$) or ($B\psi \in \mathcal{L}^{LA}(X)$ and $\mathcal{M}, s' \not\models B\phi$)
  iff $B\psi \in \mathcal{L}^{LA}(X)$

- If $\phi := A\psi$, then $\mathcal{M}, s' \models \phi$ or $\mathcal{M}, s' \models \neg\phi$
  iff ($\mathcal{M}, s' \Downarrow \psi$) or ($A\psi \in \mathcal{L}^{LA}(X)$ and not $\mathcal{M}, s' \Downarrow \psi$, impossible given the induction hypothesis)
  iff ($\mathcal{M}, s' \Downarrow \psi$)
  iff $\psi \in \mathcal{L}^{LA}(X)$ (by the induction hypothesis)
  iff $A\psi \in \mathcal{L}^{LA}(X)$

□

## References


1. Fagin, R.; Halpern, J.; Moses, Y.; Vardi, M. *Reasoning about Knowledge*; MIT Press: Cambridge, MA, USA, 1995.
2. Hintikka, J. Impossible Worlds Vindicated. *J. Philos. Log.* **1975**, *4*, 475–484.
3. Wansing, H. A General Possible Worlds Framework for Reasoning about Knowledge and Belief. *Stud. Log.* **1990**, *49*, 523–539.
4. Fagin, R.; Halpern, J. Belief, Awareness, and Limited Reasoning. *Artif. Intell.* **1988**, *34*, 39–76.
5. Modica, S.; Rustichini, A. Unawareness and Partitional Information Structures. *Games Econ. Behav.* **1999**, *27*, 265–298.
6. Dekel, E.; Lipman, B.; Rustichini, A. Standard State-Space Models Preclude Unawareness. *Econometrica* **1998**, *66*, 159–173.
7. Halpern, J. Plausibility Measures : A General Approach for Representing Uncertainty. In Proceeddings of the 17th International Joint Conference on AI, Seattle, WA, USA, 4–10 August 2001; pp. 1474–1483.
8. Heifetz, A.; Meier, M.; Schipper, B. Interactive Unawareness. *J. Econ. Theory* **2006**, *130*, 78–94.
9. Li, J. Information Structures with Unawareness. *J. Econ. Theory* **2009**, *144*, 977–993.
10. Galanis, S. Unawareness of theorems. *Econ. Theory* **2013**, 52, 41–73.
11. Feinberg, Y. *Games with Unawareness*; Working paper; Stanford Graduate School of Business: Stanford, CA, USA, 2009.
12. Heifetz, A.; Meier, M.; Schipper, B. Dynamic Unawareness and Rationalizable Behavior. *Games Econ. Behav.* **2013**, *81*, 50–68.
13. Rêgo, L.; Halpern, J. Generalized Solution Concepts in Games with Possibly Unaware Players. *Int. J. Game Theory* **2012**, *41*, 131–155.
14. Schipper, B. Awareness. In *Handbook of Epistemic Logic*; van Ditmarsch, H., Halpern, J.Y., van der Hoek, W., Kooi, B., Eds.; College Publications: London, UK, 2015; pp. 147–201.
15. Aumann, R. Interactive Knowledge. *Int. J. Game Theory* **1999**, *28*, 263–300.
16. Harsanyi, J. Games with Incomplete Information Played by 'Bayesian' Players. *Manag. Sci.* **1967**, *14*, 159–182.
17. Fagin, R.; Halpern, J.; Megiddo, N. A Logic for Reasoning About Probabilities. *Inf. Comput.* **1990**, *87*, 78–128.
18. Lorenz, D.; Kooi, B. Logic and probabilistic update. In *Johan van Benthem on Logic and Information Dynamics*; Baltag, A., Smets, S., Eds.; Springer: Cham, Switzerland, 2014; pp. 381–404.
19. Heifetz, A.; Mongin, P. Probability Logic for Type Spaces. *Games Econ. Behav.* **2001**, *35*, 34–53.
20. Cozic, M. Logical Omniscience and Rational Choice. In *Cognitive Economics*; Topol, R., Walliser, B., Eds.; Elsevier/North Holland: Amsterdam, Holland, 2007; pp. 47–68.
21. Heifetz, A.; Meier, M.; Schipper, B. Unawareness, Beliefs and Speculative Trade. *Games Econ. Behav.* **2013**, *77*, 100–121.
22. Schipper, B. Impossible Worlds Vindicated. University of California, Davis. Personal communication, 2013.
23. Sadzik, T. *Knowledge, Awareness and Probabilistic Beliefs*; Stanford Graduate School of Business: Stanford, CA, USA, 2005.

24. Savage, L. *The Foundations of Statistics*, 2nd ed.; Dover: New York, NY, USA, 1954.
25. Board, O.; Chung, K. *Object-Based Unawareness*; Working Paper; University of Minnesota: Minneapolis, MN, USA, 2009.
26. Board, O.; Chung, K.; Schipper, B. Two Models of Unawareness: Comparing the object-based and the subjective-state-space approaches. *Synthese* **2011**, *179*, 13–34.
27. Chen, Y.C.; Ely, J.; Luo, X. Note on Unawareness: Negative Introspection versus AU Introspection (and KU Introspection). *Int. J. Game Theory* **2012**, *41*, 325–329.
28. Fagin, R.; Halpern, J. Uncertainty, Belief and Probability. *Comput. Intell.* **1991**, *7*, 160–173.
29. Halpern, J. *Reasoning about Uncertainty*; MIT Press: Cambridge, MA, USA, 2003.
30. Vickers, J.M. *Belief and Probability*; Synthese Library, Reidel: Dordrecht, Holland, 1976; Volume 104.
31. Aumann, R.; Heifetz, A. Incomplete Information. In *Handbook of Game Theory*; Aumann, R., Hart, S., Eds.; Elsevier/North Holland: New York, NY, USA, 2002; Volume 3, pp. 1665–1686.
32. Kraft, C.; Pratt, J.; Seidenberg, A. Intuitive Probability on Finite Set. *Ann. Math. Stat.* **1959**, *30*, 408–419.
33. Gärdenfors, P. Qualitative Probability as an Intensional Logic. *J. Philos. Log.* **1975**, *4*, 171–185.
34. Van Benthem, J.; Velazquez-Quesada, F. The Dynamics of Awareness. *Synthese* **2010**, *177*, 5–27.
35. Velazquez-Quesada, F. Dynamic Epistemic Logic for Implicit and Explicit Beliefs. *J. Log. Lang. Inf.* **2014**, *23*, 107–140.
36. Hill, B. Awareness Dynamics. *J. Philos. Log.* **2010**, *39*, 113–137.
37. Halpern, J. Alternative Semantics for Unawareness. *Games Econ. Behav.* **2001**, *37*, 321–339.
38. Blackburn, P.; de Rijke, M.; Venema, Y. *Modal Logic*; Cambridge UP: Cambridge, UK, 2001.

*Article*

# Strategy Constrained by Cognitive Limits, and the Rationality of Belief-Revision Policies

**Ashton T. Sperry-Taylor**

National Coalition of Independent Scholars; Ronin Institute; Lititz, PA 17543, USA; ashton.sperry@ncis.org or ashton.sperry@ronininstitute.org; Tel.: +1-717-951-9585

Academic Editors: Paul Weirich and Ulrich Berger
Received: 1 September 2016; Accepted: 22 December 2016; Published: 3 January 2017

**Abstract:** Strategy is formally defined as a complete plan of action for every contingency in a game. Ideal agents can evaluate every contingency. But real people cannot do so, and require a belief-revision policy to guide their choices in unforeseen contingencies. The objects of belief-revision policies are beliefs, not strategies and acts. Thus, the rationality of belief-revision policies is subject to Bayesian epistemology. The components of any belief-revision policy are credences constrained by the probability axioms, by conditionalization, and by the principles of indifference and of regularity. The principle of indifference states that an agent updates his credences proportionally to the evidence, and no more. The principle of regularity states that an agent assigns contingent propositions a positive (but uncertain) credence. The result is rational constraints on real people's credences that account for their uncertainty. Nonetheless, there is the open problem of non-evidential components that affect people's credence distributions, despite the rational constraint on those credences. One non-evidential component is people's temperaments, which affect people's evaluation of evidence. The result is there might not be a proper recommendation of a strategy profile for a game (in terms of a solution concept), despite agents' beliefs and corresponding acts being rational.

**Keywords:** backward induction; bayesian epistemology; belief-revision policy; epistemic game theory; evolutionary game theory; naturalistic game theory; strategy

---

## 1. Introduction

Ken Binmore [1] demarcates two modes of analysis in game theory. The eductive mode (comprised of deduction, induction, and adduction) analyzes social behavior according to the strict idealizations of rational choice theory. The evolutive mode analyzes social behavior according to dynamic models that relax said idealizations. I substitute the disciplines of epistemic and evolutionary game theory for Binmore's modes of analysis. Epistemic game theory provides the epistemic foundations for rational justification of social behavior. Conversely, evolutionary game theory establishes social behavior that is evolutionarily fit and stable. It is descriptive and provides for the causal explanation of social behavior.

A recent discipline—I denote it 'naturalistic game theory'—splits the difference between epistemic and evolutionary game theory. The disciplines of neuroeconomics and behavioral game theory [2], the application of heuristics from an adaptive toolbox to real-world uncertainty [3,4], and the study of optimization under cognitive constraints [5,6] are three converging approaches to a more realistic basis for game theory. All foundational concepts are analyzed vis-à-vis real people with cognitive limits. Work has started on common knowledge, convention, and salience; I contribute to this discipline with my analysis of strategy and the rationality of belief-revision policies.

I argue that the very concept of strategy is best understood vis-à-vis real people's cognitive limits. A strategy is traditionally defined as a complete plan of action, which evaluates every contingency in a game of strategy. However, only ideal agents can evaluate every contingency. Real people require policies to revise their beliefs and guide their choices when confronted with unforeseen contingencies. The very concept of strategy requires belief-revision policies. Yet, in epistemic game theory, belief-revision policies are part of those idealizations (the game's structure, the agents' rationality, and the game's epistemic context) stipulated to formally prove conditional results about the game's outcomes. The rationality of strategies or acts depends on belief-revision policies, but the rationality of belief-revision policies remains unanalyzed.

The rationality of belief-revision policies is the subject of Bayesian epistemology. The objects of belief-revision policies are doxastic or epistemic states (I, henceforth, use 'epistemic' to refer to both types of states). Hence, the objects of analysis of belief-revision policies are epistemic states, not strategies and acts. Whereas traditional epistemology studies three basic judgments (belief, disbelief, or suspension of belief), Bayesian epistemology studies degrees of belief as credences, which indicate the confidence that is appropriate to have in various propositions. I extend the epistemic foundations of game theory to issues in Bayesian epistemology. I construct an arbitrary belief-revision policy by establishing its components: agents' credences are constrained by the probability axioms, by conditionalization, and by the principles of indifference and of regularity. I also consider non-evidential factors that affect credence distributions, despite those credences being rationally constrained.

I explain the general program of naturalistic game theory in Section 2. The main claim is that foundational concepts are best understood vis-à-vis real people's cognitive limits, which explains why analyzing said concepts in highly abstract and artificial circumstances is problematic. I then argue the same case for strategy and belief-revision policies in Section 3. Indeed, the charge of incoherence to backward induction is best explained by naturalism. However, this leaves open the question of how we analyze the rationality of belief-revision policies. I suggest doing so through Bayesian epistemology in Sections 4 and 5. I place the epistemic foundations of game theory into a more foundational epistemology.

## 2. Naturalistic Game Theory

Naturalistic game theory can be encapsulated in one claim: concepts foundational to game theory require realism. Realism provides understanding of said concepts that better matches our common sense and intuitions, while keeping in rigor. Remove realism and the foundational concepts are confronted by potentially irredeemable difficulties. Common knowledge, convention, rationality, and salience are excellent candidates for requiring realism.

I use Zachary Ernst's [5] argument form: an inference to the best explanation. The literature has recorded that the aforementioned concepts succumb to conceptual difficulties. One explanation is that the proper formal analysis has yet to be developed to account for said difficulties. But the favored explanation for naturalism is that these concepts are heuristics optimized for a range of realistic circumstances. The conceptual difficulties recorded in the literature are actually markers for the concepts' limits of application, and it is unsurprising they fail in extremely abstract and artificial circumstances. I provide two examples below: (Section 2.1) principles of rationality fail to prescribe the proper choice of action in specific abstract or artificial circumstances; and (Section 2.2) ideal agents cannot act in circumstances in which real people easily do so. This is counterintuitive. Ideal agents are *ideal*, because they have all of the information and cognitive ability to evaluate any choice and act on it in any circumstance.

### 2.1. Newcomb's Problem

Newcomb's Problem [7] is specifically constructed to contrast recommendations between the principles of expected-utility maximization and of dominance. There are two boxes, one opaque and the other transparent. An agent chooses between the contents of the opaque box (one-boxing) or the contents of *both* the opaque and transparent boxes (two-boxing). The transparent box contains one-thousand dollars. The opaque box is either empty or contains one-million dollars, depending on a perfect prediction made about the agent's choice. If the prediction is that the agent will one-box, the opaque box contains one-million dollars. If the prediction is that the agent will two-box, the opaque box is empty. If the agent chooses one-boxing, the prediction of one-boxing is guaranteed. If the agent chooses two-boxing, the prediction of two-boxing is guaranteed. Hence, one-boxing's expected utility is greater than two-boxing's expected utility. The principle of expected-utility maximization prescribes one-boxing. However, the outcome of two-boxing is better by one-thousand dollars than the outcome of one-boxing, regardless of the prediction. The principle of dominance prescribes two-boxing ([8], Section 2.1).

The principles of expected-utility maximization and of dominance are equally reasonable, but individually prescribe conflicting choices of action. There are two proposals ([8], Section 2.2). One proposal is to argue for a particular choice and its corresponding principle. Simon Burgess [9] recommends one-boxing by partitioning the problem into two stages. The agent firmly commits to one-boxing in the first stage before the prediction. In the second stage, the agent does not waver from his choice and acts on it. The other proposal is to reconcile both principles of rationality. Allan Gibbard and William Harper [10] demarcate between causal and evidential decision theory. Causal decision theory shows that two-boxing's expected utility is greater than one-boxing's expected utility. Hence, both the principles of expected-utility maximization and of dominance prescribe two-boxing. However, this depends on whether the agent's choice causes the prediction (perhaps there is retro-causality).

There is ongoing debate over the viability of these proposals. Naturalistic game theory does not argue that Newcomb's Problem can never be solved. Perhaps there is a proper formal analysis that further and further refines decision theory to provide a consistent solution. However, the including of more and more 'machinery' into an analysis can count against the reasonability of said analysis. The history of science is replete with moments where including more and more machinery into explanations has counted against the reasonability of said explanations. Naturalistic game theory instead infers that principles of expected-utility maximization and of dominance evolved in real-time as heuristics to optimize choices in a range of realistic circumstances. Newcomb's Problem marks the limits of those principles.

### 2.2. Common Knowledge

Robert Aumann's [11] common knowledge is standard in epistemic game theory: (every agent in a group knows that)$^n$ $p$, where $n$ is the degree of 'every agent in a group knows that'. The group has mutual knowledge of $p$ just in case $n = 1$. The group has common knowledge of $p$ just in case $n$ is arbitrary. Aumann's common knowledge imposes strong epistemic conditions on agents, conditions that real people cannot satisfy. Common knowledge thus requires ideal agents with no cognitive limits. However, ideal agents are beset with difficulties. For example, there are many versions of the Email Game in which agents with common knowledge cannot coordinate their actions [12]. The original game is a coordination problem where two agents attempt to coordinate their actions by sending each other emails with instructions, but each email has a small probability of not reaching its intended target. Other versions involve two generals coordinating their attacks through messengers, who might die traveling to their intended target; and two spies coordinating their risky package-exchange with clandestine signals. The agents require some level of mutual knowledge to successfully coordinate their actions. Yet the reasoning that justifies one level of mutual knowledge is equally required to justify an additional level of mutual knowledge, and an additional level of mutual knowledge, ad infinitum. The result is an impossibility for ideal agents to coordinate their risky actions, because they never obtain justification. This is counterintuitive. Ideal agents ought to obtain any level of justification—hence, the idealization.

Zachary Ernst [5] and Robin Cubit and Robert Sugden [13,14] explain that David Lewis's common knowledge is radically different from Aumann's. Lewis's book *Convention* [15] is, first and foremost, a defense of analyticity (a sentence is true in virtue of its words' meaning) against regress arguments.

Suppose language is true by convention. Then speakers must know how the convention applies to that language. However, preexisting knowledge of the convention is required to know how the convention applies to the language. This creates a vicious regress, and language is not true by convention.

Lewis defends analyticity by developing a concept of convention that allows for meaning of words to be conventional, and for agents to be part of the convention, without requiring those agents to have explicit knowledge of the convention. He develops an account of common knowledge with minimal epistemic conditions so agents can be part of a convention. The agents know how to apply the convention but this knowledge does not require inordinate levels of mutual knowledge. Lewis instead emphasizes states of affairs as sources of common knowledge: the occurrence of a state of affairs about $p$ indicates to agents in a group that they have a reasonable degree of mutual knowledge about $p$. Lewis's common knowledge emphasizes agents' environments much more than their cognitive capacities.

Again, naturalistic game theory does not argue that Aumann's common knowledge can never be reconciled in coordination problems such as the Email Game. However, Lewis provides a realistic account of common knowledge that real people can obtain and regularly act on, despite being unable to reason to inordinate levels of mutual knowledge. Real people do not require strong justification, and can obtain common knowledge with little reasoning. Lewis's realism also explains the fragility of common knowledge and convention [16].

## 3. Strategy for Ideal Agents and for Real People

The concept of strategy has the same fate as rationality and common knowledge: strategy is best conceived vis-à-vis real people's cognitive limits. Indeed, I argue that the charge of incoherence to backward induction is an excellent example of the limits of the traditional concept of strategy.

The charge of incoherence is presented as either one of two problems. The first problem is that backward induction requires agents to evaluate all contingencies in a game to settle on the backward induction profile, but certain contingencies cannot be evaluated without violating backward induction [17–23]. The second problem is that backward induction 'seems logically inescapable, but at the same time it is intuitively implausible' ([24], p. 171).

A game theorist analyzes a game from an outside perspective to provide a game's complete analysis, which is its theory [18]. The game theorist specifies the game's structure, the set of agents' strategies, and their payoffs. He defines the agents' rationality, and then specifies the game's epistemic context, which is the agents' epistemic states concerning their circumstances. The game theorist then proves conditional results about the game's outcomes as profiles of agents' strategies [25–28]. Backward induction applies to the formation of strategies.

For any sequential game with perfect (and complete) information, a strategy is a function that assigns an action at every stage the agent can possibly reach, including those stages not reached during gameplay ([29], p. 108). Consider Robert Stalnaker's [30,31] sequential game in Figure 1:

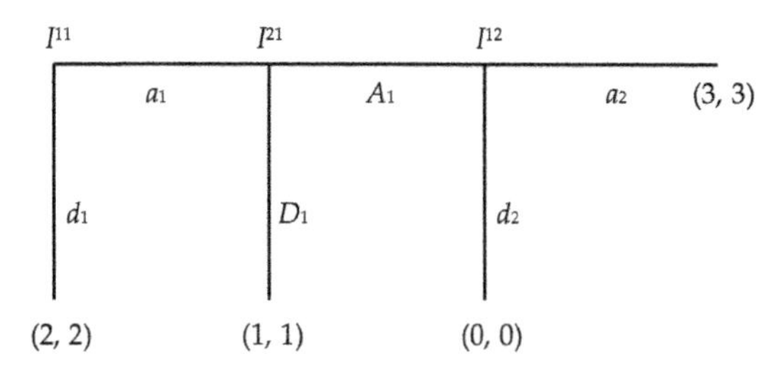


**Figure 1.** $I^{ij}$ is the $j$-th information set ($j \geq 1$) of agent $i$ ($i$ = 1, 2). Robert Stalnaker's game has two pure strategies at each information set: Either 'Down' ending the game, or 'Across' allowing the other agent to choose an action. The agents' payoffs are represented at the tree's endpoints, the left member being the first agent's payoff, the right member being the second agent's payoff. The backward induction profile is $(a_1a_2, A_1)$.

Suppose we differentially distribute the agents' epistemic states across the game's stages and consider each stage as an agent's information state such that:

(1) The first agent is rational at $I^{12}$.
(2) The second agent is rational at $I^{21}$ and knows (1).
(3) The first agent is rational at $I^{11}$ and knows (2).

The agents are rational insofar as they maximize their expected payoffs. Then the distributed epistemic states are all that are required to construct the agents' strategies through backward induction. Beginning at the third stage and reasoning our way backwards to the first stage, backward induction settles the first agent's strategy $(a_1, a_2)$ as the best response to the second agent's strategy $(A_1)$, and vice versa. The outcome $(a_1a_2, A_1)$ is the sub-game perfect equilibrium and the game's solution entailed by (1), (2) and (3). Though it is not the case in this game, an important condition of the formal definition of strategy is that agents assign actions for those stages that will not be reached by the agent during gameplay.

Robert Aumann argues that common knowledge of rationality implies the backward induction profile in sequential games with perfect information [32–34]. Backward induction requires that there are no stages where it is unclear whether a choice maximizes an agent's expected payoff. Aumann introduces clarity by providing agents with the right sort of rationality throughout the game. An agent is substantively rational just in case every choice maximizes his expected payoffs at all stages the agent can possibly reach. Substantive rationality includes those stages that an agent will not actually reach during gameplay. This is the sort of rationality Aumann associates with backward induction. This is different from evaluating acts that have occurred. An agent is materially rational just in case every choice the agent actually acts on maximizes his expected payoffs. Notice the trivial case that if an agent never acts in a game, that agent is materially rational.

Consider the following case from Stalnaker [31] with the profile $(d_1a_2, D_1)$. This profile is a deviation from the backward induction profile, and Stalnaker uses it to show the difference between substantive and material rationality. Suppose the first agent knows the second agent is following strategy $D_1$. The first agent's act $d_1$ maximizes his expected payoff, given $D_1$. And the first agent's act $a_2$ maximizes his expected payoff, given the game's structure and payoffs. The first agent is substantively rational. Now suppose the second agent knows the first agent is following strategy $d_1a_2$. Thus, the second agent knows that the first agent initially chooses $d_1$, and, if the first agent reaches the third stage, the first agent chooses $a_2$. Stalnaker explains that the second agent is materially rational, since the second agent never acts on his strategy. His material rationality is preserved through inaction. Is the second agent, nonetheless, substantively rational? Aumann argues that the second agent is not substantively rational, because his choice $D_1$ does not maximize his payoff at the second stage.

However, Stalnaker argues that there is insufficient information about what the second agent knows about the first agent, to assess the second agent's choice $D_1$ at the second stage.

Stalnaker explains that substantive rationality conflates two types of conditionals, given deviations from the backward induction profile. The first conditional is causal: if the second stage were reached, the second agent only evaluates subsequent stages and believes the first agent would choose $a_2$, since the first agent is rational. The causal conditional expresses a belief about an opponent's disposition to act in a contingency that will not arise. The second conditional is epistemic: suppose there is a deviation from the backward induction profile. The second agent revises his beliefs about his opponent, *given* that the second stage *is reached*. The epistemic conditional concerns the second agent's belief-revision policy. The second agent is not required to maintain his current beliefs about his opponent upon learning there is a deviation from the backward induction profile. Therefore, if the formal definition of strategy requires agents to evaluate all contingencies in a game to settle on the backward induction profile, and if deviations cannot be evaluated without violating backward induction, then a belief-revision policy is required to evaluate deviations without violating backward induction. The epistemic conditional is the conditional relevant to evaluating deviations and circumventing the charge of incoherence to backward induction.

I argue, however, that the charge of incoherence to backward induction presumes a concept of strategy that requires an *explicit* choice of action for every contingency, which further presumes an ideal agent. Dixit, et al. [35] provide a simple test to determine whether a strategy is a complete plan of action. A strategy:

> ... should specify how you would play a game in such full detail—describing your action in every contingency—that, if you were to write it all down, hand it to someone else, and go on vacation, this other person acting as your representative could play the game just as you would have played it. He would know what to do on each occasion that could conceivably arise in the course of play, without ever needing to disturb your vacation for instructions on how to deal with some situation that you have not foreseen ([35], p. 27).

The test is whether your strategy is sufficient for your representative to play the game just as you would without requiring further instruction. If so, your strategy is a complete plan of action. The representative test requires an ideal agent. An ideal agent has no cognitive limits, can perform any deduction, knows all logical and mathematical truths, and has all information (traditionally in the form of common knowledge) required to choose rationally. Only an ideal agent can evaluate and then assign a choice of action for every contingency.

However, does an ideal agent have *use* for a belief-revision policy to account for deviations from the backward induction profile? I argue 'no'. An ideal agent is provided with all the information about his opponent to properly strategize. The ideal agent knows his opponent. An ideal agent can consider counterfactual circumstances about his opponent (supposing he were mistaken about his opponent or circumstances). However, the ideal agent does not need to strategize based on possible mistakes, because those mistakes are not matter-of-fact. Indeed, an ideal agent maintains his current beliefs about his opponent. Hence, strategizing precludes any deviations from expected gameplay. If an ideal agent has little need for a belief-revision policy, then an ideal agent has little need for strategizing with epistemic conditionals. The ideal agent will take his circumstances and evaluate them as they are.

Stalnaker anticipates my objection.

> Now it may be that if I am absolutely certain of something, and have never even imagined the possibility that it is false, then I will not have an explicit policy in mind for how to revise my beliefs upon discovering that I am wrong. But I think one should think of belief-revision policies as dispositions to respond, and not necessarily as consciously articulated policies. Suppose Bob is absolutely certain that his wife Alice is faithful to him—the possibility that she is not never entered his mind. And suppose he is right—she is faithful, and he really knows that she is. We can still ask how Bob would revise his beliefs if he walked

> in one day and found his wife is bed with another man. We need not assume that Bob's absolute certainty implies that he would continue to believe in his wife's faithfulness in these circumstances, and we need not believe that these circumstances are logically impossible. We can make sense of them, as a logical possibility, even if they never occur to Bob. And even if Bob never thought of this possibility, it might be true that he would react to the situation in a certain way, revising his beliefs in one way rather than another ([31], pp. 49–50).

However, what Bob knows about Alice is crucial to how he evaluates his choices in this case. What is relevant to Bob is how he will *actually* interact with Alice, and how he will actually interact with Alice depends on what he knows about her. Bob can consider a hypothetical case of his wife's infidelity and still know Alice is faithful—they are not logically incompatible. However, Bob will choose how he interacts with his wife based on his knowledge of her faithfulness, not on a hypothetical case of infidelity. Bob can even devise the most sophisticated belief-revision policy to acount for myriad hypothetical cases of Alice's infidelity. But that policy is based on nothing more than fictions: epistemic conditionals with false antecedents. Bob does not include his belief-revision policy in his decision-making when interacting with his wife, because of what he knows about her. Indeed, I can make sophisticated decisions about how I will interact in Harry Potter's world. I can devise a belief-revision policy to account for myriad magical circumstances, a policy that considers that I am mistaken about not existing in Harry Potter's world, but it is a fictional world and it does not affect how I will act in this world.

Choices track efficacy, not auspiciousness ([8], Section 1). Evidential conditionals have false antecedents. Hence, an ideal agent will not consider deviations, because they are not efficacious, regardless of their auspiciousness. The information provided to an ideal agent is information that tracks efficacy. Hence, an ideal agent evaluates his choices' causal influence on an outcome. An ideal agent's belief-revision policy just accounts for a fictional world.

Therefore, ideal agents do not require belief-revision policies. If strategy does require belief-revision policies, it is for real people with cognitive limits. Real people can only foresee some contingencies and require a strategy with a belief-revision policy to address unforeseen contingencies. Their plan of action is provisional and open-ended. Real people do not know their choice's efficacy. Moreover, the representative test is not intelligible to real people. It is not simply a listing of instructions of the form: 'If contingency C occurs, choose action A'. A strategy includes the complete reasoning behind the forming of instructions, and that reasoning includes what is known about every contingency. A complete plan of action is incomprehensible and too complex for real people. It takes an ideal agent to make sense of the representative test.

I propose a more realistic understanding of the representative test. A realistic strategy can pass the test without the need of providing a complete plan of action. A representative might play the game in your favor by following a simple strategy in the form of an algorithm that can still evaluate any unforeseen contingency. The strategy Tit-For-Tat passes the representative test. The instructions might be written as:

**Step 1:** Cooperate at the first stage of the game.
**Step 2:** If the opponent cooperates at stage $n-1$ (for $n > 2$), cooperate at stage $n$. Otherwise, defect at stage n.

The strategy is simple, yet provides sufficient information for a representative to play the game exactly as you specify, knowing what to do in any contingency without needing to contact you for further instructions given an unforeseen situation. Notice the strategy does not assign *specific* actions for every contingency in a sequential game. There is no need for such specification. This simple strategy ably passes the representative test, because it has an implicit belief-revision policy.

The implicit belief-revision policy (without formally stating it for the moment) includes epistemic dependence between an opponent's prior acts and an agent's future choice of action. I shall address this more in the next section. Belief-revision polices are crucial to the evaluation of real people's strategies and acts, but only if belief-revision policies themselves are rational. How do we assess a policy's rationality?

## 4. The Fitness of Belief-Revision Policies

I shall briefly discuss the potential role of evolutionary game theory for investigating the rationality of belief-revision policies. It can evaluate belief-revision policies by testing which polices are 'better' than others through evolutionary fitness. Robert Axelrod's tournaments [36] are well-known for directly testing strategies for evolutionary fitness. These same tournaments can indirectly test belief-revision policies for fitness, because strategies in said tournaments have corresponding belief-revision policies.

The strategy Tit-For-Tat has the following implicit policy:

**Tit-For-Tat Policy:** Cooperate at the first stage of the game. If the opponent cooperates at stage $n - 1$ (for $n > 2$), treat the opponent as a cooperator in future stages and choose cooperation at stage $n$. Otherwise, treat the opponent as a defector in future stages and choose defection at stage $n$.

The strategy calls for an agent to initially cooperate, and then consider his opponent's previous act, and then revise his belief about his opponent's subsequent act. It is a policy that has epistemic dependence between an opponent's prior acts and the agent's future choice of action.

There is nothing preventing us from devising strategies and corresponding policies with as much or as little detail as we desire. We simply include Robert Stalnaker's condition of epistemic dependence between prior acts and future choice of action. Any belief-revision policy can be represented as follows:

**Generic Belief-Revision Policy:** If an opponent does action $a_j$ at stage $n - 1$ (for $n > 2$), treat the opponent with disposition $D$ and respond with action $A_j$ at stage $n$. Otherwise, continue with current set of beliefs and choice of action at stage $n$.

The response simply depends on the agent's belief-revision policy, which attributes a disposition to the opponent (for example, the opponent's hand trembled, is irrational, is a cooperator, etc.).

This program is fruitful and necessary: it explains the fitness of belief-revision policies. However, it is limited. It provides only as much information about which policy is fittest, as it does about which strategy is fittest. Kristian Lindgren and Mats Nordahl [37] have studied numerous strategies in Axelrod tournaments and found that there is no 'optimal', 'best', or outright 'winning' strategy. A strategy such as Tit-For-Tat performs better than other strategies for a period of time, before another strategy replaces it. The same might be said about corresponding belief-revision policies.

## 5. Constructing Belief-Revision Policies

An ideal agent with no cognitive limits can settle on a strategy for every contingency in a game. He has no need for a belief-revision policy as Robert Stalnaker envisions. However, real people with cognitive limits can only consider some contingencies. They do require a belief-revision policy for contingencies they have not considered. Thus, if a strategy requires a belief-revision policy, the concept is best understood vis-à-vis real people's cognitive limits. I shall focus on this realism in this section.

Epistemic game theory has developed Bayesian analyses of agents revising their beliefs during gameplay. However, the discipline has not fully considered some foundational issues in Bayesian epistemology. The most significant issue is that virtually any rational credence distribution can be assigned to a set of propositions (without changing the evidential circumstances). The aim is to compute a unique credence distribution for a set of propositions, given the evidential circumstances. I provide insight into the rationality of belief-revision policies by introducing their components through an example. The aim is for a unique credence distribution that is actionable. belief-revision policies have evidential and non-evidential components. The evidential components are Bayesian principles that guide how agents ought to revise their beliefs during gameplay. The non-evidential component that I consider is agents' temperaments. I show that evidential components can pare down an agent's credences into a unique distribution, but temperaments change credence distributions between agents—a concern that might have to be accepted about belief-revision policies.

Traditional epistemology treats three sorts of judgments: Belief, disbelief, and suspension of belief [38]. Bayesian epistemology instead treats credences, which indicate the differences in an agent's confidence in a set of propositions. Bayesian epistemology provides rational constraints on credences. The two fundamental rational constraints are synchronic and diachronic [39]. The synchronic thesis states that an agent's credences are rational at a given moment just in case those credences comply with Andrey Kolmogorov's [40] probability axioms. The diachronic thesis states that an agent's credences are rational across time just in case those credences update through conditionalization in response to evidence. Both the synchronic and diachronic theses are typically justified by Dutch Book arguments.

The credence function $Cr$ maps a set of propositions to real numbers. Conditionalization applies Bayes's theorem to a conditional credence $Cr(H \mid E) = (Cr(E \mid H)Cr(H))/Cr(E)$, given that $Cr(E) > 0$. $Cr(H)$ is the unconditional credence in the hypothesis. $Cr(E \mid H)$ is the likelihood of the hypothesis on the total evidence. An unconditional credence distribution is computed for a set of propositions from the conditional credence distribution for those propositions, given the total evidence. Conditionalization provides rational closure. If an unconditional credence distribution is constrained by the probability axioms and updates by conditionalization, the resulting credence distribution is constrained by the probability axioms, too.

An arbitrary belief-revision policy is thus far composed of credences constrained by the probability axioms and by conditionalization. In general, I apply credences to finitely repeated and sequential (two-agent) games with perfect (but incomplete) information. Propositions are about agents' strategies. Agents have cognitive limits, but they know certain features of a game. They have perfect information: they know all acts that occur during gameplay. This might not be a fair specification for real people (real people have imperfect memories), but it tracks conditionalization. Agents also know the structure of the game, its payoffs, and the set of available strategies. However, they do not know each other's chosen strategy or immediate acts. Indeed, I specify that agents do not settle on a complete plan of action before gameplay. They instead settle on an immediate choice of action. An agent applies conditionalization on his opponent's acts during gameplay. Furthermore, they do not know each other's rationality. I specify that each agent is rational insofar as they each maximize their expected payoffs, but they do not know about each other's rationality. Each agent has some initial credence that his opponent is irrational, given that some of his opponent's available strategies might not maximize expected payoffs.

I use the Centipede Game in Figure 2 as an example. Agents are not endowed with any evidence about each other's strategies before gameplay. The first agent has $2^3$ available strategies: $(a_1a_2a_3)$, $(a_1a_2d_3)$, $(a_1d_2a_3)$, $(a_1d_2d_3)$, $(d_1a_2a_3)$, $(d_1a_2d_3)$, $(d_1d_2a_3)$, and $(d_1d_2d_3)$. The second agent has $2^2$ available strategies: $(A_1A_2)$, $(A_1D_2)$, $(D_1A_2)$, and $(D_1D_2)$.

Agents are thus far not precluded from assigning $Cr = 1$ or $Cr = 0$ to contingent propositions. Agents might have certainty or complete unbelief about contingent propositions. But it is more plausible that real people with cognitive limits are uncertain about opponents' strategies. Any evidence about opponents' strategies accumulated during gameplay will provide some credence about those strategies. Thus, a credence distribution assigns each strategy some credence. I thus include the principle of regularity into any belief-revision policy. It states that a contingent proposition in a rational credence distribution receives $1 > Cr > 0$ [41]. The principle of regularity is an additional rational constraint to the probability axioms and conditionalization. It pares down credence distributions that can be assigned to a set of contingent propositions.

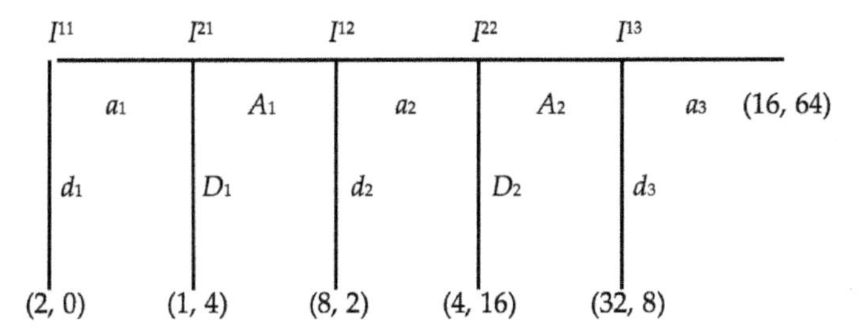


**Figure 2.** $I^{ij}$ is the $j$-th information set ($j \geq 1$) of agent $i$ ($i = 1, 2$). This Centipede Game has two pure strategies at each information set: Either 'Down' ending the game, or 'Across' allowing the other agent to choose an action. The agents' payoffs are represented at the tree's endpoints, the left member being the first agent's payoff, the right member being the second agent's payoff. The backward induction profile is $(d_1d_2d_3, D_1D_2)$.

However, it does not, together with the probability axioms and conditionalization, compute a unique credence distribution. Suppose each agent begins the game with a $Cr = 1/n$ for a number $n$ of an opponent's available strategies. The second agent is assigned $Cr = 1/8$, while the first agent is assigned $Cr = 1/4$ for each of his opponent's available strategies. Suppose the first agent chooses and acts on $a_1$ at $I^{11}$, and the second agent must decide whether to choose $D_1$ or $A_1$ at $I^{21}$, which depends on whether the first agent will choose $d_2$ or $a_2$ at $I^{12}$. The second agent's choice is between his opponent's four remaining available strategies: $(a_1a_2a_3)$, $(a_1a_2d_3)$, $(a_1d_2a_3)$, $(a_1d_2d_3)$. It is important to recognize that the act $a_1$ at $I^{11}$ is evidence that alone does not favor any particular strategy over another in the remaining set of available strategies. The second agent, thus, applies conditionalization on said evidence and evenly assigns credences to those strategies (the evidence does not favor one strategy over another). This is a reasonable application of the principle of indifference. It states that an agent updates his credences proportionally to the evidence, and no more [42]. The second agent applies conditionalization upon the current evidence of $a_1$ at $I^{11}$, and has $Cr = 1/4$ in strategies $(a_1a_2a_3)$, $(a_1a_2d_3)$, $(a_1d_2a_3)$, $(a_1d_2d_3)$. Removing acts that have occurred, he has $Cr = 1/4$ in $(a_2a_3)$, $(a_2d_3)$, $(d_2a_3)$, $(d_2d_3)$. The principle of indifference guides the second agent's judgment of his opponent's future choice of action. It is an additional rational constraint to regularity, to the probability axioms, and to conditionalization.

How does the second agent choose an action, given his current credence distribution? The second agent has a simplified decision problem at $I^{21}$. He has two choices between $D_1$ for the immediate payoff of 4, or $A_1$ for the chance of a payoff of 16 at $I^{22}$. This depends on the likelihood of the first agent choosing $a_2$ at $I^{12}$. We can compute a credence threshold for the second agent. His expected payoff for $A_1$ at $I^{21}$ is $Cr(a_2 \text{ at } I^{12})(16) + (1 - Cr(a_2 \text{ at } I^{12}))(2)$, which is set equal to 4. Apply some algebra and the second agent has a threshold $Cr(a_2 \text{ at } I^{12}) = 1/7$. If the second agent's credence at $I^{21}$ surpasses his threshold, he maximizes his expected payoff by choosing $A_1$ at $I^{22}$. The second agent's credence in my example is $1/2$ that $a_2$ at $I^{12}$.

Now, suppose the first agent chooses $a_2$ at $I^{12}$, and the second agent must decide whether to choose $D_2$ or $A_2$ at $I^{22}$. This depends on whether the first agent will choose $d_3$ or $a_3$ at $I^{13}$. Again, the act $a_2$ at $I^{12}$ is evidence that does not alone favor any particular strategy over another in the remaining set of available strategies. The second agent thus updates his credences proportionally to the evidence, and no more. The second agent applies conditionalization upon the current evidence of $a_2$ at $I^{12}$, and has $Cr = 1/2$ in strategies $(a_1a_2a_3)$, $(a_1a_2d_3)$. Removing acts that have occurred, he has $Cr = 1/2$ in $a_3$ and $d_3$. The second agent's credence threshold that $a_3$ at $I^{13}$ is $1/7$, as computed above. He maximizes his expected payoff by choosing $A_2$ at $I^{22}$. The reader might find this result surprising, because it is stipulated that each agent is rational—the first agent will choose $d_3$ at $I^{13}$. However, remember that agents only know about their own rationality, not about each other's.

The principle of indifference is an additional rational constraint to regularity, to the probability axioms, and to conditionalization. It pares down credence distributions that can be assigned to a set of contingent propositions. However, there are objections to the principle that it recommends different credences for the same evidence, depending on how we partition a set of propositions. Here is a standard example. I explain that my ugly sweater is some color. You consider whether my ugly sweater is red, and partition the choices to red and not-red. The principle of indifference recommends that $Cr(\text{R}) = 1/2$, and that $Cr(\sim\text{R}) = 1/2$. However, you then consider whether my ugly sweater is one of the primary colors. Then the principle recommends that $Cr(\text{R}) = 1/3$, and that $\text{Cr}(\sim\text{R}) = 2/3$. The principle recommends different credences depending on how you partition the choice of colors, given the same evidence.

I argue, however, that the problem is not with the principle of indifference. Evidence is both subject to evidential and non-evidential components, and this objection to the principle of indifference is an example of the influence of non-evidential components. Consider that epistemology has two comprehensive goals. One goal is to minimize the amount of falsehoods an agent obtains. The other goal is to maximize the amount of truth an agent obtains. No goal is guaranteed, and realistically an agent balances between both in evaluating evidence. Thus, some agents have a skeptical temperament. They cautiously accept evidence, which is reflected by low credences in particular propositions. Some agents have a credulous temperament. They too willingly accept evidence, which is reflected by high credences in particular propositions. Temperaments affect how agents partition a set of propositions, or agents' confidence in the evidence they evaluate. This non-evidential component combines with evidential components to establish credence distributions. Non-evidential components make the difference between two agents applying conditionalization with the same evidence, resulting in different credence distributions, despite their rational constraint.

How is epistemic game theory to account for non-evidential components? Epistemic game theory is traditionally concerned with the use of epistemic states to rationally justify a game's outcome as a profile of agents' strategies—the solution to the game. A Bayesian approach to belief-revision policies (for agents with cognitive limits, where temperaments affect their credences) might have to concede that there is no rational prescription for games as an outcome with a unique profile of strategies—no ultimate solution. It might prescribe a set of profiles where many of a game's outcomes will satisfy that set of profiles. But this is simply saying that a game can have many outcomes real people might rationally settle on. This is not to say their beliefs and corresponding actions are not rational. If there is a lesson to learn from Bayesianism, agents' beliefs and their corresponding acts can be rational while simultaneously not settling on an optimal outcome.

As a side note, naturalistic game theory does investigate non-evidential components. The study of heuristics [3,4] is a study of temperaments, or, more specifically, cognitive biases. The term 'cognitive bias' has a negative connotation, but biases can provide better outcomes in specific circumstances than more sophisticated tools, especially in uncertainty. The study of biases as heuristics is a study of the environment in which the biases evolved; and there is ongoing investigation into how agents' temperaments (non-evidential components) are environmentally constrained.

## 6. Conclusions

The motivation behind the recent formation of naturalistic game theory does stem from dissatisfaction with the current debates over foundational concepts. However, the intent is not to replace epistemic and evolutionary game theory. How then does naturalistic game theory split the difference between said disciplines? Naturalistic game theory purports that foundational concepts are confronted by potentially irredeemable difficulties. Epistemic game theory provides a valuable role by testing the limits of those concepts foundational to the discipline. Once those limits are established, the principal aim for naturalistic game theory is to provide guidance for real people's judgments in their realistic circumstances. Thus, I suggest that epistemic game theory has two aims. One aim is to derive conditional results based on the game theorist's desired idealizations and stipulations. The second aim is to clarify the conceptual limits of those idealizations and stipulations. This latter aim extends to more traditional epistemological concerns about those idealizations foundational to analysis. Some idealizations are innocuous, such as a game's structure and its payoffs. However, some idealizations are not innocuous—the concepts of strategy and of belief-revision policies are two examples.

**Acknowledgments:** I thank Paul Weirich and two anonymous referees for very helpful comments on earlier drafts of this article.

**Conflicts of Interest:** The author declares no conflict of interest.

## References

1. Binmore, K. Equilibria in Extensive Games. *Econ. J.* **1985**, *95*, 51–59. [CrossRef]
2. Camerer, C. *Behavioral Game Theory: Experiments in Strategic Interaction*; Princeton University Press: Princeton, NJ, USA, 2003.
3. Gigerenzer, G.; Selten, R. *Bounded Rationality: The Adaptive Toolbox*; MIT Press: Cambridge, MA, USA, 2002.
4. Gigerenzer, G.; Gaissmaier, W. Heuristic Decision Making. *Annu. Rev. Psychol.* **2011**, *62*, 451–482. [CrossRef] [PubMed]
5. Ernst, Z. What is Common Knowledge? *Episteme* **2011**, *8*, 209–226. [CrossRef]
6. Sperry-Taylor, A.T. Bounded Rationality in the Centipede Game. *Episteme* **2011**, *8*, 262–280. [CrossRef]
7. Nozick, R. Newcomb's Problem and Two Principles of Choice. In *Essays in Honor of Carl G. Hempel*; Rescher, N., Ed.; D Reidel: Dordrecht, The Netherlands, 1969.
8. Weirich, P. *Causal Decision Theory*; Stanford Encyclopedia of Philosophy: Stanford, CA, USA, 2016.
9. Burgess, S. Newcomb's Problem: An Unqualified Resolution. *Synthese* **2004**, *138*, 261–287. [CrossRef]
10. Gibbard, A.; Harper, W. Counterfactuals and Two Kinds of Expected Utility. In *Foundations and Applications of Decision Theory*; Hooker, C.A., Leach, J.L., McClennan, E.F., Eds.; D Reidel: Dordrecht, The Netherlands, 1978; pp. 125–162.
11. Aumann, R.J. Agreeing to Disagree. *Ann. Stat.* **1976**, *4*, 1236–1239. [CrossRef]
12. Rubinstein, A. The Electronic Mail Game: Strategic Behavior under "Almost Common Knowledge". *Am. Econ. Rev.* **1989**, *79*, 385–391.
13. Cubitt, P.R.; Sugden, R. Common Knowledge, Salience and Convention: A Reconstruction of David Lewis' Game Theory'. *Econ. Philos.* **2003**, *19*, 175–210. [CrossRef]
14. Cubitt, P.R.; Sugden, R. Common Reasoning in Games: A Lewisian Analysis of Common Knowledge of Rationality. *Econ. Philos.* **2014**, *30*, 285–329. [CrossRef]
15. Lewis, D. *Convention*; Harvard University Press: Cambridge, MA, USA, 1969.
16. Paternotte, C. Fragility of Common Knowledge. *Erkenntnis* **2016**, 1–22. [CrossRef]
17. Reny, P. Rationality, Common Knowledge, and the Theory of Games. Ph.D. Thesis, Princeton University, Princeton, NJ, USA, June 1988.
18. Bicchieri, C. Self-Refuting Theories of Strategic Interaction: A Paradox of Common Knowledge. *Erkenntnis* **1989**, *30*, 69–85. [CrossRef]
19. Bicchieri, C. *Rationality and Coordination*; Cambridge University Press: Cambridge, UK, 1993.
20. Bicchieri, C.; Antonelli, A.G. Game-Theoretic Axioms for Local Rationality and Bounded Knowledge. *J. Logic Lang. Inf.* **1995**, *4*, 145–167. [CrossRef]
21. Binmore, K. Modeling Rational Players: Part 1. *Econ. Philos.* **1987**, *3*, 179–214. [CrossRef]
22. Binmore, K. Rationality and Backward Induction. *J. Econ. Methodol.* **1997**, *4*, 23–41. [CrossRef]
23. Binmore, K. Interpreting Knowledge in the Backward Induction Problem. *Episteme* **2011**, *8*, 248–261. [CrossRef]
24. Pettit, P.; Sugden, R. The Backward Induction Paradox. *J. Philos.* **1989**, *86*, 169–182. [CrossRef]
25. Pacuit, E. Dynamic Epistemic Logic I: Modeling Knowledge and Belief. *Philos. Compass* **2013**, *8*, 798–814. [CrossRef]
26. Pacuit, E. Dynamic Epistemic Logic II: Logics of Information Change. *Philos. Compass* **2013**, *8*, 815–833. [CrossRef]
27. Pacuit, E.; Roy, O. *Epistemic Foundations of Game Theory*; Stanford Encyclopedia of Philosophy: Stanford, CA, USA, 2015.

28. Roy, O.; Pacuit, E. Substantive Assumptions in Interaction: A Logical Perspective. *Synthese* **2013**, *190*, 891–908. [CrossRef]
29. Ritzberger, K. *Foundations of Non-Cooperative Game Theory*; Oxford University Press: Oxford, UK, 2002.
30. Stalnaker, R. Knowledge, Belief and Counterfactual Reasoning in Games. *Econ. Philos.* **1996**, *12*, 133–163. [CrossRef]
31. Stalnaker, R. Belief Revision in Games: Forward and Backward Induction. *Math. Soc. Sci.* **1998**, *36*, 31–56. [CrossRef]
32. Aumann, R.J. Backward induction and common knowledge of rationality. *Games Econ. Behav.* **1995**, *8*, 6–19. [CrossRef]
33. Aumann, R.J. Reply to binmore. *Games Econ. Behav.* **1996**, *17*, 138–146. [CrossRef]
34. Aumann, R.J. On the centipede game. *Games Econ. Behav.* **1998**, *23*, 97–105. [CrossRef]
35. Dixit, A.; Skeath, S.; Reiley, D.H. *Games of Strategy*; W. W. Norton and Company: New York, NY, USA, 2009.
36. Axelrod, R. *The Evolution of Cooperation*; Basic Books: New York, NY, USA, 1984.
37. Lindgren, K.; Nordahl, M.G. Evolutionary Dynamics of Spatial Games. *Phys. D Nonlinear Phenom.* **1994**, *75*, 292–309. [CrossRef]
38. McGrath, M. Probabilistic Epistemology. In *Epistemology: A Contemporary Introduction*; Goldman, A.I., McGrath, M., Eds.; Oxford University Press: Oxford, UK, 2015.
39. Hajek, A.; Hartmann, S. Bayesian Epistemology. In *A Companion to Epistemology*; Dancy, J., Sosa, E., Steup, M., Eds.; Blackwell Publishing: Oxford, UK, 2010.
40. Kolmogorov, A.N. *Foundations of the Theory of Probability*; Chelsea Publishing Company: New York, NY, USA, 1950.
41. Titelbaum, M. *Quitting Certainties: A Bayesian Framework Modeling Degrees of Belief*; Oxford University Press: Oxford, UK, 2013.
42. Jaynes, E.T. *Probability Theory: The Logic of Science*; Cambridge University Press: Cambridge, UK, 2003.

# Section 4:
# Psychology

*Article*

# Economic Harmony: An Epistemic Theory of Economic Interactions

**Ramzi Suleiman** [1,2]

[1] Department of Psychology, University of Haifa, Abba Khoushy Avenue 199, Haifa 3498838, Israel; suleiman@psy.haifa.ac.il; Tel.: +972-(0)50-5474-215

[2] Department of Philosophy, Al Quds University, East Jerusalem and Abu Dies, P.O.B. 51000, Palestine

Academic Editors: Paul Weirich and Ulrich Berger
Received: 16 September 2016; Accepted: 18 December 2016; Published: 3 January 2017

**Abstract:** We propose an epistemic theory of micro-economic interactions, termed Economic Harmony. In the theory, we modify the standard utility, by changing its argument from the player's actual payoff, to the ratio between the player's actual payoff and his or her aspired payoff. We show that the aforementioned minor epistemic modification of the concept of utility is quite powerful in generating plausible and successful predictions of experimental results, obtained in the standard ultimatum game, and the sequential common pool resource dilemma (CPR) game. Notably, the cooperation and fairness observed in the studied games are accounted for without adding an other-regarding component in the players' utility functions. For the standard ultimatum game, the theory predicts a division of $\phi$ and $1 - \phi$, for the proposer and responder, respectively, where $\phi$ is the famous Golden Ratio ($\approx$0.618), most known for its aesthetically pleasing properties. We discuss possible extensions of the proposed theory to repeated and evolutionary ultimatum games.

**Keywords:** epistemic; aspiration level; fairness; ultimatum game; common pool resource dilemma; golden ratio; Fibonacci numbers

---

## 1. Introduction

The game theoretic approach to human and animal interactions relies on the economic rationality assumption, which prescribes that in any interaction, all players are utility maximizers, and that their utilities are non-decreasing functions of their payoffs. For the case of risk-neutral players, the theory prescribes that rational players will strive to maximize their payoffs. Despite being self-consistent and mathematically sound, standard game theory is, in many cases, at odds with experimental and real-life data. Examples of strategic interactions in which the standard game theoretic predictions fail to account for human behavior, include experimental and real-life data on the provision of public goods, the management of common pool resources, bargaining situations, and situations involving trust.

From a philosophical perspective, the assumption of risk neutrality, or even of a homogeneous population with respect to risk behavior, deprives game theory from an important epistemic attribute. The fact that game theory fails to consider the players' cognitions, motivations, aspirations, and other individual attributes, reduces the analysis of economic interactions to an ontological level, in which the behaviors of players involved in the interaction are thought to be solely determined by the formal game structure. According to the standard approach, the variability of data resulting from individual differences is measurement noise which should be reduced to a minimum. Proponents of such an approach advise experimentalists to strive for lowering the data variability, by putting strict constraints on players' behavior [1]. Psychologists, on the other hand, view the data variability as the "bread and butter" of their research [2].

A major shortcoming of standard game theory, resulting from ignoring epistemic factors, is manifest in its complete failure to predict the significant levels of cooperation and fairness observed in a variety of non-cooperative and zero-sum games. For example, for the dictator game [3,4], standard game theory prescribes that the player in the role of the dictator should keep the entire amount minus an infinitesimally small portion, and the same prediction holds for the ultimatum game, in which the responder has veto power [4,5]. In experimental studies, we find that dictators behave altruistically, and transfer, on average, about 20%–25% of the entire amount, while in the ultimatum game the amount transferred, on average, is about 40% of the total amount [4–7]. Non-negligible levels of cooperation are also observed in Public Goods, Common Pool Resource (CPR), and other social dilemma games [8–10].

Several modifications of standard game theory have been proposed to account for the cooperation and fairness observed in short-term strategic interactions. Such modifications were usually accomplished by incorporating an other-regarding component in the players' utility functions. This type of modification includes the theory of Equity, Reciprocity and Competition (or ERC) [11], and Inequality Aversion theory (IA) [12]. ERC posits that along with pecuniary gain, people are motivated by the ratio of their own payoffs to the payoff of others. IA assumes that in addition to the motivation for maximizing own payoffs, individuals are motivated to reduce the difference in payoffs between themselves and others, although with greater distaste for having lower, rather than higher earnings. Another interesting model is the "fairness equilibrium" model [13] in which the rationality assumption is modified by assuming that players make decisions based on their perceptions of others' intentions. Although such epistemic modifications yield superior predictions to the standard game theoretic predictions, they do not qualify as general models of interactive behavior. For the ultimatum game discussed hereafter, the predictions of ERC and IA are uninformative. ERC predicts that the proposer should offer any amount that is larger than zero and less or equal to 50% of the entire amount, while IA's prediction is nonspecific, as it requires an estimation of the relative weight of the fairness component in the proposer's utility function. Moreover, the two theories are strongly refuted by a three-person ultimatum game [14] designed specifically to test their predictions. In this game, Player X offers to split a sum of money \$m, allocating (m − (y + z), y, z) for herself, player Y, and player Z, respectively. One of the latter players is chosen at random (with probability p) to accept or reject the offer. If the responder accepts, then the proposed allocation is binding, as in the standard ultimatum game. However, if the responder rejects the offer, then she and X receive zero payoffs, and the non-responder receives a "consolation prize" of \$c. The consolation prize was varied across four conditions with (\$)c = 0, 1, 3, 12. The probability of designating Y or Z as responder was $p = 1/2$ and the amount to be allocated in all conditions was m = \$15. The findings of [14] contradicted the predictions of both ERC and IA. Frequent rejections were detected, when both theories call for acceptance. In addition, the effect of the consolation prize for the non-responder on the probability of the responders' acceptance rate was insignificant, and did not increase monotonically with the size of the consolation prize, as both theories predict.

More recently, epistemic game theoretic models were proposed as alternatives to standard game theory, including static models based on Aumann's interactive epistemology [15,16] and dynamic models of interactive reasoning [17,18]. In the present paper, we propose a new epistemic theory of economic interactions termed Economic Harmony. Instead of looking at cold cognitive variables characterizing human players, we focus on their aspiration levels. As we shall demonstrate hereafter, accounting for the interacting players' aspiration levels proves successful in generating impressively good predictions of players' behavior in well-known experimental games.

## 2. Economic Harmony Theory

The proposed Economic Harmony theory postulates that instead of maximizing utilities, defined as functions of their own payoffs, rational players strive to maximize utilities, which are functions of their actual payoffs relative to their aspired payoffs. Moreover, while game theory looks at points of equilibrium in the game, in which no player can increase his or her utility by unilaterally changing his or her equilibrium strategy, economic harmony theory solves for a point of harmony, at which the intersection of the strategies played by the individuals yields equal utilities for all, or in psychological terms, the point at which the satisfaction levels of all players are equal.

The view that in economic decisions, individuals compare their actual payoffs with their aspired ones, has a long tradition in psychology [19–21]. It has been also well studied in the domain of individual decision-making under risk, especially in relation to Security-Potential/Aspiration (CP/A) theory [22–25]. Several studies have also incorporated individual models of aspirations in predicting behaviors in interactive situations [26–32]. As examples, study [27] investigated the role of aspiration levels in players' choices in a repeated prisoner's dilemma game, and study [30] investigated the role of aspirations in coalition formation. Common to CP/A theory and similar models, is the use of Herbert Simon's conceptualization of satisficing vs. non-satisficing outcomes [21].

The proposed model resembles the aforementioned individual choice models in assuming that individuals make choices based on their levels of satisfaction from possible outcomes, as well as in assuming that individuals evaluate their actual outcomes by comparing them with their aspired outcomes. However, the new model differs from previous aspiration level models in two significant aspects: (1) it treats interactive situations rather than individual choice ones; (2) the levels of satisfaction are defined as the ratio of the individuals' actual outcomes to their aspired outcomes, rather than the common definition as the difference between the actual and aspired outcomes. We contend that the ratio scale is more fit for defining outcome satisfaction, than the difference scale. First, the ratio scale is dimensionless and does not depend on the measurement units of the divided goods; second, it is the standard scale in psychophysics, starting with Fechner's law [33] and Steven's power law [34,35] to more recent theories of audio and visual perception [36] and signal detection [37]; third, the ratio scale is very common in physics, biology, evolution, and other exact sciences; fourth, all types of statistical measures are applicable to ratio scales, and only with these scales may we properly indulge in logarithmic transformations [34].

*Theory Predictions*

For deriving predictions based on the proposed theory, consider an economic interaction involving $n$ players. Let $\mathbf{S}^i$ denote the vector of player $i$'s admissible strategies. $\mathbf{S}^i = \{s_j^i\}_{j=1}^{J^i}$, where $s_j^i$ is strategy $j$ of player $i$ ($j = 1, 2, .. J^i$, $i = 1, 2, \ldots, n$). In Economic Harmony theory, we preserve the rationality principle, while introducing a plausible modification of the players' utility functions. Specifically, we define the subjective utility of each player $i$ as:

$$u_i(..) = u_i\left(\frac{r_i}{a_i}\right) \quad (1)$$

where $r_i$ is player $i$'s actual payoff, $a_i$ is his or her aspired payoff, and $u(..)$ is a bounded non-decreasing utility function with its argument. For simplicity, we assume that $u(0) = 0$ and $u(1) = 1$. Note that in social-psychological terms, the aforementioned definition implies that each player puts an upper limit to his or her greed.

A point of harmony in the interaction is defined as an intersection point of strategies played by the $n$ interacting individuals, at which the utilities of all players, as defined above, are equal.

In formal notation, a point of harmony in the interaction is a vector of outcomes $r^* = r_1^*, r_2^*, r_3^*, \ldots . r_n^*)$ for which the subjective utilities of all $n$ players' outcomes satisfy:

$$u_i\left(\frac{r_i^*}{a_i}\right) = u_j\left(\frac{r_j^*}{a_j}\right) \text{ For all } i \text{ and } j \quad (2)$$

Assuming linear utilities, Equation (2) becomes:

$$\frac{r_i^*}{a_i} = \frac{r_j^*}{a_j} \text{ For all } i \text{ and } j \quad (3)$$

Two remarks are in order: First, harmony points are not equilibrium points. In equilibrium, no player can increase his or her utility by changing his or her equilibrium strategy unilaterally. In contrast, if a harmony point is reached, a player may increase his or her utility (and decrease the utilities of other players) by switching to another strategy. For a harmonious solution to emerge and stabilize, it should be supported by an efficient social or institutional mechanism. In the ultimatum and the sequential CPR games, discussed hereafter, the supporting mechanism is a second-party punishment. However, other supporting mechanisms can be effective in sustaining harmony, such as a third-party punishment [38], group punishment [39], punishment by a central authority [40], reciprocity [41], reputation [42] and religious and civic moralizing [43]. Second, as in the case of the Nash equilibrium, players are not expected to "solve" the game and play their harmonious strategies. Rather, it is conjectured that harmonious strategies can emerge through processes of learning and adaptation. In our epistemic theory, adaptation processes are expected to act interdependently on each individual's decisions and aspirations, according to criteria of success or failure [21,44,45]. In fact, valuable insights into the co-evolution of strategies and aspiration levels have been provided by recent simulation studies of both evolutionary, and reinforcement learning, games, which we shall discuss in the closing section.

In the following sections, we shall demonstrate that despite its evident simplicity, Economic Harmony theory is highly successful in accounting for the cooperation and fairness reported in several experiments on the standard ultimatum, and sequential CPR games.

## 3. Predicting Behavior in Experimental Games

### 3.1. Predicting Offers in the Ultimatum Game

In the standard two-person ultimatum game [4,5], one player is designated the role of "proposer", and the second player is designated the role of "responder". The proposer receives an amount of monetary units and must decide how much to keep for himself or herself, and how much to transfer to the responder. The responder replies either by accepting the proposed allocation, in which case both players receive their shares, or by rejecting it, in which case the two players receive nothing. Thus, whereas the proposer has complete entitlement to make an offer, the responder can inflict a harsh, although costly, punishment on an unfair proposer.

Game theory predicts that a rational proposer, who believes that the responder is also rational, should offer the smallest amount possible, since the responder, being rational, will accept any positive offer. Experimental findings of numerous ultimatum studies refute this prediction. The modal offer in most experiments is the equal split, and the mean offer is about 40% of the entire amount. In the first ultimatum experimental [5], the mean offer was 41.9%. Numerous studies have repeatedly replicated these results. For example, despite differences in culture, socio-economic background, and type of currency, Kahneman et al. [6] reported a mean offer of 42.1% (for commerce students in an American university), and Suleiman [7] reported a mean offer of 41.8% for Israeli students. A more recent meta-analysis of ultimatum experiments conducted in twenty-six countries with different cultural backgrounds [46] reported a mean offer of 41.5% (S.D. (standard deviation) = 5.7%), and yet another large cross-cultural study conducted in 15 small-scale societies [47] reported a mean offer of 40.5% (S.D. = 8.3). In stark difference with the prediction of game theory, in the above cited studies, and in many other studies, a division of about 60–40 (%), for the proposer and responder, respectively, seems to be robust across countries, cultures, socio-economic levels, monetary stakes, etc.

It is important to note that altruism alone cannot account for the relatively fair divisions observed in ultimatum bargaining. This conclusion is supported by several experiments [7,48,49]. As examples, studies [48] and [49] compared the mean offers made by the allocators in the ultimatum and dictator games. It was hypothesized that if genuine concern for the recipients' well-being is the major factor behind equitable allocations, then the mean allocation to them should be similar in the two games. Results from the studies cited above refuted the above hypothesis, by showing that the mean allocations to recipients were significantly lower in the dictator game, than in the ultimatum game. For example, in [49], 70% of the dictator games played under a double anonymity condition ended with the first mover demanding the whole 'cake'.

3.1.1. Economic Harmony Prediction

For simplicity, but without loss of generality, we set the sum to be divided by the proposer to be one monetary unit (1 MU). If he or she demands to keep $x$ MUs (and transfer $1 - x$ MUs to the responder), then using Equation (3) we can write:

$$\frac{x}{a_p} = \frac{1-x}{a_r} \tag{4}$$

where $a_p$ and $a_r$ are the aspiration levels of the proposer and the responder, respectively. Solving for $x$ yields:

$$x = \frac{a_p}{a_p + a_r} \tag{5}$$

And the amount offered to the responder is:

$$x_r = 1 - x = 1 - \frac{a_p}{a_p + a_r} = \frac{a_r}{a_p + a_r} \tag{6}$$

Determining $x$, which guarantees a harmonious allocation, in the sense of equal levels of satisfaction, requires the measurement of the players' aspiration levels. Probing the players' aspiration levels by self-reports before, or after, they make their decisions in the game is somewhat problematic due to the notorious interdependence between attitudes and behavior. However, this problem could be eliminated or at least minimized by using suitable experimental designs. A possible design is to utilize the "one–one" ultimatum game treatment played according to the strategy protocol [50]. In this game, proposers are instructed to divide a pie of a given size between themselves and their designated responders. If the amount proposed is equal or larger than the responder's demand, the two players receive their shares and the game ends; but if the proposer's offer is less than the responder's demand, then the proposer is given a second chance to make an offer. Under this treatment, rational proposers and responders are expected to utilize the first round of the game to signal their aspiration levels.

In the absence of empirical data about the players' aspiration levels, the assumption of rational players, who are motivated to maximize their individual utilities, enables us to make a first-order estimate about their aspiration levels. We conjecture that a self-interested proposer would aspire for the entire sum (i.e., $a_p = 1$). On the other hand, a rational responder, who believes that the proposer is also rational, cannot aspire that the proposer gives him or her the entire amount or even any amount larger than the amount that the proposer would keep for himself or herself. Embedded in this line of reasoning is the view that in contemplating their aspiration levels, proposers and responders use qualitatively different reference points. Proposers are assumed to adhere to a fixed reference-point, by comparing their demands to the complete sum, which they were given the right to divide. Their highest aspiration would be to keep the entire amount for themselves. The responders are assumed to adhere to a social comparison rule, using the amount demanded by the proposer as their reference-point. Another possibility, which we shall consider hereafter, is that responders might base their aspiration levels on the equality norm, and aspire to receive half of the total amount. Under the "social comparison" assumption, we have $a_r = x$, while under the "equality norm" assumption we have $a_r = \frac{1}{2}$. For the latter case, substituting $a_p = 1$ and $a_r = \frac{1}{2}$ in Equation (5) yields:

$$x = \frac{a_p}{a_p + a_r} = \frac{1}{1 + \frac{1}{2}} = \frac{2}{3}\,\text{MU} \tag{7}$$

And the amount offered to the responder is:

$$x_r = 1 - x = 1 - \frac{2}{3} = \frac{1}{3}\,\text{MU} \tag{8}$$

Under the social comparison assumption, we have $a_p = 1$ and $a_r = x$. Substitution in Equation (5) yields:

$$x = \frac{1}{1 + x} \tag{9}$$

Solving for $x$. we get:

$$x^2 + x - 1 = 0 \tag{10}$$

Which yields the positive solution:

$$x = \frac{\sqrt{5} - 1}{2} = \phi \approx 0.618\,\text{MU} \tag{11}$$

where $\phi$ is the famous Golden Ratio, which plays a key role in many fields of science and aesthetics [51,52]. This ratio is known to be equal to $\lim_{n \to \infty} \left( \frac{f_n}{f_{n+1}} \right)$, where $f_n$ is the $n$th term of the Fibonacci Series: 0, 1, 1, 2, 3, 5, 8, 13, 21, 34, 55, 89, 144, .., in which each term is equal to the sum of the two preceding terms, or: $f_n = f_{n-1} + f_{n-2}$. The corresponding portion for the responder is equal to:

$$x_r = 1 - x = 1 - \phi \approx 0.38\,\text{MU} \tag{12}$$

### 3.1.2. Empirical Validation

We tested the prediction in Equation (12) using data from study [46], which reported a Meta-analysis on 75 ultimatum experiments conducted in 26 countries with different cultural backgrounds, and from study [47], which reported results from 15 small-scale societies, including three groups of foragers, six groups of slash-and-burn horticulturalists, four groups of nomadic herders, and two groups of small-scale agriculturalists. The reported mean proportional offers in the two aforementioned studies were 0.395 and 0.405 for studies [46,47], respectively, which are quite close to the Golden Ratio prediction of $\approx$0.382. A Two one-sided test of equivalence (TOST) [53] validates this conjecture. A rule of thumb in testing for equivalence using TOST, is to set a confidence level of $\pm$10%. For study [46], the analysis yielded significant results for the upper and lower bounds of the equivalence range (upper bound = 42.016, $p < 0.0001$; lower bound = 34.377, $p = 0.0425$; overall significance = 0.0425). For study [47], the results were also significant (upper bound = 42.016, $p = 0.012$; lower bound = 34.377, $p = 0.0255$; overall significance = 0.0255).

### 3.1.3. Relaxing the Rationality Assumption

In deriving the "harmony" points, we assumed that a rational proposer would aspire for the entire amount (of 1 MU). We relax this assumption by supposing that proposers might aspire to receive any amount between 1 MU, and $1 - \alpha$ ($0 \leq \alpha \leq 0.5$), where $\alpha$ is a "security" parameter [22]. Under the assumption that the responder aspires to receive $\frac{1}{2}$, Equation (3) becomes:

$$\frac{x}{1 - \alpha} = \frac{1 - x}{0.5} \tag{13}$$

Solving for $x$ we get:

$$x = \frac{1-\alpha}{(1.5-\alpha)} \tag{14}$$

And the amount proposed to the responder becomes:

$$x_r = 1 - x = 1 - \frac{1-\alpha}{(1.5-\alpha)} = \frac{0.5}{(1.5-\alpha)} \tag{15}$$

On the other hand, assuming that the responder aspires to be treated equally (i.e., $a_r = x$), we have:

$$\frac{x}{1-\alpha} = \frac{1-x}{x} \tag{16}$$

Solving for $x$ yields:

$$x^2 + (1-\alpha)\,x - (1-\alpha) = 0 \tag{17}$$

Which solves for:

$$x = \frac{\sqrt{(1-\alpha)^2 + 4\,(1-\alpha)} - (1-\alpha)}{2} \tag{18}$$

And the offer to the responder is:

$$x_r = 1 - x = 1 - \frac{\sqrt{(1-\alpha)^2 + 4\,(1-\alpha)} - (1-\alpha)}{2} = \frac{3 + \alpha + \sqrt{(1-\alpha)^2 + 4\,(1-\alpha)}}{2} \tag{19}$$

Figure 1 depicts the predicted offers by Equations (15) and (19) as functions of $\alpha$ in the range $\alpha = 0 - 0.5$.

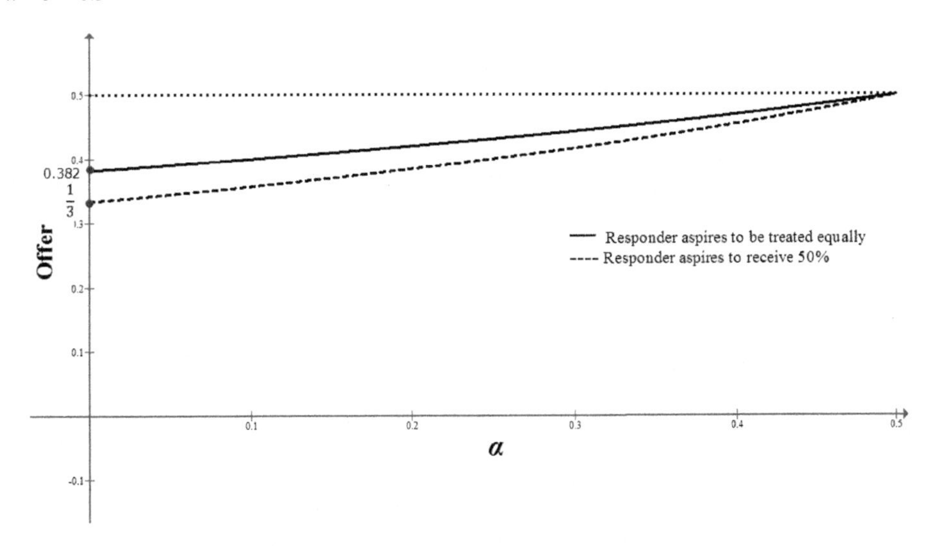


**Figure 1.** Predicted offers as a function of the proposer's security level.

If we assume that the proposers might aspire for any amount between 1 and $1 - \alpha$ with equal probability, the predicted mean offer could be calculated by averaging $x_r$ over the range $(0, \alpha)$. Under the assumption $a_r = \frac{1}{2}$, using Equation (15) we have:

$$x_r = \frac{1}{\alpha}\int_0^{\alpha} \frac{0.5}{(1.5-\alpha)}\,\mathrm{d}\alpha = \frac{1}{2\alpha}\left[\ln(3) - \ln(3-2\alpha)\right] \tag{20}$$

While under the assumption $a_r = x$, using Equation (19) we get:

$$\begin{aligned} x_r &= \tfrac{1}{\alpha}\int_0^{\alpha}\left(\tfrac{3+\alpha+\sqrt{(1-\alpha)^2+4(1-\alpha)}}{2}\right)\mathrm{d}\alpha \\ &= 2\,[3\alpha - \tfrac{\alpha^2}{2} - \tfrac{1}{2}\left(\tfrac{\alpha}{2} - \tfrac{3}{2}\right)\sqrt{\alpha^2-6\alpha+5} + 2\ln(3-\alpha \\ &\qquad - \sqrt{\alpha^2-6\alpha+5}) \end{aligned} \tag{21}$$

As an example, suppose that the proposer aspires to get any amount between 100% and 75% of the entire amount. Under the equality norm assumption, substituting $\alpha = 0.25$ in Equation (20) yields an offer of $x_r \approx 0.37$, which is only slightly higher than the offer of $\frac{1}{3}$, predicted for a completely rational proposer (i.e., for $\alpha = 0$). Under the social comparison assumption, substituting $\alpha = 0.25$ in Equation (21) yields an offer of $x_r \approx 0.40$, which is also slightly higher than 0.38, the predicted offer for $\alpha = 0$.

3.1.4. Relaxing the Linearity Assumption

For the proposer and responder with power utility functions $u_p = (\frac{x}{a_p})^a$ and $u_r = (\frac{1-x}{a_r})^b$, applying the harmony condition in Equation (2) yields:

$$\left(\frac{x}{a_p}\right)^a = \left(\frac{1-x}{a_r}\right)^b \tag{22}$$

Setting $a_p = 1 - \alpha_p$ and $a_r = x - \alpha_r$ we obtain:

$$\left(\frac{x}{1-\alpha_p}\right)^a = \left(\frac{1-x}{x-\alpha_r}\right)^b \tag{23}$$

For the case of $\alpha_p = \alpha_r = 0$ we obtain:

$$\frac{x^{a+b}}{(1-x)^b} = 1 \tag{24}$$

Which could be written as:

$$x\left(x^{\beta}+1\right) = 1 \tag{25}$$

where $\beta = \frac{a}{b}$.

Figure 2 depicts the predicted offer, $x_r = 1 - x$, as a function of $\beta$. For practical cases, it is plausible to assume that players are generally risk averse, preferring a sure thing over a gamble of equal expected value, and a gamble with low variance over a riskier one [54]. It is also plausible to conjecture that proposers who face the risk of losing a larger amount (in case or rejecting their offers), will display more risk aversion than responders who would lose a smaller amount in case they decide to reject an unfair offer. We thus may assume that $a$ and $b$ are smaller than 1, and that $a < b$, implying $0 \leq \beta \leq 1$. For $\beta$ in the range (0.5, 1), numerical integration on $x_r$ as a function of $\beta$ yields a mean offer of about 0.402, which is only about 0.02 less than the 0.382 predictions obtained under the linearity assumption.

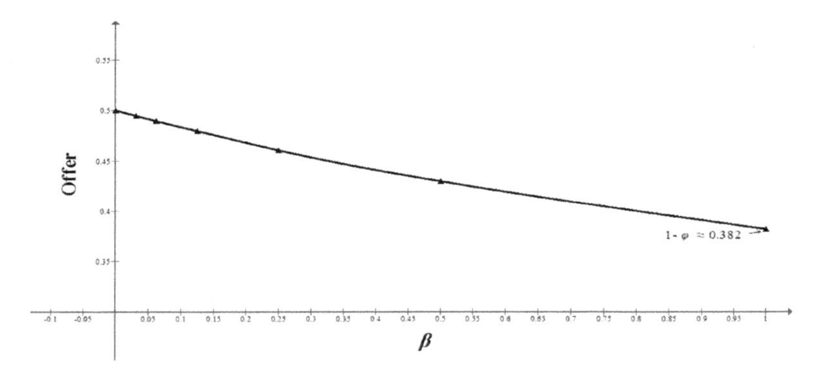


**Figure 2.** The predicted offer as a function of the ratio between the risk indices of the proposer and responder.

### 3.2. *Predicting Requests in a Sequential CPR Game*

The common pool resource dilemma (CPR) game models situations in which a group of people consume a limited shared resource. Under the sequential protocol of play [55] with a step-level rule, individual requests are made in an exogenously determined order, which is common knowledge, such that each player knows his or her position in the sequence and the total requests of the players who proceed him or her in the sequence. If the total request does not exceed the pool size, all players receive their requests. However, if the total request exceeds the pool size, all players receive nothing [56–58].

The $n$-person sequential CPR with a step-level resource has the structure of an $n$-person ultimatum game, in which by making a sufficiently large request, any player can "veto" the request decisions of the players preceding him or her in the sequence. The sub-game perfect equilibrium of the sequential CPR game prescribes that the first mover should demand almost all the amount available in the common pool, leaving an infinitesimally small portion to the other players. This prediction is strongly refuted by experimental results, showing that first movers do not exploit their advantageous position in the sequence, and that they leave a substantial portion of the resource to the other players. Moreover, these studies reveal a robust position effect: individual requests are inversely related to the players' positions in the sequence, with the first mover making the highest request and the last mover making the lowest request [58–61].

#### 3.2.1. Economic Harmony Prediction

To derive the harmony solution for the sequential CPR game, consider a game with $n$ players. Denote the request of the player occupying position $i$ in the sequence by $r_i$ $(i = 1, 2, \ldots n)$. For two successive players $r_i$ and $r_{i+1}$, the game is reduced to a two-person ultimatum game, in which harmony is achieved when $r_i = \phi$, and $r_{i+1} = 1 - \phi$, where $\phi$ is the Golden Ratio (see Section 3.1.1). Thus:

$$\frac{r_{i+1}}{r_i} = \frac{1-\varphi}{\varphi} = \frac{1}{\varphi} - 1 = 1 + \varphi - 1 = \varphi \approx 0.618 \; (i = 1, 2, \ldots n-1) \tag{26}$$

#### 3.2.2. Empirical Validation

We tested the above prediction using data reported in one study using groups of three players and three studies using groups of five players [56,59–61]. In all studies, the pool size was 500 points. The resulting predictions (using Equation (26) and a pool size of 500), together with the experimental results, are depicted Figure 3. As shown in the figure, the match between the theoretical prediction, and the experimental results is impressive, particularly for $n = 3$ (see Figure 3a). For the three, and five players' games, Kolmogorov–Smirnov tests revealed that the difference between the distributions of the theoretical and observed requests are non-significant ($p = 1$ and $p = 0.975$, for the three and five players' games, respectively).

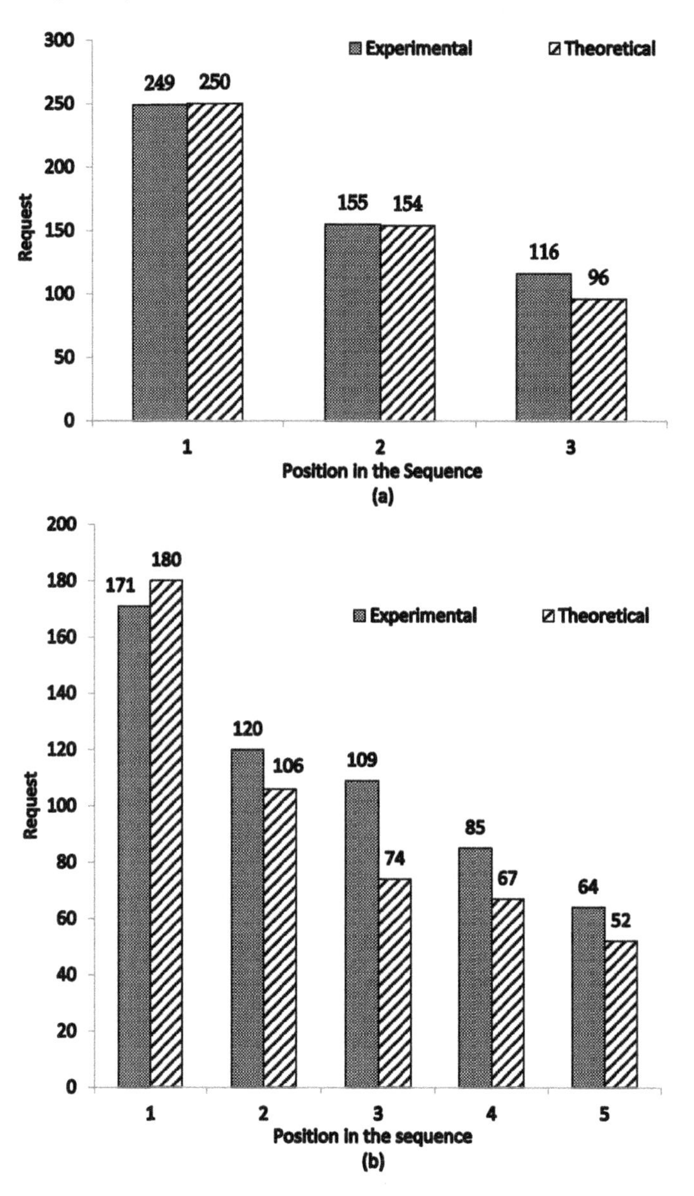


**Figure 3.** Empirical and predicted mean of requests in a sequential common pool resource (CPR) Dilemma with resource size = 500 for three players (**a**); and five players (**b**).

## 4. Summary and Concluding Remarks

Previous theories of cooperation and fairness in strategic interactions in the short term have usually attempted to account for the observed cooperation in zero-some and mixed-motive games, by adding an other-regarding component in the players' utility functions (e.g., [11,12]). In the proposed Economic Harmony theory, we retained the rationality principle, but altered the utility function by defining the players' utilities as the ratio between their actual payoffs and their aspired payoffs. Instead of looking at points of equilibrium, we looked at the point of harmony in a given interaction, at which the subjective utilities (satisfaction levels) of all players are equal.

For the two games discussed in this article, the Golden Ratio emerged as the point of harmony. Notably, for the two-person ultimatum game, the same result was obtained independently in [62] using the method of infinite continued fractions and the Fibonacci numbers. The derived solution prescribes that a rational proposer should offer a fair division of about (0.618, 0.382) to himself/herself, and the responder, respectively. The emergence of the Golden Ratio as a point of harmony, or homeostasis in economic interactions, adds to its numerous appearances in all fields of sciences [63–67]. It is not unreasonable to conjecture that the Golden Ratio is a point of homeostasis in physical, biological, and socioeconomic systems. Moreover, the fact that the Golden Ratio plays a key role in human aesthetics and beauty [51,68] suggests that humans' taste for fairness and beauty are neutrally correlated. A recent *f*MRI (functional magnetic resonance imaging) study [69] lends support to our conjecture, by demonstrating that participants who performed evaluations of facial beauty (beautiful/common), and morality in scenes describing social situations (moral/neutral), exhibited common involvement of the orbitofrontal cortex (OFC), inferior temporal gyrus, and medial superior frontal gyrus.

It is worth emphasizing that in the discussed games, fairness is predicted by the theory, and not presupposed by it, as in ERC and inequality aversion theories. Notwithstanding, the predicted fairness (at the Golden Ratio) was shown to account quite nicely for the levels of fairness observed in many ultimatum and CPR games. We have also shown elsewhere [70] that the theory is successful in predicting the results of several variants of the ultimatum game, including a three-person ultimatum game with uncertainty regarding the identity of the responder [14], a modified ultimatum game with varying veto power [7], and an ultimatum game with one-sided uncertainty about the "pie" size [71]. In addition, the theory was shown to yield good predicting of the players' decisions in the prisoner's dilemma game and in the public goods game with punishment (see [72]).

In deriving the analytic solutions for the discussed games' points of harmony, we relied on rational reasoning to conjecture about the aspiration levels of the interacting players. In reality, it is likely that different individuals might adhere to different rules and anchor their aspiration levels on different reference points. Accounting for such epistemic variability could be undertaken by an empirical assessment of each individual's aspiration level, given the game structure and the player's position in the game. When measurement of the individuals' aspiration levels is unreliable, costly, or difficult to perform, the predictability power of the proposed theory could be enhanced by assuming that the players' aspiration levels are sampled from a theoretical probability distribution. An example for using an underlying probability distribution of aspiration levels to infer about offers was briefly attended to in Section 3.1.3, in which we assumed that the proposers' aspiration levels are in the range between 1 (perfect rationality), and $1 - \alpha$ (a bounded rationality), where $\alpha$ is a uniformly distributed "security" factor.

We view the proposed theory as one step towards a more general theory of rational fairness-economics. An interesting inquiry, which we hope to address in future research, concerns the dynamic aspects of aspirations. By using repeated experimental games or computer simulations, we can test the dynamics of players' aspirations in strategic games, including the two games treated in the present paper. In particular, we would like to investigate the co-evolution process of aspirations and decisions in repeated and evolutionary games and to test whether the predicted points of harmony for each game will emerge as attractors for the behavioral dynamics. An insightful explanation of the process in which players in such games might update their aspiration levels was provided in [73]. Using analytical solutions and computer simulations, the study demonstrated that when individuals could obtain some information on which offers have been accepted by others in previous encounters, responders in the game learn to maintain high aspiration levels. As explained by the authors of the aforementioned study, responders who lower their aspiration levels (low demands), increase the chance of receiving reduced offers in subsequent encounters. In contrast, by maintaining high aspiration levels (high demands), although costly, they gain a reputation as individuals who insist on a fair offer. The authors summarize their findings by stating that "when reputation is included in the ultimatum game, adaptation favors fairness over reason ... information on the co-player fosters the emergence of strategies that are non-rational, but promote economic exchange" [73].

Several other studies on evolutionary ultimatum games [74,75], using various updating rules, reveal that, in general, the dynamics in the investigated ultimatum game were similar to the one reported in [73]. In general, proposers' demands and responders' acceptance thresholds converge to a relatively steady-state which is far away from the predicted equilibrium. Remarkably, the mean proposers' demands converged to levels that were strictly above the equal split, while the mean responders' demands converged to much lower levels, implying that above-equal demands were accepted. In [74], the rate of proposers' demands ($p$) and the rate of responders' acceptance thresholds ($q$) were co-evolved with an underlying network typology containing 1500 individuals. The results of many simulations run for two million steps showed that at the beginning of the simulation, p and q underwent a fast decrease from $p = q = 0.5$ (equal split) to about $p = 0.38$ and $q = 0.19$, after which the decrease in both demands was very slow, till the simulations' termination after 2 million steps. A similar result was reported in [75], using a learning-mutation process. The study found that the mean offer increased if it was smaller than 0.4, and oscillated between 0.4 and 0.5.

Interestingly, convergence to a stable mean offer of about 0.4 is also evident in simulated games played in the intermediate term using reinforcement learning models [76,77]. The findings in [76] show that in many simulations, starting from initial conditions drawn randomly from a uniform distribution over the integers 1–9, the simulated proposers started by demanding 5 out of 9, but increased their demand in the progression of the simulation to 6. Interestingly, the mean proposers' demands remained in the range between 6 and 7, and did not increase further, since strategies higher than 7 died out in the course of the simulated game. These dynamics are explained by the difference in the speed at which the proposer and the responder update their demands. This explanation is confirmed in [78], which demonstrated that "proposers learn not to make small offers faster than responders learn not to reject them" [78].

The findings described above share several features: (1) the proposers' mean demands in the relatively steady states reached in all the aforementioned simulation studies are much lower than the predicted (sub-game) equilibrium; (2) the mean demands increase steeply from 0.5 at the initial phase of the game, reaching a ratio of about 0.6 at the final simulation trials; (3) the proposers' mean acceptance threshold decreases from 0.5, at the initial phase of the game, to low levels of about 0.2–0.3 at the final simulation trials. While the aforementioned common features seem to agree with Economic Harmony predictions, the question of whether the mean demands in these and in similar games converge to the Golden Ratio or to a different nearby point is left to future research.

**Acknowledgments:** I am very grateful to Amnon Rapoport, David V. Budescu, Thomas Wallsten, Efrat Aharonov-Majar, and three anonymous reviewers for their helpful remarks on earlier drafts of the paper. I am also grateful to Stefan Schuster who shared with me his research on the Golden Ratio in the ultimatum game.

**Conflicts of Interest:** The author declares no conflict of interest.

## References

1. Hertwig, R.; Andreas, O. Experimental practices in economics: A methodological challenge for psychologists. *Behav. Brain Sci.* **2001**, *24*, 433–452.
2. Suleiman, R. Different perspectives of human behavior entail different experimental practices. *Behav. Brain Sci.* **2001**, *24*, 429.
3. Bardsley, N. Dictator game giving: Altruism or artefact? *Exp. Econ.* **2008**, *11*, 122–133. [CrossRef]
4. Camerer, C.; Thaler, R.H. Ultimatums, dictators and manners. *J. Econ. Perspect.* **1995**, *9*, 209–219. [CrossRef]

5. Güth, W.; Schmittberger, R.; Schwartze, B. An experimental analysis of ultimatum games. *J. Econ. Behav. Org.* **1982**, *3*, 367–388. [CrossRef]
6. Kahneman, D.; Knetsch, J.L.; Thaler, R.H. Fairness and the assumptions of economics. *J. Bus.* **1986**, *59*, 285–300. [CrossRef]
7. Suleiman, R. Expectations and fairness in a modified ultimatum game. *J. Econ. Psychol.* **1996**, *17*, 531–554. [CrossRef]
8. Dawes, R.M. Social Dilemmas. *Annu. Rev. Psychol.* **1980**, *31*, 69–193. [CrossRef]
9. Kollock, P. Social Dilemmas: The Anatomy of Cooperation. *Annu. Rev. Sociol.* **1998**, *1*, 83–214. [CrossRef]
10. Van Lange, P.A.M.; Joireman, J.; Parks, C.D.; van Dijk, E. The psychology of social dilemmas: A review. *Organ. Behav. Hum. Dec.* **2013**, *120*, 125–141. [CrossRef]
11. Bolton, G.E.; Ockenfels, A. ERC: A theory of equity, reciprocity, and competition. *Am. Econ. Rev.* **2000**, *90*, 166–193. [CrossRef]
12. Fehr, E.; Schmidt, K.M. A theory of fairness, competition, and cooperation. *Q. J. Econ.* **1999**, *114*, 817–868. [CrossRef]
13. Rabin, M. Incorporating fairness into game theory and economics. *Am. Econ. Rev.* **1993**, *83*, 1281–1302.
14. Kagel, J.H.; Wolfe, K.W. Test of fairness models based on equity consideration in a three-person ultimatum game. *Exp. Econ.* **2001**, *4*, 203–219. [CrossRef]
15. Aumann, R.J. Interactive epistemology I: Knowledge. *Int. J. Game Theory* **1999**, *28*, 263–300. [CrossRef]
16. Aumann, R.J.; Brandenburger, A. Epistemic conditions for Nash equilibrium. *Econometrica* **1995**, *63*, 1161–1180. [CrossRef]
17. Stalnaker, R. Knowledge, belief and counterfactual reasoning in games. *Econ. Philos.* **1996**, *12*, 133–163. [CrossRef]
18. Battigalli, P.; Siniscalchi, M. Hierarchies of conditional beliefs and interactive epistemology in dynamic games. *J. Econ. Theory* **1999**, *88*, 188–230. [CrossRef]
19. Hilgard, E.R.; Sait, E.M.; Margaret, G.A. Level of aspiration as affected by relative standing in an experimental social group. *J. Exp. Psychol.* **1940**, *27*, 411–421. [CrossRef]
20. Lewin, K.; Dembo, T.; Festinger, L.; Sears, P.S. Level of aspiration. In *Personality and the Behavior Disorders*; Hunt, J., Ed.; Ronald Press: Oxford, UK, 1944; pp. 333–378.
21. Simon, H.A. Theories of decision-making in economics and behavioral science. *Am. Econ. Rev.* **1959**, *49*, 253–283.
22. Lopes, L.L. Between hope and fear: The psychology of risk. *Adv. Exp. Soc. Psychol.* **1987**, *20*, 255–295.
23. Lopes, L.L. Algebra and process in the modeling of risky choice. In *Decision Making from a Cognitive Perspective*; Busemeyer, J., Hastie, R., Medin, D.L., Eds.; Academic Press: San Diego, CA, USA, 1995; Volume 32, pp. 177–220.
24. Lopes, L.L.; Oden, G.C. The role of aspiration level in risky choice: A comparison of cumulative prospect theory and SP/A theory. *J. Math. Psychol.* **1999**, *43*, 86–313. [CrossRef] [PubMed]
25. Rieger, M.O. SP/A and CPT: A reconciliation of two behavioral decision theories. *Econ. Lett.* **2010**, *108*, 327–329. [CrossRef]
26. Siegel, S. Level of aspiration in decision making. *Psychol. Rev.* **1957**, *64*, 253–262. [CrossRef] [PubMed]
27. Crowne, D.P. Family orientation, level of aspiration, and interpersonal bargaining. *J. Personal. Soc. Psychol.* **1966**, *3*, 641–664. [CrossRef]
28. Hamner, W.C.; Donald, L.H. The effects of information and aspiration level on bargaining behavior. *J. Exp. Soc. Psychol.* **1975**, *11*, 329–342. [CrossRef]
29. Rapoport, A.; Kahan, J.P. Standards of fairness in 3-quota 4-person games. In *Aspiration Levels in Bargaining and Economic Decision-Making*; Tietz, R., Ed.; Springer: New York, NY, USA, 1983; pp. 337–351.
30. Komorita, S.S.; Ellis, A.L. Level of aspiration in coalition bargaining. *J. Personal. Soc. Psychol.* **1988**, *54*, 421–431. [CrossRef]
31. Thompson, L.L.; Mannix, E.A.; Bazerman, M.H. Group negotiation: Effects of decision rule, agenda, and aspiration. *J. Personal. Soc. Psychol.* **1988**, *54*, 86–95. [CrossRef]
32. Tietz, R. Adaptation of aspiration levels—Theory and experiment. In *Understanding Strategic Interactions*; Albers, W., Güth, W., Hammerstein, P., Moldovanu, B., van Damme, E., Eds.; Essays in Honor of Reinhard Selten; Springer: Berlin/Heidelberg, Germany, 1997; pp. 345–364.

33. Fechner, G.T. *Elemente der Psychophysik (Elements of Psychophysics)*; Holt, Rinehard and Winston: New York, NY, USA, 1860.
34. Stevens, S.S. On the theory of scales of measurement. *Science* **1946**, *103*, 677–680. [CrossRef] [PubMed]
35. Stevens, S.S. The psychophysics of sensory function. *Am. Sci.* **1960**, *48*, 226–253.
36. Luce, R.D.; Steingrimsson, R.; Narens, L. Are psychophysical scales of intensities the same or different when stimuli vary on other dimensions? Theory with experiments varying loudness and pitch. *Psychol. Rev.* **2010**, *117*, 1247–1258. [CrossRef] [PubMed]
37. Posch, M. Win-stay, lose-shift strategies for repeated games – memory length, aspiration levels and noise. *J. Theor. Biol.* **1999**, *198*, 183–195. [CrossRef] [PubMed]
38. Henrich, J.; McElreath, R.B.; Ensminger, A.; Barrett, J.; Bolyanatz, C.; Cardenas, A.; Gurven, J.C.; Gwako, M.; Henrich, E.; Lesorogol, N.; et al. Costly punishment across human societies. *Science* **2006**, *312*, 1767–1770. [CrossRef] [PubMed]
39. Fehr, E.; Gächter, S. Altruistic punishment in humans. *Nature* **2002**, *415*, 137–140. [CrossRef] [PubMed]
40. Samid, Y.; Suleiman, R. Effectiveness of coercive and voluntary institutional solutions to social dilemmas. In *New Issues and Paradigms in Research on Social Dilemmas*; Biel, A., Eek, D., Garling, T., Gustafsson, M., Eds.; Springer: Berlin, Germany, 2008; pp. 124–141.
41. Trivers, R. The Evolution of Reciprocal Altruism. *Q. Rev. Biol.* **1971**, *46*, 35–57. [CrossRef]
42. Haley, K.J.; Fessler, D.M.T. Nobody's watching? *Evol. Hum. Behav.* **2005**, *26*, 245–256. [CrossRef]
43. Shariff, A.F.; Norenzayan, S.A. God is watching you: Priming God concepts increases prosocial behavior in an anonymous economic game. *Psychol. Sci.* **2007**, *18*, 803–809. [CrossRef] [PubMed]
44. Lant, T.K.; Mezias, S.J. Managing discontinuous change: A simulation study of organizational learning and entrepreneurship. *Strateg. Manag. J.* **1990**, *11*, 147–179.
45. Lant, T.K. Aspiration Level Adaptation: An Empirical Exploration. *Manag. Sci.* **1992**, *38*, 623–644. [CrossRef]
46. Oosterbeek, H.; Sloof, R.; van de Kuilen, G. Cultural differences in ultimatum game experiments: Evidence from a meta-analysis. *J. Exp. Econ.* **2004**, *7*, 171–188. [CrossRef]
47. Henrich, J.; Boyd, R.; Bowles, S.; Camerer, C.; Fehr, E.; Gintis, H.; McElreath, R.; Alvard, M.; Barr, A.; Ensminger, J.; et al. Behavioral and brain sciences "Economic man" in cross-cultural perspective: Behavioral experiments in 15 small-scale societies. *Behav. Brain Sci.* **2005**, *28*, 795–855. [CrossRef] [PubMed]
48. Forsythe, R.; Horowitz, J.L.; Savin, N.E.; Sefton, M. Fairness in Simple Bargaining Experiments. *Game Econ. Behav.* **1994**, *6*, 347–369. [CrossRef]
49. Hoffman, E.; McCabe, K.; Shachat, K.; Smith, V. Preferences, property rights, and anonymity in bargaining games. *Game Econ. Behav.* **1994**, *7*, 346–380. [CrossRef]
50. Weg, E.; Smith, V. On the failure to induce meager offers in ultimatum game. *J. Econ. Psychol.* **1993**, *14*, 17–32. [CrossRef]
51. Livio, M. *The Golden Ratio: The Story of Phi, the World's Most Astonishing Number*; Broadway Books: New York, NY, USA, 2000.
52. Posamentier, A.S.; Lehmann, I. *The Fabulous Fibonacci Numbers*; Prometheus Books: Amherst, NY, USA, 2007.
53. Liua, J.P.; Chow, S.C. A two one-sided tests procedure for assessment of individual bioequivalence. *J. Biopharm. Stat.* **1997**, *7*, 49–61. [CrossRef] [PubMed]
54. Kahneman, D.; Lovallo, D. Timid choices and bold forecasts: A cognitive perspective on risk taking. *Manag. Sci.* **1993**, *39*, 17–31. [CrossRef]
55. Harrison, G.; Hirshleifer, J. An experimental evaluation of weakest-link/best-shot models of public goods. *J. Public Econ.* **1989**, *97*, 201–225. [CrossRef]
56. Rapoport, A.; Budescu, D.V.; Suleiman, R. Sequential requests from randomly distributed Shared resources. *J. Math. Psychol.* **1993**, *37*, 241–265. [CrossRef]
57. Budescu, D.V.; Au, W.; Chen, X. Effects of protocol of play and social orientation in resource dilemmas. *J. Organ. Behav. Hum. Decis.* **1997**, *69*, 179–193. [CrossRef]
58. Budescu, D.V.; Au, W. A model of sequential effects in common pool resource dilemmas. *J. Behav. Decis. Mak.* **2002**, *15*, 37–63. [CrossRef]
59. Suleiman, R.; Budescu, D.V. Common Pool Resource (CPR) dilemmas with incomplete information. In *Game and Human Behavior*; Budescu, D.V., Erev, I., Zwig, R., Eds.; Lawrence Erlbaum Associates, Inc. Publishers: Mahwah, NJ, USA, 1999; pp. 387–410.

60. Budescu, D.V.; Rapoport, A.; Suleiman, R. Common pool resource dilemmas under uncertainty: Qualitative tests of equilibrium solutions. *Game Econ. Behav.* **1995**, *10*, 171–201. [CrossRef]
61. Budescu, D.V.; Suleiman, R.; Rapoport, A. Positional order and group size effects in resource dilemmas with uncertain resources. *J. Organ. Behav. Hum. Dec.* **1995**, *61*, 225–238. [CrossRef]
62. Schuster, S. The Golden Ratio as a Proposed Solution of the Ultimatum Game: An Epistemic Explanation by Continued Fractions. Available online: https://arxiv.org/abs/1502.02552v1 (accessed on 21 December 2016).
63. Klar, A.J.S. Fibonacci's flowers. *Nature* **2000**, *417*, 595. [CrossRef] [PubMed]
64. Shechtman, D.; Blech, I.; Gratias, D.; Cahn, J.W. Metallic phase with long-range orientational order and no translational symmetry. *Phys. Rev. Lett.* **1984**, *53*, 1951–1954. [CrossRef]
65. Coldea, R.; Tennant, D.A.; Wheeler, E.M.; Wawrzynska, E.; Prabhakaran, D.; Telling, M.; Habicht, K.; Smeibidl, P.; Kiefer, K. Quantum criticality in an Ising chain: Experimental evidence for emergent E8 symmetry. *Science* **2010**, *327*, 177–180. [CrossRef] [PubMed]
66. Weiss, H.; Weiss, V. The golden mean as clock cycle of brain waves. *Chaos Solitons Fract.* **2003**, *18*, 643–652. [CrossRef]
67. Roopun, A.K.; Kramer, M.A.; Carracedo, L.M.; Kaiser, M.; Davies, C.H.; Traub, R.D.; Kopell, N.J.; Whittington, M.A. Temporal interactions between cortical rhythms. *Front. Neurosci.* **2008**, *2*, 145–154. [CrossRef] [PubMed]
68. Pittard, N.; Ewing, M.; Jevons, C. Aesthetic theory and logo design: Examining consumer response to proportion across cultures. *Int. Mark. Rev.* **2007**, *24*, 457–473. [CrossRef]
69. Wang, T.; Mo, L.; Mo, C.; Tan, L.H.; Cant, J.S.; Zhong, L.; Cupchik, G. Is moral beauty different from facial beauty? Evidence from an fMRI study. *Soc. Cogn. Affect. Neurosci.* **2015**, *10*, 814–823. [CrossRef] [PubMed]
70. Suleiman, R. An Aspirations Model of Decisions in a Class of Ultimatum Games. Unpublished Manuscript. 2014. Available online: http://vixra.org/pdf/1412.0147v1.pdf (accessed on 21 December 2016).
71. Rapoport, A.; Sundali, J.A.; Seale, D.A. Ultimatums in two-person bargaining with one-sided uncertainty: Demand games. *J. Econ. Behav. Organ.* **1996**, *30*, 173–196. [CrossRef]
72. Suleiman, R. Economic Harmony: An Epistemic Theory of Economic Interactions. In Proceedings of the International Conference on Social Interaction and Society, ETH Zurich, Switzerland, 26–28 May 2016.
73. Nowak, M.A.; Page, K.M.; Sigmund, K. Fairness versus reason in the ultimatum game. *Science* **2000**, *289*, 1773–1775. [CrossRef] [PubMed]
74. Miyaji, K.; Wang, Z.; Tanimoto, J.; Hagishima, A.; Kokubo, S. The evolution of fairness in the coevolutionary ultimatum games. *Chaos Soliton Fract.* **2013**, *56*, 13–18. [CrossRef]
75. Zhang, B. Social Learning in the Ultimatum Game. *PLoS ONE* **2013**, *8*, e74540. [CrossRef] [PubMed]
76. Erev, I.; Roth, A.E. Learning in extensive-form games: Experimental data and simple dynamic models in the intermediate term. *Am. Econ. Rev.* **1998**, *88*, 848–881.
77. Roth, A.E.; Erev, I. Learning in extensive-form games: Experimental data and simple dynamic models in the intermediate term. *Game Econ. Behav.* **1995**, *8*, 164–212. [CrossRef]
78. Cooper, D.J.; Feltovich, N.; Roth, A.E.; Zwick, R. Relative versus absolute speed of adjustment in strategic environments: Responder behavior in ultimatum games. *Exp. Econ.* **2003**, *6*, 181–207. [CrossRef]

MDPI AG
St. Alban-Anlage 66
4052 Basel, Switzerland
Tel. +41 61 683 77 34
Fax +41 61 302 89 18
http://www.mdpi.com

*Games* Editorial Office
E-mail: games@mdpi.com
http://www.mdpi.com/journal/games

Lightning Source UK Ltd.
Milton Keynes UK
UKOW07f0059171117
312853UK00004B/70/P

9 783038 424222